KU-548-154

Scientific Writing for Young Astronomers
Part 2

EAS Publications Series, Volume 50, 2011

Scientific Writing for Young Astronomers

A Collection of Papers on Scientific Writing

Part 2

UNIVERSITY OF NOTTINGHAM
GEORGE GREEN LIBRARY OF
SCIENCE AND ENGINEERING
WITHDRAWN
FROM THE LIBRARY

Edited by: *Christiaan Sterken*

EDP SCIENCES

17 avenue du Hoggar, PA de Courtabœuf, B.P. 112, 91944 Les Ulis cedex A, France

First pages of all issues in the series and full-text articles
in PDF format are available to registered users at:
http://www.eas-journal.org

Cover Figure

The tapestry *Astronomie* depicting the Muse *Astronomia*.
Linen, wool and silk, 240 × 340 cm, made in Paris, *circa* 1500–1510.
Photographed by Swedish photographer Håkan Berg.
© Reproduced with permission of the *Röhsska Museum of Fashion, Design and Decorative Arts* (*Röhsska Museet för Konsthantverk och Design*), Göteborg, Sweden.

Indexed in: ADS, Current Contents Proceedings – Engineering & Physical Sciences, ISTP®/ISI Proceedings, ISTP/ISI CDROM Proceedings.

ISBN 978-2-7598-0639-3 EDP Sciences Les Ulis
ISSN 1633-4760
e-ISSN 1638-1963

1006588499

The copyright of each paper in this volume resides with its author under the terms of the Creative Commons Attribution-Noncommercial License 3.0, which permits unrestricted use, distribution, and reproduction, provided the original work is properly cited and that it is not reproduced for commercial purposes.

© The authors, published by EAS, EDP Sciences, 2011
Printed in France

Table of Contents

Expanded Table of Contents VII

Preface XV

Acronyms XVII

About the Authors XIX

Writing a Scientific Paper I. The writing process
C. Sterken 1

Writing a Scientific Paper II. Communication by Graphics
C. Sterken 65

Writing a Scientific Paper III. Ethical Aspects
C. Sterken 173

Subject Index 285

Name Index 297

Expanded Table of Contents

Preface XV

Acronyms XVII

About the Authors XIX

WRITING A SCIENTIFIC PAPER I. THE WRITING PROCESS 1

Preamble **1**

1 Introduction **2**

2 Why do scientists write? **2**

3 Why do scientists publish? **3**

4 Printed paper *versus* oral communication **4**

5 Categories of scientific papers **5**

5.1 Regular research paper in a refereed journal 5
5.2 Letter to the Editor 6
5.3 Research Note 6
5.4 Review paper 6
5.5 Errata and Comments 6
5.6 Information bulletins and telegrams 7
5.7 Data papers 7
5.8 Papers on the design of instruments 7
5.9 PhD thesis book 7
5.10 Essays 8
5.11 Invited talk at a conference 8
5.12 Contributed paper at a conference 8
5.13 Conference posters and ticket papers 8
5.14 Book reviews 9
5.15 Publicity papers 9
5.16 Salami papers 9
5.17 Hoaxes 9
5.18 Duplicate papers 10
5.19 Gray literature 10

6 About language **11**

7 The reading process **13**

8 Writing a research paper **15**

9 The structure and organization of a scientific paper **18**
9.1 Title, subtitle and running title 19
9.2 Authors 20
9.3 Affiliations 22
9.4 Abstract 22
9.5 Key Words 23
9.6 Introduction 23
9.7 Methods, observations, computations and theory 24
9.8 Results 24
9.9 Analysis, discussion and conclusions 24
9.10 Acknowledgments 25
9.11 Dedications 26
9.12 References and citations 27
9.13 Postscript and Appendix 28
9.14 Index and glossary 28

10 Submitting your manuscript **28**
10.1 The very last step: deep proofreading 28
10.2 Etiquette and netiquette 30

11 What to avoid at all price **31**
11.1 Writing things that you do not understand 31
11.2 Drowning in acronyms 31
11.3 Mixing American & British styles 32
11.4 Clichés 32
11.5 Inconsistent capitalization 32
11.6 Footnotes and unnecessary emphasis 32
11.7 Sexism 33
11.8 Non-standard nomenclature 33
11.9 Misquoting, misciting and quoting out of context 34

12 Instructions and guidelines **37**
12.1 Editorial instructions 37
12.2 Copyrights 39

13 FAQs about the editorial process **40**
13.1 Who should be coauthor? 40
13.2 What is *honorary authorship?* 41
13.3 How to settle the order of author names? 42
13.4 What if a coauthor passes away? 43
13.5 Can the list of authors change during the review process? 43

13.6 Who submits the manuscript to the Editor? 43
13.7 Must I wait for a coauthor's consent to submit? 43
13.8 When to submit a manuscript to `astro-ph`? 44
13.9 How to communicate with the referee? 45
13.10 When to thank the referee? 46
13.11 Can a student be referee? 47
13.12 With whom can I share a referee report? 47
13.13 May I reproduce complete sentences? 47
13.14 How to cite or quote a straying paper? 49
13.15 How to sort words that incorporate diacritics? 49
13.16 To cite the URL or to cite the complete bibliometric reference? . 50
13.17 Why publish your data? 50
13.18 Why do journals engage in publishing data? 51
13.19 How to cope with writer's block? 51
13.20 Should I allow someone else to completely rewrite my manuscript? 52
13.21 What to do when disagreeing with editorial changes? 52
13.22 Is it too much to expect authors to read the papers they cite? . . 54
13.23 How to review a paper? 54
13.24 How to revise a multi-authored paper? 56
13.25 When can I ask for another referee? 57
13.26 Why do Editors edit, and why do referees review? 58

14 Summary **59**

Postscript **59**

WRITING A SCIENTIFIC PAPER II. COMMUNICATION BY GRAPHICS 65

1 Introduction **65**

2 Data graphics and visual communication **67**
2.1 Written communication 67
2.2 Processing visual information 69
2.3 Why use graphs at all? 70
2.4 Images, graphs and tabular structures 70

3 Text without graphics and graphics without text **70**
3.1 Consequences of non-graphical communication: an example 71
3.2 Non-textual communication: an example 72

4 Design types of graphs **75**
4.1 Points and lines . . . 75
4.1.1 Scatter plots . . . 75
4.1.2 Vector plots . . . 76
4.1.3 Line graphs . . . 76
4.1.4 Time series and phase diagrams . . . 76
4.1.5 Timelines . . . 77
4.2 Box-whisker plots . . . 77
4.3 Surface plots . . . 78
4.3.1 Histograms . . . 78
4.3.2 Bar graphs . . . 80
4.3.3 Pie charts . . . 80
4.3.4 Bubble graphs and pictographs . . . 82
4.3.5 Contour maps . . . 82
4.3.6 Sky maps . . . 82
4.3.7 Ternary diagrams . . . 82
4.4 Genuine 3-D surfaces . . . 83
4.5 Drawings and cartoons . . . 87
4.6 Which design type of graph? . . . 88

5 Computer graphics **89**
5.1 Vector images . . . 90
5.2 Raster or bitmap images . . . 90
5.3 Converting bitmap to vector data and *vice versa* . . . 91
5.4 File formats . . . 91

6 Photographs **94**
6.1 Halftones . . . 94
6.2 Halftones and Moiré patterns . . . 95

7 Elements of a graphical image **96**
7.1 Aspects of textual components in graphics . . . 97
7.1.1 Typefaces and fonts . . . 97
7.1.2 Typesetting units of length: pixels, points and picas . . . 98
7.1.3 Which typeface and which font to use in graphics? . . . 98
7.1.4 Font size . . . 99
7.2 Graph title . . . 101
7.3 Axes . . . 102
7.3.1 Axis orientation, and origin . . . 102
7.3.2 Axis tickmarks . . . 102
7.3.3 Axis title and units . . . 104
7.3.4 Axis type . . . 106
7.3.5 Axis labels . . . 108
7.3.6 Axis scale . . . 108

7.3.7 Axes and framing 110
7.3.8 Background colors and lighting conditions 110
7.4 Plotting 112
7.4.1 Plot symbols and lines 112
7.4.2 Autoclipping 112
7.4.3 Error bars 112
7.5 Graphics and texts as image components 112
7.5.1 Legend or key, and labels 113
7.5.2 Scale factors and orientation 114
7.5.3 Insets 114
7.5.4 Grid lines 115
7.5.5 Decoration 116
7.6 Aspects of composition 117
7.6.1 Galleries 117
7.6.2 Size ratios 118
7.6.3 Figure captions 119
7.6.4 Figure numbering and placement 119

8 Image quality: high resolution and low resolution 120
8.1 Image resolution 120
8.2 Sampling frequency and scanning 122

9 The complexities of handling color 124
9.1 Why use color in graphics communication? 124
9.2 Different colors for everyone 125
9.3 An early generic mathematical color model 126
9.4 Color models, color spaces and color palettes 130
9.4.1 The CIE XYZ color space 131
9.4.2 The RGB and $sRGB$ color spaces 132
9.4.3 The $CMYK$ color space 132
9.4.4 Device-independent colors: spot colors 135
9.4.5 Degenerate color spaces 135
9.5 Color workflow 138
9.5.1 Color management 138
9.5.2 Proper color balance and gray tonality 139
9.5.3 Color transformations 140
9.5.4 When to change color space? 142

10 Principles of graphical excellence 143
10.1 Managing ink 143
10.2 Revise, edit, revise, edit, and revise, edit 144
10.3 Layout: the typeset page as an image 146
10.4 Style 147

11 Lying with graphs **148**
11.1 The copyright 148
11.2 Retracing graphs 149
11.3 The lie factor 150
11.4 Outliers 151
11.5 Dot joining, or "helping" the eye by connecting points 153
11.6 The perception of linearity 154
11.7 Curve fitting and overshoots 159
11.8 Models and calibrations 160

12 Tables and lists **162**

13 What to avoid at all price **164**

14 Summary **166**

Selected color graphics **171**

WRITING A SCIENTIFIC PAPER III. ETHICAL ASPECTS 173

1 Introduction **173**

2 On truth **174**

3 On error **176**

4 On quality and value **179**
Quality. 179
Value. 180

5 On bibliometry and bibliometrics **182**
5.1 The measure of journals 182
5.1.1 The impact factor 182
5.1.2 The cited half-life 184
5.1.3 The immediacy index 184
5.2 The measure of scientists 184
5.2.1 Paper counts 184
5.2.2 Citation rates and citation index 187
5.2.3 The Hirsch h index 188
5.3 Should bibliometrics guide you in the decision where to publish? . 193

6 Researcher misconduct **194**
6.1 Amplification of work 194
6.2 On secrecy and secretiveness 195
6.3 Forging 195

6.4 Trimming 197
6.5 Cooking 197
6.6 Data manipulation 199
6.7 Image manipulation 200

7 Author misconduct **201**
7.1 Lying about publication status 201
7.2 Cascade submissions 202
7.3 Omitting author names 202
7.4 Incorrect authorship order 202
7.5 Adding noncontributing authors 203
7.6 Refusal to accept responsibility 203
7.7 Abusing the referee 204
7.8 Misquoting, improper attribution and wrongful citation practice . 204
7.9 Hoaxes 205
7.10 Copyright infringement and plagiarism 205
7.11 Self-plagiarism, text recycling and paraphrasing 208
7.12 Dual and redundant publications 209

8 Referee misconduct **210**
8.1 Breaching confidentiality 211
8.2 Hasty or superficial revision 212
8.3 Long revision times 212
8.4 Professional arm wrestling and taking revenge 212
8.5 Silencing dissent 212
8.6 Anonymity and its effects 213
8.7 Conflicts of interest 214

9 Grant-reviewer misconduct **216**

10 Editorial misconduct and editorial mismanagement **217**
10.1 Non-responsive or inactive Editors 217
10.2 Inappropriate modifications of manuscripts 218
10.3 Proximity between authors and journals 219
10.4 Forging, bypassing or shortcutting the refereeing process 219
10.5 Manipulation of the journal impact factor 220
10.6 Breaching confidentiality 221

11 Publisher misconduct and mismanagement **222**
11.1 Manipulation of journal content 222
11.2 Manipulation of journal impact factors 223

12 Research supervisor misbehavior **223**
12.1 Creativity or routine? 225
12.2 Authorship ethics 226

13 Dealing with the Press **228**

14 The signatures of scientific misconduct **231**

15 The causes of scientific misconduct **232**
15.1 Increased administrative burdens 232
15.2 Career pressure and competition 232
15.3 Misunderstanding bibliometry 233

16 Cures for misconduct **234**
16.1 Accepting the limits of discovery 234
16.2 Research misconduct policies 234
16.3 Mending the business model of journals 235
16.4 Establishing policies on research misconduct: an example 237
16.5 Detecting dual publication and plagiarism: an example 239

17 Where to go for help: whistleblowing **241**
17.1 Motivations for whistleblowing 241
17.2 Whistleblowing: how to 242
17.3 Whistleblowing and its aftermath 244

18 On the biases of bibliometric indices **246**
18.1 Errors of confusion by the database construction 247
18.2 Errors of confusion by the database user 248
18.3 Errors of omission and inclusion 249
18.4 Inherent bias 252
18.5 Errors of procedure 255
18.6 On rankings based on one single quantity 259
18.7 On the use and abuse of bibliometric indices 263
18.8 Some serious warnings 264
18.9 Summary of bias factors 265
18.10 Repairing bibliometrics 267
18.11 On the true meaning of bibliometric indicators 270

19 How can truth and value ever survive? **271**

Epilog **273**

Subject Index 285
Name Index 297

Preface

The *Astronomy & Astrophysics* (A&A) Board of Directors and EDP Sciences, the Publisher of A&A, decided in 2007 to organize a School on the various aspects of scientific writing and publishing, with two aims: (a) to teach young PhD students how to express their scientific thoughts and results through adequate and efficient written communication, and (b) to discuss the operation of A&A as an example of an international peer-reviewed journal in astronomy. To this end, a first *Scientific Writing for Young Astronomers* (SWYA) School was organized in 2008, and a second one in 2009. The Schools were financed by A&A and by EDP Sciences, who also provided all logistic support.

This book, together with Volume 49 of the EAS Publications Series, is the outcome of these Schools. Whereas Volume 49 presents guidelines and examples for publishing in academic journals (like A&A), the three papers in this Volume are aimed at supplying guidelines to PhD students and postdoctoral fellows to help them compose scientific papers for different forums (journals, proceedings, thesis manuscripts, etc.). These papers address the writing process, graphics, ethics, and bibliometry, and cover the information that was presented in about a dozen lectures. These chapters feature several examples and case studies in astronomical context, in addition to many examples from scientific enquiry in a broader sense – even from art and from business life.

Paper I copes with the preparation of manuscripts, with the handling of copyrights and permissions to reproduce, with communicating with editors and referees, and with avoiding common errors. More than two dozen FAQs (on authorship, on refereeing, on revising multi-authored papers, etc.) are answered.

Paper II is entirely dedicated to communication with graphics, *i.e.*, to all facets of visual communication by way of images, graphs, diagrams and tabular material. Design types of graphs are explicated, as well as the major components of graphical images. The basic features of computer graphics are explained, as well as concepts of color models and of color spaces (with emphasis on color graphics for viewers suffering from color-vision deficiencies). Special attention is given to the verity of graphical content, and to misrepresentations and errors in graphics and in associated basic statistics. Dangers of dot joining and curve fitting are discussed, with emphasis on the perception of linearity, the issue of nonsense correlations, and the handling of outliers. The remainder of the chapter illustrates the distinction between data, fits and models.

The main theme of Paper III is truthful communication of scientific results, and hinges on the pillar that every scientist daily comes across: **ethics** – involving two major aspects of research, *viz.*, the measurement of scientific value, and the enforcement of proper conduct in research and in scientific writing. The following bibliometric parameters are explained: the journal impact factor, the journal cited half-life, and the journal immediacy index, as well as paper counts, citation rates, citation index and the Hirsch index. These bibliometric indices and indicators are illustrated with real examples derived from bibliometric analyses of the astronomical literature. The biases of bibliometric indices, and the use and abuse of bibliometrics are worked out. Moreover, suggestions for remediating the present system of bibliometric measurement and evaluation are made. Scientific misconduct in the broadest sense is discussed by category: researcher misconduct, author misconduct, referee and grant-reviewer misconduct. But also publisher misconduct, editorial misconduct and mismanagement, and research supervisor misbehavior are dealt with. The overall signatures of scientific misconduct are focussed on, as well as the causes and the cures. This is followed by a Section devoted to whistleblowing. This chapter illustrates the complex endeavor that education in science really is, and also comprises passages that directly deal with the hopes and concerns of our students, either expressed during and after the lectures, or through private contacts later on.

The best way to use this book is to read it twice: a first reading for skimming its content, and then a deeper reading of the passages that are relevant to what the reader wants to know. Or use this book as a reference, and read the Sections corresponding to items listed in the index or in the expanded Table of Contents.

The opinions expressed in these papers are those of the author, and those of the authors of the cited papers, and do not necessarily reflect the views of the Editors of A&A. None of these papers were submitted for language editing to a professional language editor.

The Schools took place in Hotel and Conference Center Aazaert in the city of Blankenberge on the North Sea coast of Belgium. The meetings were very rewarding, and happened in a very pleasant atmosphere, thanks to the dedication and professionalism of the complete staff: from the barkeeper to the waiters, and from the administrative staff to the chefs in the kitchen – they all contributed greatly in making us feel at home, and in giving us an unforgettable time. In the name of all participants, I express deep appreciation to Mr. Peter Kamoen, Manager of Hotel Aazaert, and to his entire staff.

The Manager of the Schools (on behalf of EDP) was Mrs. Isabelle Turpin, assisted by Mrs. Aurélie Chastaingt. They did a professional job preparing each event, and their presence during the complete duration of each School was a blessing for anyone who needed help. In the name of all lecturers and students, I extend a warm thanks to Isabelle and to Aurélie.

Dr. Agnès Henri, Mrs. Sophie Hosotte, Mrs. Isabelle Houlbert and Mrs. Marie-Louise Chaix (EDP) are given thanks for their most adequate communication and helpful support during the preparation of this book.

My special thanks go to those students who – either during or after the lectures, or through subsequent private contacts – shared with me their hopes and concerns.

I am very much indebted to Dr. Hilmar W. Duerbeck, Dr. Chris Koen, and Dr. Laszlo Szabados for deep proofreading of, and criticism, discussion and advice on the three papers included in this book.

I acknowledge financial support from the Vrije Universiteit Brussel, and I express gratitude to my Employer, the Belgian Fund for Scientific Research (FWO), for supporting several research projects that, over the years, gave me the opportunity to acquire the specialized knowledge presented in this book.

Christiaan Sterken
Editor
University of Brussels, Belgium

DOI: 10.1051/eas/1150000

Acronyms

AAS	American Astronomical Society
AD	Anno Domini
ADS	NASA Astrophysics Data System
AJ	Astronomical Journal
ASCII	American Standard Code for Information Interchange
ASP	Astronomical Society of the Pacific
A&A	Astronomy & Astrophysics
ApJ	Astrophysical Journal
ApJS	Astrophysical Journal Supplement Series
BC	Before Christ
BCE	Before the Common Era
BMI	Body Mass Index
BMJ	The British Medical Journal
BMP	Windows bitmap format
Caltech	California Institute of Technology
CCC	Copyright Clearance Center
CCD	Charge Coupled Device
CDS	Centre de Données astronomiques de Strasbourg
CIE	Commission Internationale de l'Éclairage
CMMI	Computer Modern Math Italic
COPE	Committee on Publication Ethics
CRT	Cathode Ray Tube
CV	Curriculum Vitae
DOI	Digital Object Identifier
dpi	dots per inch
DSS	Digital Sky Survey
EDP	EDP Sciences
EPS	Encapsulated Postscript
ESA	European Space Agency
ESO	European Southern Observatory
FAQ	Frequently Asked Questions
FFP	Fabrication, Falsification and Plagiarism
FITS	Flexible Image Transport System
GCPD	General Catalogue of Photometric Data
GIF	Graphics Interchange Format
HJD	Heliocentric Julian Date
HST	Hubble Space Telescope
IAU	International Astronomical Union
IBVS	Information Bulletin on Variable Stars
ICMJE	International Committee of Medical Journal Editors

IF	Journal Impact Factor
IMRAD	Introduction, Methods, Results and Discussion
IQ	Intelligence Quotient
IRAS	Infrared Astronomical Satellite
ISI	Institute for Scientific Information
ISO	Infrared Space Observatory
JCR	Journal Citation Report
JPEG	Joint Photographic Experts Group
LBV	Luminous Blue Variable
LISA	Library and Information Services in Astronomy
LMC	Large Magellanic Cloud
lpi	lines per inch
MJD	Modified Julian Date
MMS	Manuscript Management System
MNRAS	Monthly Notices of the Royal Astronomical Society
NASA	National Aeronautics and Space Administration
OA	Open Access
OCR	Optical Character Recognition
$O-C$	Observed minus Calculated
ORCID	Open Researcher & Contributor ID
PASP	Publications of the Astronomical Society of the Pacific
PDF	Portable Document Format
PICT	Apple Picture Format
PMT	Photomultiplier Tube
PNG	Portable Network Graphics
ppi	points per inch
pt	printer's point
QPO	Quasi Periodic Oscillation
RAS	Royal Astronomical Society
SAO	Smithsonian Astrophysical Observatory
SDSS	Sloan Digital Sky Survey
SI	International System of Units
SIMBAD	Set of Identifications, Measurements, and Bibliography for Astronomical Data
SPIRES-HEP	High Energy Physics database, Stanford
S/N	Signal to noise ratio
SWYA	Scientific Writing for Young Astronomers
TAI	International Atomic Time
TIFF	Tagged Image File Format
URL	Uniform Resource Locator
USNO	United States Naval Observatory
UT	Universal Time
UTC	Coordinated Universal Time
VLT	Very Large Telescope
WAME	World Association of Medical Editors
WET	Whole Earth Telescope
WoS	Web of Science
ZAMS	Zero-age main sequence

About the Authors

Christiaan Sterken is Research Director of the *Belgian Fund for Scientific Research* (Fonds Wetenschappelijk Onderzoek), and works at the Department of Physics at the University of Brussels (Vrije Universiteit Brussel). His principal field of research is the photometry of variable stars (luminous blue variables, massive binaries, pulsating main sequence stars, and cataclysmic variables), and comets. He also teaches courses in observational astronomy, and on the history of natural sciences at the University of Brussels.

Scientific Writing for Young Astronomers
C. Sterken (ed)
EAS Publications Series, **50** *(2011) 1–63*
www.eas-journal.org

WRITING A SCIENTIFIC PAPER I. THE WRITING PROCESS

C. Sterken[1]

Abstract. Based on my direct personal experience as an author, referee, editor, and PhD thesis supervisor, I present in this paper some advice and guidelines on writing scientific papers. This article copes with the preparation of manuscripts, with the handling of copyrights and permissions to reproduce, with communicating with editors and referees, and with avoiding common errors. More than two dozen FAQs are dealt with, among others: who should be coauthor, what is honorary authorship, how to settle the order of author names, may the list of authors change during the review process, who submits the manuscript to the Editor, must one wait for coauthors' consent to submit, when to submit a manuscript to `astro-ph`, how to communicate with the referee, with whom to share a referee report, can students be referee, how to cite and quote a straying paper, how to cope with writer's block, should others be allowed to completely rewrite a manuscript, how to review a paper, how to revise a multi-author paper, when to ask for another referee, why do Editors edit, and why do referees review?

Preamble

This paper is the first of three expanded sequels to *Advice on Writing a Scientific Paper*[1], Sterken (2006), exploring key aspects of the scientific writing process. In particular, advice is given on how to prepare manuscripts, handle copyrights and permissions to reproduce, how to communicate with Editors and referees, how to avoid common errors, what are the requirements for demanding authorship, and which guidelines to follow when determining a fair authorship sequence. The second paper handles graphics (images, graphs and Tables), and the third paper deals with ethics in scientific communication.

As these papers (hereafter referred to as Paper I, Paper II and Paper III, respectively) deal with tutoring on writing and publishing, the reader will notice some overlap

[1]Complete text freely available at `http://www.aspbooks.org/publications/349-0445.pdf`

©The authors, published by EAS, EDP Sciences, 2011
DOI: 10.1051/eas/1150001

among these three articles, and also between these papers and the other papers in these Proceedings. Such repetition is unavoidable, since every lecturer's approach is from a different viewpoint (Editor, referee, author, thesis supervisor, ...), but the redundancy is also included on purpose since repetition is an innate element of the practice of teaching. Papers I–III also present ample advice on research practice through teaching by example.

These papers are written in the first and second person singular – a not so common approach in scientific communication that commonly utilizes *passive voice* – but since they deal with somewhat personal advice based on hard-won experience, this approach is much more direct and persuasive than when speaking in the third person. The reader is thus addressed in the second person, the target audience being the young astronomer at the beginning of her or his life as an author (though senior researchers may occasionally take profit of the advice presented here). As the Scientific Writing for Young Astronomers (SWYA) schools were organized on behalf of the journal *Astronomy & Astrophysics* (A&A), some of the practical advice given here is tuned for publishing in this specific journal.

In these three papers, I proceed on the premise that the reader is strongly committed to the core business of science: *verity* and *truth*, even more so to the *truthful communication* of scientific findings (see also Sect. 5.17). Paper II, and the Epilog of Paper III, address an additional element of science and scientific communication: *beauty* and *aesthetics*.

1 Introduction

The structure of this paper is as follows. The next three Sections deal with the question why we write and why we publish, and what the principal differences are between the spoken word (lecture) and written (published) communication. Then comes a Section describing the different categories of research papers, followed by three Sections dealing with language, and with the reading and the writing process. A complete Section is devoted to the structure of a scientific paper. Finally, a selection of FAQs on practical matters is presented.

2 Why do scientists write?

We are so used to see facts and theories as the only building blocks of science, that we often forget that there is much more, as is so well conveyed by Shermer (2007):

> To be of true service to humanity, science must be an exquisite blend of data, theory and narrative.

Data and theory are complemented and completed by *narration*. Any narration or storytelling consists of three elements: *i)* the **content** of the message or story, *ii)* the **form** of the story, and *iii)* the **direction** of the communication (author to reader, author to Editor, author to referee). Moreover, the writing phase of a research project not seldomly modifies the research conclusions, as is testified by Perkowitz (1989):

> I have learned that when I write a research paper I do far more than summarize conclusions already neatly stored in my mind. Rather, the writing

> PROCESS IS WHERE I CARRY OUT THE FINAL COMPREHENSION, ANALYSIS, AND SYNTHESIS OF MY RESULTS.

Thus, without publication, research results remain incomplete. This cognitive process is known as *advancing insight*: ideas that arise as the work evolves. Umberto Eco also says[2]

> WRITING IS AN ACT OF COMMUNICATION, IT'S AN ACT OF LOVE, IT'S SOMETHING YOU DO IN ORDER TO BE UNDERSTOOD.

Perkowitz's and Eco's statements underline that it is very important to learn how to properly produce narrative (*i.e.*, write texts, or make good graphics), but also how to consume narrative (*i.e.*, how to read papers, and how to understand referee reports). The main aim of writing is communication: above all, **we write to be read** – and not to be cited as a first purpose.

Before embarking on the writing of papers, it is useful to first deal with the question of why we publish, then to concentrate on the crucial difference between the printed word *versus* the oral message, and finally to explain the reading process.

3 Why do scientists publish?

Scientific research is a collective undertaking, and involves the distribution of knowledge, and the sharing of results. This dissemination of knowledge, using established rules and standards of scientific reporting, is the prime reason for publishing. Evidently, we do not always and only publish with this motive in mind; in real practice there are several gradations in the incentive for writing. The following short paragraphs list, in order of diminishing importance, some motivations for publishing.

1. BECAUSE I WANT TO REPORT NEW SCIENTIFIC RESULTS AND GET THE CREDIT. Besides the wish and need to communicate results, it is very important to publish as soon as new results and discoveries are considered solid. Credit is usually (but not always) attributed to the person who first makes a result public. Early publication allows another necessary process to commence: *experimental verification* of the observations or theories – and, perhaps, *refutation*.

2. BECAUSE I NEED A JOB, A PROMOTION, OR A GRANT. Any successful application for a job or a research grant requires proof of scientific creativity and productivity. This evidence indirectly emanates from the publication record. A colleague once wrote me that he sought a contract renewal and, therefore, needed to produce *"a paper, quick and dirty"*. The resulting quick paper may have been effective in his pursuit of a job, but needless to say, the paper cannot have been on a high standard of creativity if the job hunt was the only motive for writing it.

3. BECAUSE I WANT TO ACHIEVE SOCIAL CLIMBING BY BEING VISIBLE IN ADS[3]. This ambition to gain esteem by fellow scientists is somehow related to the previous one, in fact to the

[2]Quoted by Mahtot Teka, Time Magazine, 10 December 2007.

[3]The NASA Astrophysics Data System, `http://www.adsabs.harvard.edu/`

complete mechanism of "measuring" scientific stature by "counting" a person's papers and citations through bibliometric tools like ADS (see Sect. 5.2 in Paper III for a more profound discussion).

4. BECAUSE I AM TRAVELING, AND PAPERING IS THE ONLY WAY TO COVER MY TRAVEL COSTS. This is unfortunate but realistic: many conference attendees get their expenses covered only if they present a paper (in practice, a poster is displayed most of the time).

The first motive for publishing is the most earnest: announce new results, describe new findings, or report significant errors in published works. I could not say it clearer than the Chilean artist Violeta Parra (1917–1967) does in *Yo canto a la diferencia*:[4]

> ...I DO NOT PLAY THE GUITAR FOR APPLAUSE. I SING OF THE DIFFERENCE BETWEEN WHAT IS TRUE AND WHAT IS FALSE. OTHERWISE, I DO NOT SING.

It all boils down to the amount of new-information content of your paper: if, for example, the information content of a poster paper is insignificant – either because of poor content, or because the poster is just a duplicate of another paper (going to be) published elsewhere – then the poster paper simply is irrelevant, and should not be published in the first place.

4 Printed paper *versus* oral communication

There are fundamental differences between oral and literate modes of communication – between a talk and a written paper – such as:

- the printed message is not instantaneous: whereas the speaking voice is a direct address to an audience, the printing and distribution process takes months of time;
- speaking is continuous communication, the printed message is discontinuous;
- printed text does not render any intentional or inadvertent body language – bodily movement, posture or facial expression, but also vocal stress and changes in pitch – that underlines and even amplifies the message;[5]
- a printed message does not provide any non-verbal "leakage", *i.e.*, hesitations and signs of poor preparation – though printed text may carry equivalent signs of sloppiness, such as typographical errors;
- a written paper calls for careful wording in order not to be misunderstood, or in order to avoid being cited out of context, (see also Sect. 11.9);
- a printed text allows the author to present a very different degree of detail in the description of the research procedures and results;

---

[4]*...no tomo la guitarra por conseguir un aplauso. Yo canto a la diferencia que hay de lo cierto a lo falso. De lo contrario, no canto.*

[5]There are many languages of gesture, and facial expressions are not always intelligible to those raised in another culture, see Thomas (1991).

- because the spoken word is straight communication, it provides for instantaneous feedback;
- whereas a confusing idea – whether ambiguously described or inherently complex – may get across the attentive reader's mind after a second reading of a same phrase, the time frame of the spoken word does not allow such a second pass: even simple messages do not always penetrate the receiver's mind in real time, because attention occasionally drops off to the level that *you may have heard the message, though without really listening*[6] *to it*;
- published phrases remain forever, spoken words often persist only for a very short time;
- whereas sitting through a bad lecture may possibly yield a minor bonus (*e.g.*, when the lecturer uses well-designed slides and attractive animations), a poorly written paper has no value whatsoever, since such papers *always* convey the message that *"what is poorly written is also of doubtful scientific value"* (Stein 2006);
- a lecture – an invited lecture above all – is a social event, with a direct relationship between speaker and audience, and with considerable emphasis on the pedagogic form. Reading a written text is a more solitary experience.

A published paper is quite often the written record of a talk, like some of the papers in these Proceedings. Whereas one just cannot publish a *verbatim* record of a lecture, the written article normally corresponds to the talk (at least in content). In this set of three papers, however, I deliberately deviate from this principle, because the matter covered was spread out over almost a dozen lectures in both SWYA schools.

5 Categories of scientific papers

There are many categories of scientific papers. This Section describes the most common types, as roughly ranked by importance (see also the A&A website `www.aanda.org`).

5.1 Regular research paper in a refereed journal

Submission of a paper to a refereed journal (like A&A) mostly implies that the paper is the result of original research not previously published, nor submitted for publication elsewhere, nor considered for later reproduction without the consent of the copyright holder. Note that some refereed journals will not even submit to the referee a paper that is "too light". Bertout & Schneider (2004), for instance, give a typical example of the type of topic that they deem unsuitable for submission to A&A: *"... an unremarkable binary together with a standard interpretation of its light curve"*. It is thus advisable to check the editorial policies of a journal, and verify what types of papers are published by that journal.

[6]The verb *to hear* means to perceive or apprehend by the ear, the verb *to listen* means to hear something with thoughtful attention (*Merriam–Webster Collegiate Dictionary*).

Do not forget that papers that are not refereed are discounted for career promotions and job applications. Most refereed journals require that manuscripts be submitted in the so-called referee format (double spacing: full page instead of two-column format), so that the referee can make annotations.

5.2 Letter to the Editor

A *Letter* is a research paper (or a short opinion paper) that requires rapid publication. Some types of Letters are esteemed higher than regular refereed papers, others have a much lower esteem. A&A publishes Letters to communicate important new results that require rapid publication. They are restricted in length to four printed pages, and are usually published within 4–8 weeks of acceptance (see the paper by Claude Bertout in Part 1 of these Proceedings). Whether a manuscript is accepted as a Letter, or is diverted for publication as a regular paper, is the decision of the Editor, though the opinion of the referee not seldomly is the determining factor.

5.3 Research Note

A&A accepts *Research Notes*, *i.e.*, short papers that contain either new results as an extension of work reported in a previous paper, or limited observations not urgent enough to be published as a Letter, or useful calculations that have no definite immediate astrophysical applications. A Research Note can also present a preliminary analysis of a large dataset deemed important for timely communication to the readership of the journal.

5.4 Review paper

A review paper is a wide retrospective survey of a specific field, including a critical evaluation of the scientific subdiscipline dealt with. A straightforward compilation of literature sources or cataloged data is by no means a review paper – especially not in our times when compilations can so easily be carried out. In particular, when composing overview Tables of astrophysical parameters[7] from different sources, one should always make sure that the resulting compilation lists data that are *commensurable* and *homogeneous*, so that their use in statistical studies is permitted. A&A does not publish review papers.

5.5 Errata and Comments

Errata refer to errors in previously published papers, and are normally signed by all coauthors of the original paper. *Comments* are usually not published by A&A, except in very special cases. Three conditions are necessary for a *Comment* to be considered for publication: *i*) it refers to a paper published by A&A, *ii*) it does unambiguously solve the problem or question it raises, and *iii*) its publication will be useful to the community. Some journals publish comments as *Letter to the Editor*.

[7]For example, photometric indices, effective temperatures, abundances, ...

5.6 Information bulletins and telegrams

A bulletin is a brief news item intended for immediate publication. An example of a bulletin journal in astronomy is the *Information Bulletin on Variable Stars*[8] (IBVS), published by the Konkoly Observatory in Hungary. IBVS publishes short but significant news on variable-star astronomy.

The *Central Bureau for Astronomical Telegrams*[9] (CBAT) issues a telegram journal that is responsible for the dissemination of information on transient astronomical events via the International Astronomical Union Circulars (IAUCs).

The basic difference between these two communication media is the speed (IAUCs being faster) and the publication cost (IBVS not imposing page charges). Telegrams get some editorial attention, but are not necessarily refereed: the degree of refereeing in such short papers depends on the editorial policies of the journal.

5.7 Data papers

Standard journals require concise reporting, hence few journals can accommodate large volumes of experimental data. Some major journals have (or had) associated supplement series, where results of experiments (theoretical or observational) are published. Most data are nowadays published in machine-readable form, and quite some papers in A&A have the associated data (time series, comprehensive Tables, large illustrations, etc.) deposited at the CDS (Centre de Données astronomiques de Strasbourg), see also the paper by Laurent Cambrésy in Part 1 of these Proceedings. Section 14 of A&A, *Online catalogs and data*, offers online publication of large sets of data.

Very few refereed journals accept "data only" papers, *i.e.*, the data without an accompanying analysis. Still, there is a substantial difference between a *data archive* and a *data paper*: whereas the former merely is a set of tabular material with an accompanying `readme` file, the latter provides the space to discuss the data calibration, to examine all aspects of the error budget, and even to include some scientific results.

5.8 Papers on the design of instruments

Publications on the conception and design of instruments belong to a somewhat similar category as data papers, have a relatively short life time (see also Sect. 5.1.2 in Paper III), and tend to get buried in the archives of observatory libraries. A&A has a Section *Astronomical instrumentation*, with refereed papers on instrumentation. This Section is published online only.

5.9 PhD thesis book

Most universities require the PhD candidate to submit a PhD manuscript. As the pressure to publish forces the student to publish new findings as fast as possible, PhD manuscripts,

[8] `http://www.konkoly.hu/IBVS/IBVS.html`
[9] `http://cfa-www.harvard.edu/iau/cbat.html`

as a rule, are compilations of published papers, with an introductory chapter that gives the outline of the thesis, and its most important conclusions. As most journals require the transfer of copyrights from the authors (see Sect. 12.2, and Papers II and III), the reformatting of papers in a thesis manuscript could, in principle, be seen as a breach of the copyright agreement. But most journals do not object to reproducing published papers in the PhD manuscripts of their authors.

5.10 Essays

The definition of this form of paper is rather vague, and generally refers to a short interpretive work usually dealing with its subject from a personal point of view of the author.

5.11 Invited talk at a conference

In principle on invitation, such papers give a review of one of the themes of the meeting. Good meetings frequently have several such landmark papers, which are quite often very informative, and certainly are recommended literature for every beginner in the field. My experience is that the term *invited* should not be taken too literally: organizers of meetings often receive an explicit request from a potential participant demanding to be "invited" for a talk.

5.12 Contributed paper at a conference

Such papers, as a rule, are short, and quite often bring a preview of new results that are not yet published. In principle, every conference paper should be able to pass mild refereeing, but in practice, such papers are of widely variable quality, and some would never pass a strict refereeing test. There is now an increasing tendency to include so-called *refereed conference papers* in Proceedings of meetings. The refereeing is then done by the Scientific Organizing Committee of the conference, or by other participants. Needless to say, this approach is a far cry from genuine refereeing, since only the really very bad papers are rejected. The Editor(s), moreover, are in a difficult position to refuse such papers, simply because their content (oral or poster presentation) has been revealed, and the authors risk losing credit if an attendee rushes to print with similar results.

5.13 Conference posters and ticket papers

Poster sessions were created to allow participants to present their results, even if the meeting format only allows invited talks and reviews. The advantage of posters is that participants can view and discuss the material at their own pace, in an atmosphere that encourages direct discussion. But I have seen cases of reproduction or duplication of talks and posters by the same authors at the same meeting – even people postering decade-old stuff. Such duplicates have no value in a curriculum, and such duplication should be avoided. After all, why duplicate a paper if you can come up with other interesting aspects of your research as a *variation* on previously published results? This reminds me of one of my PhD students who proudly told me about his Master-thesis supervisor who was so "clever" that he succeeded in getting the same paper published three times. Frankly, this is more a sign of skill in deception than of cleverness. Section 7.12 in Paper III discusses the problem of paper duplication.

The term *ticket paper* is slang for a conference paper that only serves to get your travel covered. University administrations often require participants to give a talk when they apply for travel support. As such, a short paper or poster is often presented for the sole purpose of acquiring the travel subsidy.

5.14 Book reviews

Books are not really reviewed by a referee, although publishers do ask an external expert to advise them on whether or not to publish a submitted book manuscript. But once a book is published, the publishing company seeks to have the book reviewed by several independent readers, and such reviews are then published in scientific journals. Copies of new books are quite often sent to Editors of journals, who in turn look for an appropriate reviewer. A&A does not publish book reviews.

The principal issues to present in such reviews are content-related criticisms, technical aspects of presentation (such as quality of graphics, occurrence of typographical errors), but also comments on the price and on whether value for money is delivered. There seems to be a clear-cut difference between one- and two-author books (textbooks, monographs) and multi-authored books (mainly proceedings): the latter quite often receive unfavorable judgment because of untidy editing, or as Duerbeck (2005) puts it

> ... VERY SLOPPILY EDITED – OR IS EDITING NOWADAYS NOTHING MORE THAN THE UNCONTROLLED PROCESSING OF LaTeX FILES?

5.15 Publicity papers

Some papers do communicate results, though only for the purpose of bringing to public attention some achievements of a large team or of a scientific consortium. Such papers are typically published in annual reports of institutes, or in observatory magazines.

5.16 Salami papers

The term refers to the not so uncommon habit of reporting results in slices, *i.e.*, instead of presenting all results from a single study in one cohesive paper, the work is partitioned in multiple papers, and submitted to different journals. This technique, for sure, increases visibility in a bibliographic database, but the dilution takes its toll at the moment when a peer-review body asks your *"three most important papers of the last three years"*. More on paper slicing in my contribution on ethics in Paper III.

5.17 Hoaxes

A scientific *hoax* is a fabrication. If the intention of the author is to dupe the reader, we speak of a *fraud* (see also Sect. 7.9 in Paper III). There is a very special type of hoax or *parody* that leads to exposure of corrupt research practices or customs. A most interesting example is Alan Sokal's *Transgressing the Boundaries: Toward a Transformative Hermeneutics of Quantum Gravity*, Sokal (1996). There is a more recent case, where an *Open Access* journal agreed to publish a nonsensical article written by a computer program, claiming that the manuscript was peer reviewed and even requesting that the authors pay 800 US$ in page charges.[10] The debacle led to the resignation of the Editor-in-chief of the journal.[11]

Believe it or not, fabrication of false coauthor names also occurs. One famous example is the most interesting case of immunologist Polly Matzinger, who refused to use passive voice in her papers, and considered that if she is forced to refer to herself as "we", she consequently needs a coauthor. Hence she added the name of her Afghan hound as *Galadriel Mirkwood*[12] in Matzinger & Mirkwood (1978).

Although hoax papers are made to expose misconduct, they are not appreciated, since the form and method of the protest basically violates the principle of mutual trust, as laid out in the last paragraph of the Preamble (page 2). This also applies to inventing coauthors: such fabrications may harm the fabricator in the first place, as Polly Matzinger may have found out.

5.18 Duplicate papers

Paper duplication occurs, not seldomly for commercial or promotional reasons. Sets of papers on one single project (for example, space experiments: see Fig. 22 in Paper III) are often bundled in a special issue of a journal. For celebrating the 40th anniversary of A&A, the Editors published a special anniversary issue as a compilation of 40 papers, which they esteem to have a strong impact in the community at large. This collection appeared as Volume 500 (2009), and contains facsimile duplicates of the original papers, and every such paper was accompanied by a 1- or 2-page commentary by a reviewer. The good thing about this compilation is that an ADS search on any author's name will return only the original paper's bibcode, so that citing these papers will not generate any bibliometric bias. A confusion that may arise when such precautions are not taken is discussed in Paper II (see also Sect. 13.16).

Note also that book Publishers even rename books to fit their markets: for example, Paul Davies' book *The Goldilocks Enigma: Why is the Universe Just Right for Life?* is sold in the US as *Cosmic Jackpot: Why Our Universe Is Just Right for Life.*

5.19 Gray literature

Gray literature refers to instrument and software manuals, applications for observing or computing time, grant applications and grant reports (progress and final reports): in principle all such manuscripts and papers should follow the same guidelines of rigor and

[10] `http://www.the-scientist.com/blog/display/55756/`

[11] `http://www.outsidethebeltway.com/archives/science_editor_quits_after_hoax/`

[12] Galadriel is a fictional character, and Mirkwood is the name of a fictional forest, in the legends created by J.R.R. Tolkien (1892–1973).

clarity as any of the above-mentioned classes of earnest papers. Dishonest attitudes in grant applications and reports are often severely punished: as a jury member for grant allocations I have witnessed prompt disqualification of applicants who had moved forward the position of their own name in the list of their papers (see also Sect. 11.9). Even upgrading the class of paper (for example selling a poster abstract as a refereed paper) may lead to downgrading an application: some jury members do check!

Press Releases also belong to the category of gray literature, since such texts have, strictly spoken, no author, but a composer.[13]

6 About language

A most important element of the narrative is the language wherein the scientific communication occurs. **Language is the link from the mind of the writer to the mind of the reader** – either instantaneously, or over time. If the world had only one common language, we would have an information-sharing network with formidable power. But there exist approximately 7000 languages worldwide (and several thousand are expected to disappear in the next century), so which language to select for scientific communication?

> LANGUAGE IS NOT JUST ANY CULTURAL INVENTION, BUT THE PRODUCT OF A SPECIAL HUMAN INSTINCT,

says Pinker (1994): humans possess this very special common basic biological skill to learn the processing of information through language.

As Pinker conveys, *"There are Stone Age societies, but there is no such thing as a Stone Age language ... "*. So, basically, almost any language could be used for the communication of science,[14] since all languages consist of a finite list of elements (words or characters) together with a program to construct blocks of information: the syntax for building sentences. But very considerable culture-specific differences exist in communication through language: for example, some cultures express thoughts in an indirect, verbose, affective and tentative way, others communicate in a more direct way. Cultural-specific aspects are evident when someone uses his or her native language, but these aspects are even more detectable when a person expresses thoughts in a language different from the native language.

Today, English happens to be the *lingua franca* in science, and Joli Adams documents this fact in her paper on language editing in Part 1 of these Proceedings. Already in its first year of existence, the Editors of A&A urged authors to use English for communicating their results (see Fig. 1). They point out two prime advantages of using one common language:

1. the work becomes more widely known, and
2. the pool of **active** and **competent** referees becomes larger.

[13] See Paper II (Section 4.4) for an example of a paper that has a composer, but no author.

[14] Provided it possesses an extensive technical vocabulary, accompanied by a proper dictionary.

1. We wish to advise our French-speaking and German-speaking authors to explore possibilities to write in English in order that the results of their work become more widely known. We now have Editorial help to revise the language of their papers in English if they need it. May we remind the authors that is it becoming increasingly difficult to find active and competent referees capable of reading French or German.

Fig. 1. "First important point" in the very first A&A Editorial (Pottasch & Steinberg 1969).

The use of one common language foreign to so many countries and cultures unavoidably leads to many variants of English, for example Chinese English, Dutch English, French English, American English, etc.: variants[15] where the correct English words are used[16], but that are dominated by the native-language syntax that also permeates the English at the reader's end. And there are also variants of style: scientific style is direct, plain, assertive, context-independent, and brief – in contrast to prosaic or almost any other style that we use on a daily basis. Modern scientific style, however, mostly uses the passive and impersonal voice, and shuns active voice verbs and first-person pronouns.[17]

There are two basic types or classes of languages: *non-tonal languages* (stress languages, like English) that use one dominant *syllable*[18] in the word, and *tonal languages*, which use tone to distinguish words (like Mandarin and Vietnamese).

A tricky situation arises with transliterations from one class to another, like the *Hanyu Pinyin* scheme, which is the most common Standard Mandarin romanization scheme (*Hanyu* refers to the Chinese language, *pin* stands for connection and *yin* for sound): foreign words are phonetically split in syllables that have a meaning in Chinese. Figure 2 shows, as an example, two Pinyin transliterations for the name of the city of Blankenberge, kindly (and independently) provided by two participant students who had never heard about that place before going there. The word is read as *Bù lán kěn bei ge*, and the syllables have meanings of cloth, blue (or orchid), be willing to (or consent[19]), uncle, and grid (or meter). As one can see, the fourth syllable is transposed differently. I have also encountered a kind of "inverse pinyin" in the western transliteration of the first name of Taiwanese student Hsiang-Wen Hsu, who uses the phonetically similar "Sean" as his *callname*.[20] Callnames should never be used in citations or as part of an author name, so this person's bibliometric records in ADS list Hsu, H. as well as Hsu, Hsiang-Wen.

But it really gets catchy when dealing with Asian names of persons. For example, entering the family name Zhang in an ADS search yields almost 14 000 entries, whereas Zhang, L. leads to a list of about 600. The reason is that this name not only is very common in China, but that the Pinyin transliteration for Zhang originates from at least

[15] With *variants* I do not mean the national dialects of English.

[16] Though not without confusion: a billion, for example, means 10^9 to some people, and 10^{12} to others.

[17] Notable exceptions are to be found in the weekly journal *Nature*.

[18] A *syllable* is a natural unit in which a word can be analyzed.

[19] The third *grapheme* (a fundamental unit in a written language), and its meaning, are also used for the second syllable in my own surname – referring to a consenting person, perhaps?

[20] A *callname* is a name given in place of the official name, for example Bob for Robert, Liz for Elizabeth, Bill for William, Dick for Richard, and Ted for Edward. A callname is not necessarily a *nickname*.

布兰肯贝格
布兰肯伯格

Fig. 2. Pinyin transliterations for *Blankenberge* (courtesy Wu Yue and Zhang Li Yun).

four different root names. And that is not the only difficulty: Chinese names, traditionally, are formatted differently from Western names. Whereas we write our names in the order given name(s) (first name and middle names) followed by the last name (*i.e.*, family name, or surname), the traditional format in Asian cultures is the other way around, with the family name first (mostly a one-character name – Wu and Zhang in the caption of Fig. 2), followed by a one or two-character given name (sometimes linked by a dash). Nowadays there develops a strong tendency to switch to western format in publications, and even in daily life. This situation greatly adds confusion in reference lists and citations (see also Sect. 9.12), as Jia Wei[21] points out in Qiu (2008):

> I REALLY HAVE A HARD TIME SORTING OUT WHO HAS PUBLISHED WHAT.

There are many ways to confuse western names, though. For example, surnames in Spanish-speaking countries come from both father and mother: the first part of the surname is the father's family name, the second surname is the family name from the mother's side (and often only the first letter is mentioned, like in Violeta Parra S.[22]). This format should not be confused with the case of people who have truly composite surnames, nor with cases where people change name by marriage (see also Sect. 9.2). Names transliterated from Cyrillic (Russian and a number of other languages of eastern Europe and Asia) also pose a problem for correct identification, since they do not all use the same Cyrillic spelling: for example, all combinations of Karamazov, Karamazoff, Dostoevsky, Dostojewski, Dostoyevsky, ... all appear in book translations.

7 The reading process

When reading a text, you are assisted by a kind of mental *daemon*[23] that – in computer jargon – is called a *parser*, a process that **unconsciously captures the hierarchy of the input text and transforms it into a form suitable for further processing**. In linguistics, the parser is a mental program that analyzes sentence structure during reading and, normally, checks for syntax errors at the same time (Pinker 1994).

Let us look at an example from one of my students' manuscripts: *"through linear least-squares fitting an equation that best fitted them was extracted ... "*. When I read that sentence for the first time, I was mentally slowed down or even stopped after "equation", and again after "them" (*i.e.*, the data). The first part of that sentence contains

[21] Associate Dean at the Pharmacy School of Shanghai Jiao Tong University.
[22] Her complete name is Violeta del Carmen Parra Sandoval.
[23] A *daemon* is a computer program that runs in the background.

tautologies: "linear least-squares fitting an equation" and "that best fitted" are useless repetitions, since "linear" already specifies the equation, and "least-squares" always implies a best fit. Moreover, "them" is confusing, and can only be understood by remembering the previous sentence. So, a much shorter way of saying the same would be to refer to a linear least-squares fit of two specified variables.

A similar slowing down of the reading pace occurs when previously undefined acronyms are used: for example in the same text the acronyms QPO (quasi-periodic oscillation) and AD were used, and the reviewer commented *"It took me a while to associate AD (Anno Domini) with accretion disk... "*. Note that the parser keeps track of bits and pieces of previous sentences (especially acronyms), and that a memory stack of only 6 to 8 positions can be properly maintained. Fortunately, an ambiguity like QPO can be minimized through the context in which it appears – though the writer of a scientific text should always avoid to communicate in context-sensitive statements that imply guesswork by the reader. Note that the parser – just like any other semi-automatic skill – develops with time and, therefore, **intense reading is a prime requirement for adequate development of your writing skills**.

When I read a research paper, for example as a referee, I first read it cover to cover in one fling, and make some brief annotations if needed. This pass is very useful to cool the parser. In a second approach, I concentrate more on the Abstract, the graphs, the Table and Figure captions, and the Conclusions.

One very specialized form of reading is *proofreading* (see also Sect. 10.1), a most important step in the publication process. Proofreading occurs at two phases: several times before submitting a manuscript to the Editor, and then one last time after receiving the proofs from the Publisher. Proofreading occurs already at the earliest stages of writing: in fact, every time you read a few pages of your self-composed text (or text produced by a coauthor), you engage in proofreading, although the proper term for this activity phase rather is *revizing* and *editing*. You can easily put the parser to sleep by too many passes, especially of your own compositions. That is why proofreading is best done by another pair of eyes as well. It also happens that the first Sections are subject to more proofreading passes than the last ones, and that should be avoided, since the last Sections are not the least important!

Proofreading is so important that Scholastic, the company that publishes the Harry Potter books, has a full-time *continuity editor* who doublechecks the complete proofreading. She explains:[24]

> For instance, with Bertie Bott's Every Flavor Beans, I make sure it's always B-o-t-t-apostrophe-s. Every Flavor is not hyphenated, and Flavor does not have a u.

In other words, tight standards of English, as well as of the Harry Potter constructed magic language, are rigidly observed throughout all Volumes of that series.

[24]Quoted by Cheryl Klein, Time Magazine, 9 July 2007.

8 Writing a research paper

Writing a paper is a *process*, a chain of gradual steps that lead to an acceptable and pleasing end product. As such, learning to write is a continuous process that demands a lot of time for acquiring the necessary writing skills. The best procedure is to start drafting a paper while the research work is still in progress, recording first the preliminary structure, and then completing those parts that are rather easy to handle: description of the methods, observations, references, Figure and Table captions, etc. One advantage of starting early is that you become familiar with the journal's house style, and almost immediately can see some of the fruits of your labor in typeset format, and that you avoid *writer's block*, *i.e.*, staring at blank pages or at sloppy notes (see Sect. 13.19). Another, more important, bonus is that starting early may also lead you to follow possible sidelines, even to a spin-off paper derived from your main work. Such a by-product paper is an excellent item to present at a meeting as a poster, to show new findings in the context of the core work, without needless duplication of material that has already been published elsewhere. In such a way, you can **make publicity for your core work, without violating copyright rules, nor creating copy-and-paste duplicates**.

There are four indispensable basic requirements for producing a good paper:

1. write clearly,
2. write accurately,
3. be brief (avoid verbose and pompous styles), and
4. build a logical structure: the train of thought should be logical, and avoid a winding and repetitive course in the suite of ideas.

Besides these basic requirements, there are a number of issues to absolutely avoid, see Section 11.

A very important point is to learn mastering *the exact word in the exact place*. Table 1 comprises sets of nouns, verbs and adjectives that look like synonyms, but do not always have exactly the same meaning in context, by which these words are often incorrectly used. It is very good practice (also for native English-speaking authors) to consult a *thesaurus* – a list of words grouped together according to similarity of meaning – rather than always fall back to translation by means of a dictionary.

A guaranteed way to use the wrong word in the wrong place is grandiloquence, *i.e.*, the excessive use of verbal ornamentation and pompous language. Or giving impressive numbers with incomplete information content, like the digital-camera manufacturers' marketing hype of playing with zoom factors and resolution specifications. I recently reviewed a paper where the authors stated that they used a 390 150-pixels CCD camera, whereas the more modest-looking 715×510 gives the reader much more information on the chip's dimensions and size.

Some verbs like *look, believe, seem* and *think* should be used with great care: avoid writing that "X thinks ...", because you cannot really know what X thinks or believes. One of my postdocs liked to state "the data look good ...": *good* is not a generic synonym for *excellent*, and covers *pleasing* as well as *dependable*, hence a more quantitative and

Table 1. Groups of words that look like synonyms, but are often incorrectly used.

error, uncertainty, mistake, defect, flaw, blunder	scatter, noise
dimension, size, extent, extension	trend, pattern
regression, correlation, fit	chance, probability
confidence, significance	amplitude, range, power
standard star, comparison star	flux, intensity, luminosity
seeing, scintillation, scattering	precision, accuracy
color index, filter, passband, color	robust, stable
standard, classic, normal, default	parameter, variable, estimate
misbehavior, misconduct, misdemeanor	symmetry, isotropy
define, postulate, speculate	bias, residual
parameter, observable, factor	minimum, lower bound
colloquium, symposium, conference	maximum, upper bound
presentation, talk, lecture, seminar, class	typeface, font, type
supervisor, mentor, coach, patron, boss, sponsor	retention, secretiveness
do, perform, conduct, carry out	great, large, big
period, frequency, mode, harmonic	simple, elegant, elementary
fact, observation, measurement	randomness, entropy, chaos
model, doctrine, theory, hypothesis, mechanism	duplication, redundancy
supposition, proposition, assumption, premiss	mean, average
example, metaphor, conjecture	consideration, thought
prove, claim, maintain, demonstrate, verify, contend	invention, discovery, insight
assert, avow, support, advise, suggest, establish	significant, relevant
predict, anticipate, forecast, foresee	whole, complete, entire
instructions, guidelines, requirements, rules	mention, cite, quote, refer
calculate, compute, reckon, count	emulate, simulate
authenticity, integrity, honesty	code, algorithm, program
expect, believe, suppose, estimate, guess, think	intercept, zero point
excellent, good, satisfactory, acceptable, sound	pleasing, dependable
tutorial, guide, manual, treatise, memoir	manuscript, paper, text
absolute, relative, differential	ask, suggest, recommend
explain, understand, apprehend, comprehend	inconceivable, impossible
confirm, affirm, corroborate, validate	induce, conclude
entail, implicate, impose, imply	luck, chance, serendipity
enormous, immense, indefinite, infinite, innumerable	refute, overthrow
copyright infringement, (self-)plagiarism, paraphrase	repeat, replicate
jargon, terminology, nomenclature	construe, interpret
capture, acquire, sample	analyze, reduce, process
discern, distinguish, recognize, scrutinize	utilize, use, apply
invent, contrive, formulate, imagine, devise	revise, review, referee

less emotional expression is to be preferred instead. Qualifying terms, such as "likely, probably, perhaps" are also not very recommended ways to make your point, unless the point is so vague that "perhaps" it is not a point at all. Modifiers like "possibly, very" weaken the point even more, as well as "it is clear" or "it is inconceivable", etc. Declarations that are made empathically, like "there can be no doubt", should also be avoided, as they suggest that no supporting evidence is necessary.

Look out when using words with a very strict meaning in mathematics and physics: sometimes, a quantity or a well-defined parameter is used in a context so vague, that the reader has no clue of what that quantity or concept really represents. One common error, for example, is to call a frequency a *mode*, before even knowing that the frequency is associated with stellar pulsation. Another one is the symbol σ, which is normally used for the standard deviation[25] of a single observation, associated with $\pm$ pointing to the standard error of the mean, identifying the range of values that encloses the true value of the measured quantity. But in past times, $\pm$ was often used to indicate probable error (p.e.). Hence, avoid writing σ, or "the sigma value", but specify what σ means. And do not assume lightly that an error distribution is *normal*[26] (or Gaussian), which strictly means that the expectation of every variable is the same as the expectation of any other in the sample (and each is equal to the mean or average), and implies uncorrelated values. Another frequently occurring confusion is *variance – i.e.*, the square of the standard deviation – where a biased or an unbiased estimate can be implicated. And in counting samples, *ten or more* is not exactly the same as *more than ten*.

An example of an ambiguous word is *bug*, meaning either an insect, or a hidden microphone, or an error in a computer program. Another example strictly belongs to computing: the symbol `NaN`[27] is mostly produced as the result of an invalid operation, and this symbol should not be used for missing values in Tables, unless it is specified that the symbol stands for a missing number and not for the result of an unsound computing operation (see also Paper II).

Non-native English-speaking authors[28] should always look up the meaning of a word whenever in doubt, especially when words with different meanings have almost the same pronunciation (so-called *homophones*, for example "whole" and "hole"). Words that are considered synonyms in your mother tongue, may have a different connotation in English (like large/great/big, intriguing/interesting, claim/maintain/prove, or listen/hear, see the footnote on page 5).

Beware also those words that are called *false friends*: pairs of words in two languages that look or sound similar, but differ in meaning. A very frequent misunderstanding involves *eventually* (meaning an unspecified later time), which matches the French *éventuellement* (possibly). Other examples are *data* and *date*, *curriculum* (syllabus) *versus* CV[29], and the mixup of *to use* and the French *user* instead of *utiliser*.

---

[25] Standard deviation is a measure of the dispersion of a data set.

[26] *Normal* is a misleading word, and has a different meaning in "normal distribution" and "normal equations".

[27] `NaN`: Not a Number.

[28] Sometimes referred to as *non-native users of English* (Salager-Meyer 2009).

[29] *Curriculum vitae*.

Beware also of constructing English words by simple logic: someone who theorizes, can be called a theorist, but someone who socializes is not necessarily a socialist. And one who organizes is an organizer, not an organist (though some organists do organize organ concerts). Do not forget that language is a continually evolving communication system with strict rules – though these rules do not lead to 100% logic consequences.

Be very careful with translation, and never write your paper first in your native language for translation afterwards (by yourself, or by a friend): **there is a real danger that the translation process changes your message**. As an anecdotic example, I refer to an instruction manual of an electric door bell, which stated that installation was easy, and that only "three sons" were needed: the automaton-like translator had mixed up the French word for cables (*fils*) and for sons (*fils*) without attention for the coherence of the message. This kind of machine translation is offered by internet search engines, and can lead to very strange sentences when applied to scientific texts, as the following example illustrates.

Li *et al.* (2007) published a paper entitled *A new study of the long-term variability of the SX Phoenicis star CY Aquarii*. The Google translation of the abstract goes: *"By phoenix SX-type variable star CY Aqr was determined by observing two new light changed a great time, ... Combining the published data points fitting reveals a new formula for calculating a great time"*. Obviously, the constellation was identified with a mythical being, and time of maximum was rendered as "a great time".

In order to avoid translational gaffes, when quoting a text in translation, it is advisable to give the original statement together with your translation (as is done in the footnotes on pages 4 and 20). Non-native English-speaking authors should also consider the advantages of using English-language versions of computer operating-systems, of viewing non-dubbed English movies, and of reading English books in their original editions.

9 The structure and organization of a scientific paper

People often mix up *manuscript* and *paper*. A manuscript[30] is a document submitted to an Editor (or to be submitted in a not too distant future). A paper is a manuscript accepted for publication – that is, as soon as it has been stamped "accepted", or has received a Digital Object Identifier[31] (DOI®) from a Publisher.

We have seen that words are the raw materials, and that language syntax gives the sentences their meaning. In a paper, sentences are grouped in paragraphs, and paragraphs become the elements of Sections. Most scientific papers have a very similar structure in a well-tried format suitable to efficiently transfer facts and interpretations of facts. Papers are mostly organized in Sections according to the so-called IMRAD model, where the acronym stands for **I**ntroduction, **M**ethods (observations, computation, theory), **R**esults **a**nd **D**iscussion (and Conclusions).[32] The A&A website states about this common model:

[30]Etymologically, the term refers to a hand-written document, and a non-manually reproduced text was called a typescript.

[31]The Digital Object Identifier System identifies objects in the digital environment (*i.e.*, publications) with an internationally standardized unique DOI name, see `http://www.doi.org/`. The footer of the title page of each paper in these Proceedings holds the DOI.

[32]This scheme is not applicable to gray literature.

> THIS IS A WELL-TRIED FORMAT; AUTHORS SHOULD HAVE GOOD REASONS FOR DEVIATING FROM IT. THE GOAL OF A SCIENTIFIC PAPER IS NOT TO IMPRESS THE READERS BY POETIC LANGUAGE BUT TO TRANSFER FACTS AND NEW INSIGHTS AS LUCIDLY AS POSSIBLE.

An example of non-IMRaD is the paper by Oort (1970) *The Density of the Universe*: this paper has an abstract, but no further structure. This is only admissible for very short, and very important papers.

Note that the IMRaD model does not dictate that each element of the acronym must necessarily be composed after the preceding one. My experience is that the **M** and **R** Sections are the easiest to start with, whereas **I** and **D** are best finalized at a later stage. In the following Subsections we deal with each structural element of the manuscript.

9.1 Title, subtitle and running title

The title should be specific, and as brief as possible, at the same time it should be attractive. If a title is long, then the author should provide an abbreviated *running title* for the page headers. Avoid neutral titles (beginning with "A study of ..."), refrain from excess words, do not use abbreviations (unless they are very well known, for example, LMC for the Large Magellanic Cloud), and prevent grammatical errors and typos in the title. Finalize the title when you finish the paper. A subtitle is used when the title is too long to be clear, or to indicate that the paper is part of a sequence of papers.

I once saw a booklet entitled *On BH's and GW's*. It took me a moment to realize that the compilation was about black holes and gravity waves, especially because the plural of BH and GW should have been written as BHs and GWs – without apostrophe[33]. Such a title, though brief, is not clear, as the reader needs reflection to understand it. The champion of brevity and wittiness, sure enough, is "M 32 ± 1" by Lauer *et al.* (1998).

Question marks in the title can also lead to confusion, for example *Stephan's Quartet?* (Allen 1970), which implicitly refers to Stephan's Quintet. Or *Mode Identification through Photometry: Full Colours?*, where the appellation "Full Colours" refers to recognition of levels of achievement or service,[34] see Sterken (2000) – an intended pun to lead a paper that critically discusses some deficiencies of mode-typing photometric methods. A similar strategic ambiguity is to be found in the title of Watson's (2007) book *Avoid Boring People*, where boring can be read in two ways: as the adjective meaning non-exciting, or as the gerund of the verb "to bore". Such confusion is mostly undesigned, and may pose a real problem for people acquainted with languages where the usage of a verb as a noun[35] is uncommon.

A famous counterexample of a very clear and most complete title is by Lemaître (1927), who formulates his hypothesis in a most explicit way already in the title of his work[36]: *A homogeneous universe of constant mass and increasing radius, validating the*

[33] An apostrophe expresses a relationship through the genitive.

[34] in athletics, schools, clubs, etc.

[35] A quite unique aspect of the English language.

[36] *Un Univers homogène de masse constante et de rayon croissant rendant compte de la vitesse radiale des nébuleuses extragalactiques.*

radial velocities of extragalactic nebulae (see also Paper III). However, beware to never promise more in the title than it delivers!

9.2 Authors

Who should be coauthor, and in which order author names must appear, are issues many young (but also some not so young) writers struggle with. Virtually nobody carries out research without any help at all: there are several classes of people who provide assistance: scientific contributors, technical contributors, administrative and financial contributors, etc., and it is evident that some of these contributions, **if substantial**, deserve coauthorship. There are no absolute rules, except for the general expectation that the order of names should be morally upright and correct, *i.e.*, **that there is a correlation with delivered effort or labor**.

Though it may sound far-fetched, typos in author names – even the same surname appearing twice in the same author list – do occur! Typos most frequently appear in names with character accents (*e.g.* č, ç, œ...) and in composite names. The many ways of using the prefixes *van*, *le*, *du*, ... can lead to problems with search engines. Take as an example van Genderen, which could bring about Van Genderen, Van genderen, and Vangenderen. Any of the first three orthographic forms leads ADS to more than 250 abstracts coauthored by Arnout van Genderen, whereas "Vangenderen" generates no hits at all. A similar, but reverse, situation occurs with Du Bois, du Bois, du bois and "Dubois": only the last format leads to the abstracts of the papers of Patrice Dubois. And surnames are not everywhere processed in the same way when it comes to alphabetic order: names with prefixes are in some countries ordered according to prefix, and in other countries according to the principal component, for example "Genderen, van" and "Zach, von", see the name index list at the end of this book.

Beware also that many people have several first names, and that some of these are composite: there is a difference between Jean Michel (J. M.) and Jean-Michel (J.-M.). To illustrate a real-life case: I have coauthored four papers with Chinese author Liu Zongli, and the ADS search supplies four different identifications: Liu Zongli, Liu Z., Liu Zong-li and Zongli, L. And to make the confusion complete, my university's library database lists several works under "Sterken C.", and one book by "Sterken Chr." – simply because the Publisher decided, without any consultation, to adopt Chr. for the initial of my given name[37] in Sterken & Manfroid (1992).

There is an additional complication with gender: in some societies, marriage coerces females to adopt the name of their husband. So, Hillary Clinton was Hillary Diane Rodham before her marriage – Rodham was her maiden name – and she is sometimes designated as "Hillary Clinton *née* Rodham". But she is often referred to as Hillary Rodham Clinton, though some would call her Hillary Rodham-Clinton and others will refer to Hillary Clinton-Rodham, a really awful situation if she were a scientific author, and perhaps to be only surpassed by Irène Joliot-Curie[38]. Not to speak of what happens when a

[37] What prompted several colleagues to address me as Christopher.

[38] Irène Joliot-Curie was the daughter of Marie Skłodowska-Curie and Pierre Curie. In 1926 she married the physicist Jean Frédéric Joliot, and they both changed their surnames to Joliot-Curie.

female scientist divorces or marries again, implying a complete reset of the citation count. And how to handle same-sex marriages?

The only advice I can give to female scientists is to simply retain their birth name (and initials), and take an example at the American suffragist Lucy Stone (1818–1893), who was the first to claim the right not to use husband family names. To illustrate the depth of the problem, I refer to Merton (1995) on the so-called *Thomas Theorem*,[39] and quote from Dorothy S. Thomas' letter of September 10, 1973 addressed to R. K. Merton:

> I ASSURE YOU THAT I WAS A BORN THOMAS AND THEN MARRIED A THOMAS WHO WAS NO RELATION. I WAS AN ARDENT LUCY STONER[40] AND ALSO SWORE I WOULD NEVER CHANGE MY MAIDEN NAME WHICH I DIDN'T.

The proper way for these authors to sign would be Thomas, W. I. & Thomas, D. S., and the correct citation would be Thomas & Thomas (1928)[41].

An example from astronomy is Maunder & Maunder (1908): *The heavens and their story*, by Annie Scott Dill Russell and Edward Walter Maunder. Annie Russell (1868–1947) was a solar astronomer, and one of the very few 19th-century professionally-employed female astronomers in Britain (Brück 1994). In 1895 she married Edward Walter Maunder, and their book was coauthored "Annie S. D. Maunder & E. Walter Maunder" – though the front page gives "A. & W. Maunder". ADS always lists his initials as E. W. (both initials referring to his given names), whereas she is either referred to as[42] "Maunder, W., Mrs." or "Maunder, A. S. D.", where the last initial does not refer to a given name, but to her mother's surname Dill. This brings up another aspect of quoting given names: though born Edward Walter Maunder, he never used the first given name, and always signed as E. Walter Maunder. Whatever the reason for dropping the first name, if it is done when signing a paper, **it should always be done in the same way** – unless you do not mind bibliometric mix-up.

This brings me to abbreviations of given names, for example Thos. for Thomas and Wm. for William – an old tradition that was introduced for saving a few characters in print, but that causes more confusion than savings. And, as said already on page 12, common callnames, nicknames and diminutives all contribute to confusion in reference lists, and should thus not be used.

Finally, a last word on gender and given names: it is often far from obvious to guess the gender of a person on the basis of one or more first names, especially when dealing with names in other cultures. Some boys' names look deceptively similar to girls' names in another language (for example Cecil and Cécile), whereas Chris, Scott, Kim and Andrea are unisex names that serve males as well as females.[43]

---

[39] Sociologists William I. Thomas and his wife, Dorothy S. Thomas, coauthored on an axiom in the sociology of knowledge: *"If men define situations as real, they are real in their consequences"* – though for a long time all credit solely went his way.

[40] Females who insist on keeping their maiden name are called *Lucy Stoners* ever since.

[41] This reference is not included in the reference list because I had no access to this work.

[42] Maunder (1902, 1904).

[43] As recorded by the US Census Bureau `http://www.census.gov/genealogy/names/`

9.3 Affiliations

Author affiliation is not just a matter of recording contact information: affiliation information is increasingly used to rate universities and journals, and the outcome of an affiliation index can be decisive for an institute's financing. It is, therefore, very important to use correct and unequivocal affiliations for all coauthors. The following points need special attention:

1. make sure to use the same format for all coauthors;
2. beware of abbreviations and acronyms;
3. be consistent and use only one language inside the address (for example Brussel-Bruxelles-Belgium would involve three languages in one address line);
4. verify if the English translation of your institute's name is recognized by the search engines, and if it is not, then stick to the original;
5. if you mention "on leave", then be specific and correct.

The last point may need some clarification. In my editorial practice, I once encountered a PhD student stating his affiliation with an additional *"On sabbattical leave from ... "*. A *sabbattical*[44] is a prolonged absence in order to fulfill a very specific goal, *e.g.*, writing a book, and sabbaticals are typically for one year following six years of full-time teaching or administrative activities, and certainly do not belong to the scope of a regular PhD training.

9.4 Abstract

The abstract should provide maximum information with minimum words, and cover the following elements:

1. *WHY* was this research undertaken, and what is the objective of this study;
2. *HOW* did you do the research (observations, theory, calculations), and
3. *WHAT* are the new results, and what do these new results mean,

and should be comprehensible without reading the paper. The abstract is not part of the paper and should not make reference to it. It is preferably written in the third person, and draws attention to all new facts in the paper. Avoid using abbreviations and acronyms,[45] or listing references in the abstract: if referring seems inescapable, then simply give the complete reference, including journal Volume and page numbers. Grammatical errors and typos in the abstract should not occur, and any kind of review matter is out of place there.

Remember that the abstract is the first thing someone reads after reading the title (especially when consulting journal archives, where reading the complete paper may involve access and download permissions), and as such, the abstract is perhaps the most important part of the paper. Some journals do not include all back volumes in their online archives, and comprehensive online collections of journals (like SPRINGERLINK[46] or

[44] From Greek *sabbatikos* and Latin *sabbaticus*.

[45] Unless they are very common, like FAQ used in the Abstract of this paper.

[46] `www.springerlink.com/`

ScienceDirect[47]) offer publisher-imposed restricted access modes,[48] and require the reader to pay for downloading the full paper. Since the abstract is mostly available, a complete and well-composed abstract will help the reader to decide whether it is worth paying for the download.

To help you optimize your abstracts, A&A encourages the use of structured abstracts (see the editorial published by Bertout & Schneider 2005). Just like a traditional abstract, a structured abstract summarizes the content of the paper, but it does make the structure of the article explicit and visible. For doing so, the structured abstract uses headings that define several short paragraphs. Three paragraphs, entitled *Aims, Methods* and *Results*, are mandatory. When appropriate, the structured abstract may use an introductory paragraph entitled *Context*, and a final paragraph entitled *Conclusions*.

The objectives of the paper are defined in *Aims*, the methods of the investigation are outlined in *Methods*, and the results are summarized in *Results*. The heading *Context* is used when needed to give background information on the research conducted in the paper, and *Conclusions* can be used to explicit the general conclusions that can be drawn from the paper. The Editors of A&A, though, suggest that authors who prefer the traditional non-structured form are invited to implicitly follow the logical structure embedded in the proposed structure. Note that the style of the abstract should be pleasing. Bertout & Schneider (2005) demonstrate how structured abstracts may look by giving an example from an observational paper, and a second one from a theoretical paper in cosmology.

9.5 Key Words

Key words are used for indexing, and A&A allows a maximum of six key words to be listed immediately after the abstract. These words are to be selected from a list that is published each year in the first issue in January,[49] a listing that is common to the major astronomical journals.

9.6 Introduction

The introduction should clearly state why the study was undertaken, and place the research in a broad context by referring to previous works of relevance. The introduction should not contain the conclusions. Some authors tend to expand an introduction into a review paper by itself; this should be avoided, it is better to refer to papers in the well-established review journals. At the end of the introduction the outline of the paper may be described as a road map of your work, but spoiling space for presenting a list of Section headings at the end of the introduction (as I do in Sect. 1) is to be avoided when space is limited.

The introduction is also the place to mention if preliminary results or related conference papers of yours were already published or submitted. Give a clear statement of the problem you study, and present a brief literature review that cites the most relevant papers

[47] `www.sciencedirect.com/`

[48] Either embargo, *e.g.* a 360 day delay of availability of full text articles, or access to only part of the archive.

[49] Keywords are subdivided into broad categories, see `www.aanda.org/content/view/136/184/lang,en/`

related to your work, but avoid irrelevant citations (especially of the few people you personally know). Only give references to papers that you have really read, or at least have seen, and never cite a reference because the paper or software manual you read also cites that paper! Wild citing changes the citation history of a work, and this helps no one (see Sect. 18 in Paper III). If you refer to someone, refer to a specific and important paper that is of direct relevance to your work, not just to a minor poster or abstract.

In the introduction, but also throughout the rest of the paper, a proper paragraph structure should be maintained. A paragraph is a set of sentences in which one and only one main point or idea is handled. It also includes directly-related sentences with details and information that support the paragraph topic.

Finalize the introduction after finishing the discussion and conclusions, and check if all acronyms have been properly defined.

9.7 Methods, observations, computations and theory

In this Section you describe your methods in such a way that the description gives all elements needed to allow a full understanding and possible experimental or computational replication or verification of your work. Do this in a concise way, list observing logs in a Table (with maximum information in a minimum of space), and if your paper discusses multiple targets, avoid repetition of redundant descriptions for every object discussed.

9.8 Results

Point out how the data look (trends, new effects, frequencies, ...), but do not give interpretation of the results at this stage: use the data to state only the bare facts, without making inferences. Describe experimental errors, and state the accuracy of the results (but avoid tabular and graphical redundancy, since Tables and Figures illustrating the same results are mostly not accepted by journals). Do not oversell nor underreport what you found, and always question your assumptions. As Richard Feynman (1985) explains:

> ... IF YOU'RE DOING AN EXPERIMENT, YOU SHOULD REPORT EVERYTHING THAT YOU THINK MIGHT MAKE IT INVALID – NOT ONLY WHAT YOU THINK IS RIGHT ABOUT IT: OTHER CAUSES THAT COULD POSSIBLY EXPLAIN YOUR RESULTS; AND THINGS YOU THOUGHT OF THAT YOU'VE ELIMINATED BY SOME OTHER EXPERIMENT, AND HOW THEY WORKED – TO MAKE SURE THE OTHER FELLOW CAN TELL THEY HAVE BEEN ELIMINATED. ... IN SUMMARY, THE IDEA IS TO TRY TO GIVE ALL OF THE INFORMATION TO HELP OTHERS TO JUDGE THE VALUE OF YOUR CONTRIBUTION; NOT JUST THE INFORMATION THAT LEADS TO JUDGMENT IN ONE PARTICULAR DIRECTION OR ANOTHER.

9.9 Analysis, discussion and conclusions

Here comes the analysis and interpretation of the results reported in the previous Section. It is very important to keep this Section separate from the foregoing. Compare your results with published work, and point out limitations and uncertainties of your own work (see Feynman's principle in Sect. 3 of Paper III). Always translate the observational and computational accuracy to error budgets *in the physical domain*, and avoid confusing the error of a numerical fit with the error on the result (see Paper II). Give suggestions for improving your results, mention competing explanations and contrary evidence. Your conclusions must absolutely stand out in this Section.

Concerning the reliability of your results, there are two concepts that are being confused all the time:

1. *precision*: how finely a result reproduces (statistical estimate or observable), and
2. *accuracy*: how close a result is to the true value.

Young (1994) puts it like this:

> By "precision" is meant the repeatability of a measurement, usually under fixed conditions. On the other hand, "accuracy" means the absence of error, as measured against some external standard, such as a set of standard stars."

See Figure 18 in Paper II for a graphical representation of the difference between precision and accuracy.

Whereas both concepts can be reasonably well described in statistics, observation and measurement, it is not evident to disentangle them when presenting models that include many input parameters, especially when these rely on calibrations. Take, for example, the light-curve modeling of a binary star that results in almost two dozen output orbital and stellar parameters: a vast number of these quantities are frequently reported with unrealistically high accuracies.

9.10 Acknowledgments

To always give credit and acknowledge the help of others, is a matter of scholarly courtesy. It is also a matter of honesty and fairness towards your supervisor, a colleague, or a referee. The most important point here is that acknowledgments need both be fair and complete. In short, **acknowledgment is an obligation**.

Avoid pages-long acknowledgments thanking everyone who crossed your path. And be sincere and do not exaggerate: if you had a not so pleasant experience with a supervisor or assistant-supervisor, then do not thank them with extreme zeal, but find a polite way to transfer the obligation (unrestrained flatter may even insult your other colleagues). By all means avoid phrasing it like *"The author wishes to acknowledge the work of Dr. X towards formulating the concepts and writing the first draft of the manuscript"*, or *"I would like to acknowledge the help of A, B, and C. They obviously bear no responsibility for the contents of this article"*,[50] which are confusing, or may very well be interpreted as mockery. And, when mentioning people, give their complete names (Jim, John and Jane will appreciate if their surnames are also given).

I have experienced an acknowledgment of the style *"The first author wishes to thank B for useful comments on the manuscript"*, at the same time omitting C who performed a critical reading of the paper long before B came on board. Such negligence can be

[50]These quotes were taken from real publications.

interpreted by C as if his/her contribution was of no impact at all, or that this is a case of an "all take and no give" thing. Such situations may result in crushing the collaboration.

As you are not supposed to forget a coauthor in the author line, you should also not forget to include the names of all persons who contributed to the writing of your paper. If a language professional (editor or translator) has helped you with the manuscript (especially with grammar, punctuation, and style), you should acknowledge that assistance. But there is a catch here: acknowledging language assistance implies that you follow the language editor's advice, and that you do not introduce new errors when correcting the manuscript, nor that you add any last-moment sentences that the language corrector has not seen. Acknowledging a language expert, while producing a poor product, is more of an annoying affront than a thanks.

Thank the provider of the equipment that you used (library, telescope, computer facilities, etc.), but never feel obliged to add all directors and associate directors of the place to the list of authors, as is explained in Section 13.1. Explicit mentioning of research and travel grants is mandatory, and some agencies require the use of standard expressions, and also the inclusion of a list of contract numbers.

The situation is not unlike the one in the performing arts, in particular movie productions, where the important contributors – major actors and the director of the project – are listed up front on the movie poster, while lesser contributors – technical assistants, figurants and supporting bit-players are listed in the trailer at the back. Spectators, of course, do not read every name in the trailer, and the same is true for the reader of a scientific paper who will skip the acknowledgment paragraph, but it is nevertheless important for those people who contributed that it is mentioned in which part they have been involved. If you publish a volume of text, you may add a sentence like "cover design by A, artwork by B, typeset by C, printed by D Press, published by E...".

Do not assume that acknowledgments are not read. I often do, and I find that it gives me an additional insight in the sociological aspects of science. For example, I was struck by the statement

> I ALSO WISH TO RECORD MY INDEBTEDNESS TO KEVIN PRENDERGAST FOR DISCUSSIONS (IN 1979) ABOUT ACCRETION DISK PROCESSES THAT FIRST SPARKED MY INTEREST IN THE PROBLEM.

(Shore 2009), that very well illustrates the long-term aspect of this very wide service community that science is, or should be.

9.11 Dedications

A dedication is mostly a tribute to one or more persons, and is rarely used in scientific reports. But it is quite often included in a PhD thesis manuscript (*"in memory of..."* or *"to my parents"*), or in a book. A dedication not necessarily needs to include the ground for the statement, but giving a reason helps the reader to understand the message. For example, Tukey (1977) lauds two colleagues in the following terms:

> DEDICATED TO THE MEMORY OF CHARLIE WINSOR, BIOMETRICIAN, AND EDGAR ANDERSON, BOTANIST, DATA ANALYSTS BOTH, FROM WHOM THE AUTHOR LEARNED MUCH THAT COULD NOT HAVE BEEN LEARNED ELSEWHERE.

9.12 References and citations

Reading and writing are intimately connected. Citation, from the author's point of view, is the practice of referring to the work of yourself and of other scientists. Sweetland (1989) distinguishes four additional functions of citation from the reader's standpoint:

1. *information retrieval*: citing and cited papers are related by subject;
2. *document retrieval*: through (inter)library services and downloads;
3. *scientometry*: the ranking of journals, and
4. *bibliometry*: the use of citation counts to determine a scientist's productivity.

These functions require that citations are accurate, complete and correct.

Each citation in the text must entail a reference at the end of the paper, and *vice versa*,[51] each reference must lead to at least one citation in the text. Cite and quote correctly and accurately, and only give references to papers that you have really used for the paper that you are composing. Sometimes it is not possible to give a formal reference to a published paper – for example when receiving information from a visiting colleague, or through email messages and conversations in electronic discussion groups – then the proper way of citing is the author's name and year, followed by *"personal communication"*. If you use unpublished graphics, always cite the original sources of the Figures, and add *"courtesy of... "* in the caption.

Two main systems for citing and referencing are in use: the HARVARD SYSTEM, with author names and year in the text, and references in alphabetic order at the end (like in all papers of this book), and the VANCOUVER STYLE, with author-numbers in the text with references in footnotes or numbered at the end. A&A uses the Harvard system. Each book or journal has its own house style for the half a dozen basic fields that comprise:

1. Name(s) of author(s)
2. Date (year)
3. Title of paper
4. Name of journal
5. Volume number
6. Page number (or range).

It is not mandatory to include the title of a cited journal paper,[52] but for books and reference works it is required. Note that with six fields, there are in principle 720 possible combinations, and allowing for some variance in the format of the author names (leading or trailing initials, etc.) one easily arrives at millions of possibilities! Hence, some standardization is imposed, for instance by using internationally recognized abbreviations for journals, by keeping punctuation to a minimum, and by avoiding underlined or embolded elements. As publishing space is tight, cited papers with more than four authors are usually listed in the reference lists as author1, author2, author3, *et al.*

[51] Latin phrase meaning *the other way around.*

[52] In this book, all titles of papers are given in the list of references.

The fact that different citing and referencing systems are in use in our field of science, leads to an unanticipated bonus for Editors and referees: I have been able to trace several duplicate papers, simply because the authors reprocessed the same manuscript using a different style or class file, but forgot to adapt the formatting in the reference lists!

Typos in the reference list should not occur, especially in the author names and in the bibcode-related parameters. Three frequently occurring aspects of sloppy typing can really be a nuisance to the reader, in particular for on-line papers: wrong coding or even omission of *diacritics* (see Sect. 13.15), inclusion of control characters in URL addresses,[53] and adding spaces inside, or alien characters (bracket, dot, comma, colon, ...) immediately before or after an URL.

9.13 Postscript and Appendix

A *postscript*[54] is a note or series of notes appended to an article or a book. It is reserved for a very special remark or statement (see the postscript on page 59). An *appendix* contains supplementary technical matter in tabular form, usually attached to the end of a paper or book, not to confuse with the Postscript described previously (for an example of an Appendix, see the paper by Laurent Cambrésy in Part 1 of these Proceedings).

9.14 Index and glossary

An *index* is a list of words or headings and associated page numbers, and appears at the end of a book (thesis manuscript, textbook, conference proceedings), and is not included in a research paper in a journal. One can have author index, subject index and name index, but also an object index (for example with celestial object identifiers).

A *glossary* is an alphabetical list of technical terms with definitions of those words, and is usually presented as one of the appendices at the end of a book or thesis manuscript.

10 Submitting your manuscript

10.1 The very last step: deep proofreading

Proofreading is not a natural talent or instinct, but an acquired skill, and you must learn to master it by developing an "eagle eye" for your own mistakes. There are, basically, two modes: *proofreading for content* and *proofreading for error*: the former can only be done in the reading equivalent of the listening mode of hearing (see the footnote on page 5), the latter can be done in a more relaxed semi-automatic background-processing mode .

Proofread very slowly, one word at a time, and read what is written on the page, and not what you think or remember is there. The very last double-check in proofreading

[53] URLs often contain non-ASCII characters, therefore the URL encoder converts the URL into a valid ASCII format by replacing the non-ASCII characters with % followed by two hexadecimal digits. Such control characters should not appear when citing URLs. ASCII stands for *American Standard Code for Information Interchange*.

[54] POSTSCRIPT is a trademark used for a programming language optimized for printing graphics and text, but *a postscript* – in the present context – refers to *post scriptum* (abbreviated as PS), a sentence or a paragraph added after the main body of a letter or text. It is a kind of afterword or epilog.

should be title and abstract. Always ask someone else to check the title page of your thesis manuscript, and especially check for omission of blank spaces and the wrong use of the hyphen, because these are the most painful mistakes to live with. Anyone experienced in proofreading knows that it is much more difficult to detect errors during on-screen reading than when working with a printout. Basically, there are various errors to be corrected while doing the final proofreading, such as:

1. mixture of American and British style (like analyse, analyze);
2. typos and mistakes of things you know, but do not see;
3. language errors unknown to you, and
4. style and syntax errors.

The first three kinds of errors can be fixed with a spelling checker, but the last variety is more tricky, especially since a mistyped word or verb may be very well correct in a different context.

After reading over your manuscript several times, you will somehow have become blind to errors, because you subconsciously dictate rather than read. Ask a close friend or a colleague who is not involved in your work, to proofread your entire manuscript. It is good practice to look for a *sparring partner*[55], and this is how I operate for some of my manuscripts (see the acknowledgment at the end of Paper III, and in the Preface of this book).

There are several beliefs about when to stop proofreading. One is the fatal *"no matter how many times you read the manuscript, there will always be undetected errors, hence proofreading is useless"*. The other is what I overheard my own supervisor saying to a novice student *"we have done enough now, and any remaining errors will be flagged by the referee"*. Both attitudes abuse the system.

You should always be aware that, when all the experimental and analytical work is done, and when the manuscript has been circulated umpteen times to all coauthors, one must stop at some point, and refrain from going on forever changing what has been changed several times already. This phenomenon is quite typical for a situation with "too many bosses", *i.e.*, the student having to take to heart the advice of too many senior coauthors. It is good to be aware of the so-called *Law of Diminishing Returns*: a general law of economics proposed by Malthus (1798)[56], which states that if one factor of production is increased (in our case the writing) while the others remain constant (the data and analysis), the overall returns will decrease relatively after a certain point. It is a matter of training and skill to find the optimal point where to stop the adjustments in your writing.

Undetected typos are embarrassing, but they are disastrous if they change the content of your message. As an example, I quote from the paper that reports the main chapter of my own PhD manuscript, Sterken (1977): *"... similar short-term variations such as those seen in HD 160529 and eventually in HD 168607 have <u>never</u> been found in α Cygni"*, which states the opposite of the fact expressed in the preceding sentence. What happened? The submitted typescript correctly stated that such variations have **ever** been

[55] *Sparring partner*: any person with whom a professional boxer practices while training.

[56] Thomas Robert Malthus (1766–1834) was a British economist and demographer.

seen, but the typesetter[57] made a typographical mistake, and the proofreader (myself) did not spot the error, and gave his *imprimatur*.[58] The message can somehow be correctly understood from the context in the same paragraph, but citations out of context will convey an unintended opposite view.

Let me conclude this Section with an example from my editorial practice. I recently completed a conference Proceedings Volume together with two co-editors (Sterken *et al.* 2010). Both colleagues were, like me, non-native English speakers, and we set up a system of proofreading in three passes: I would first correct incoming manuscripts for typos, technical mistakes and obvious syntax errors. Then my colleagues would annotate the corrected manuscripts with their own remarks, and send them back to me. The experience showed that in about two thirds of the cases, each of us would spot the most evident language errors, and suggest a same correction. But in the remaining cases, it quite often happened that we had two or even three different interpretations of "what the author possibly wanted to say", and we then corrected the text in the most logic direction, knowing very well that a fourth party – the reader – may perhaps come to even another understanding. Do not forget that a similar situation may occur with the language editor, who in general is not a trained scientist, see Section 13.21 (and Sect. 10.2 in Paper III) for more details.

10.2 Etiquette and netiquette

Etiquette is a set of rules of conduct to be observed in social or official life, and *netiquette* is the etiquette of cyberspace. Whereas etiquette, as a code of conduct, may appear obsolete these days, business companies offering almost identical products at cutting-edge prices, find out that good manners may make the difference. The same is true when dealing with referees and Editors: sloppy conduct and bad manners may harm your cause.

The first rule is: always respond or reply to Editors and referees, also when your paper is being refused. Even sending a simple message acknowledging their mail (for example saying that you will revise the manuscript and submit again) is much better than not replying at all. By all means, avoid being rude: remember that your words are written, and that they can come back to haunt you.

The second rule is to respect the Editor's time and bandwidth by avoiding voluminous mails and unnecessary attachments. I had one case where an author sent me the inutile `.dvi` file, returned the `.sty` file, and even included Windows system files with the submitted archive. And, never submit a LaTeX file that you cannot compile, never tell the Editor that *"it correctly compiles after you hit* `return` *when it halts at an error message"*, because no Publisher will ever accept such a thing. Nor use home-made macros or local typesetting packages that are not available at the commercial printer's office.

Referees and Editors are working scientists, and have their own research deadlines and teaching duties. It is a matter of education and courtesy to submit to these people manuscripts that have received the highest degree of author attention as possible. It

[57] In those days, manuscripts were outsourced and sent overseas for typesetting.

[58] *Imprimatur* (from Latin) means consent to print.

does, of course, happen that referees let pass a poor paper. But you should by no means draw wrong conclusions when seeing such a paper in a high-impact journal: mirror your manuscript to the best papers in the journal, and do not submit a poor product because another author has succeeded in getting something published that is below standard.[59]

When a referee's report contributes to a significant improvement of the manuscript, or when the referee flagged a capital mistake, it is good practice to thank the referee (see also the last paragraph of Sect. 13.1, and Sect. 13.10). But do not thank the anonymous referee for *his* help unless you are sure that the referee is male (see Sect. 11.7).

11 What to avoid at all price

11.1 Writing things that you do not understand

Never add sentences that you do not fully understand (one of the pitfalls of the *copy and paste* facility). This is especially true for Latin abbreviations, like *et al.* that stands for the Latin *et alii*, and the very confusing *i.e.* and *e.g.*: the former stands for *id est* ("that is", or "in other words") whereas the latter means *exempli gratia*, "for example". There also is *viz.* – an abbreviation of the contraction from *vidēre licet* – which means "namely", used for clarifying something, and often is accompanied by a colon when the following statements are in a list.

Foreign terms (*Gedankenexperiment*) and other Latin words (*minus, versus, in extenso, ad hoc, a fortiori, etcetera, caveat, ...*) should, in principle, be written in italics. Some Publishers adopt the editorial convention of not italicizing the phrase "et al.".

11.2 Drowning in acronyms

Science uses a technical terminology that allows for significant gain in communication efficiency by coding concepts through the use of acronyms. Space science jargon, in particular, is peppered with acronyms for space vehicles and parts thereof. Try to avoid inventing acronyms, and if you really feel that an acronym should be used, then first count the number of times the acronym appears in your text: the introduction of a new acronym does not pay off if it is only used two or three times. Make also sure that the acronym you propose has no political or obscene meaning, and when using an acronym for the first time, you also give its definition. On page 14 the case is mentioned of two quite common acronyms, but very confusing situations can occur, for example GCPD (*General Catalogue of Photometric Data*)[60] is known by movie adepts as the fictional *Gotham City Police Department*.[61] This example may seem absurd, but it is nonsensical **only** because we know the context.

Never use acronyms that you do not understand, for example some observers may not know what a PMT (Photomultiplier Tube) is, nevertheless the acronym is used. Do not add a genitive to acronyms: "WET's results" is better written as "the results obtained by

[59] See the remark on verification and validation in Section 4 of Paper III.

[60] `http://obswww.unige.ch/gcpd/gcpd.html`

[61] Gotham City in the Batman books published by DC Comics.

WET".[62] As said already, no acronyms should appear in the abstract. Scientific terminology is unavoidable, but unnecessary jargon and complicated non-scientific terminology only slow down the reading process.

11.3 Mixing American & British styles

A paper or a book should use either American or British style. British words ending on *-our, ise, -logue, etc.* are spelled differently in American (color, analyze, catalog, etc.) – but look out for the lack of logics, see the last two lines in Table 1 (*i.e.* devise, revise, scrutinize, utilize). Note also that the Americans write *acknowledgment* for *acknowledgement*, and *cesium* for *caesium*. See also the paper by Adams *et al.* in Part 1 of these Proceedings, and a very useful list of accepted spellings (following government guidelines and major dictionaries) for UK & Ireland, South Africa, New Zealand, Australia, Canada and the US compiled by WIKIPEDIA.[63]

11.4 Clichés

A *cliché* is something that has become a commonplace. The term *asteroseismology*, for example, has become such an attractive term to use, that it appears in so many introductions to papers that have very little to do with the seismology of stars. Quite often an author has determined only a couple of frequencies – not even a complete frequency spectrum – but still wants to sell the product by opening the paper with the magic word: there is just no need for such marketing using a product label that does not cover the content.

11.5 Inconsistent capitalization

Some words may have a different meaning when capitalized: the Moon is our moon, whereas moon refers to any natural satellite, and Galaxy refers to our galaxy. A figure can be a printed character or a numerical value, whereas a Figure is a graphical representation. A table is a piece of furniture, a Table is a systematic arrangement of data in rows and columns. A Section is a group of paragraphs, a section could be a part or division, a team, or an incision, or a term from geometry – such as volume, whereas Volume refers to division of a book or a journal. In this paper I capitalize Editor and Publisher when I refer to the Editor or the Publisher[64] of a book or a journal, but when not capitalized, the terms refer to editors and publishers at large. Adjectives (like "editorial") are never capitalized.

11.6 Footnotes and unnecessary emphasis

Footnotes interrupt the reading process, and render a very poor page layout in Sections with lots of mathematical formulae. Footnotes frequently occur in review papers, and

[62] *Whole Earth Telescope.*

[63] `http://en.wikipedia.org/wiki/Wikipedia:Manual_of_Style`

[64] Capitalization, in this case, has also become a symbol of respect.

the full-page layout is better adapted to footnoting than does the two-column layout in a research journal. This book contains numerous footnotes, in particular for the many URL addresses. Keeping consistency in punctuation and capitalization in footnotes is not always evident: in Papers I–III, footnotes containing a complete sentence are closed with a period, except when the last item in the sentence is a web address. Short explanatory statements are not concluded with a period, nor do they start with a majuscule.

Some authors too often use boldface or italics to give some words and sentences particular prominence. Over-emphasizing simply works the opposite way – though for tutoring (like in this book), over-emphasis is often done on purpose. Pay attention not to use superfluous words, for example in "existing calibrations": there is no point in emphasizing "existing" since non-existing calibrations simply do not exist.

11.7 Sexism

Sexism is the collection of attitudes that foster stereotypes of social roles based on sex. Traditionally, the third personal masculine pronoun "his" was used to refer to masculine and dual-gender nouns (like astronomer). In modern usage this is considered sexist and therefore "his/her" is sometimes recommended to avoid sexism. But this is awkward, especially when used repeatedly in a text, it draws attention away from the real message, and also overloads the parser. One solution is to make the sentence plural. While it is possible to draft a paper and avoid a single occurrence of such sexist elements of expression, one should be careful when reading older papers and texts from times when such concerns were not raised. In my papers I have tried to avoid the problem, but it was not always possible to unwind any sex-based differences in communication style. But all references to gender in given examples or cases, refer to genuine situations.

Take the following example. When queuing at a Brussels-airport security gate, I saw a huge cartoon with the quote *By believing in his dreams, a man turns them into reality*, which is an homage from Hergé[65] to Neil Armstrong and the Apollo 11 crew. Struck by what would now be seen as sexist language, I consulted the original French text: *A force de croire en ses rêves, l'homme en fait une réalité*. Here, *l'homme* – man – was poorly translated to "a man". This vividly illustrates the danger of inadvertent misquoting in translation, see also page 18 and Section 11.9. And this, again, demonstrates that writing is a process: I must have spotted this phrase dozens of times without noticing the point I am making now. So why, exactly on my way to that particular meeting, was I struck by this imperfection? The answer simply is: because I was composing my paper in the weeks preceding my travel – even during my trip to the meeting covered in Sterken & Aerts (2006).

11.8 Non-standard nomenclature

When previously-defined designations of celestial objects are used in your paper, they should never be altered (*e.g.*, neither truncated nor shortened). Do not invent your private

[65] Pseudonym of Belgian artist Georges Remi (1907–1983), spiritual father of *Tintin*.

celestial-object name, while referring to the original consolidated nomenclature as "other designation": this creates confusion when other authors take over your naming scheme. Be honest and do not say that you discovered 23 new variable stars when half a dozen of them were already known and labeled before you started the study, and you simply renamed them! The original bibliographical reference for a designation should be given in the first place. Laurent Cambrésy, in his paper on archives (in Part 1 of these Proceedings), stresses the importance of this guideline, and I like to refer to a personal communication by Dubois (1995), a colleague of Cambrésy at CDS:

> CONGRATULATIONS ON YOUR PREOCCUPATION WITH THE PROBLEM OF DESIGNATIONS. IT IS A MATTER THAT IS TOO OFTEN IGNORED AND THAT CAUSES A LOT OF PROBLEMS. MY REACTION IS THE SAME AS YOURS: IF THE STARS HAVE ALREADY BEEN NAMED, IT IS PREFERABLE TO USE THE ARCHIVAL NAME . . . [66]

11.9 Misquoting, misciting and quoting out of context

Misquoting and misciting occur at several levels, and most incorrect quoting and citing are due to inaccurate reading and sloppiness. The following common errors occur:

- simple typographical errors in any of the fields listed in Section 9.12;
- quoting or citing without reading, or even having seen the source paper;
- verbatim copying references from a list without checking the source paper;
- misquotation by changing content;
- not citing at all, which definitely is the worst form of misciting;
- changing the order of names when citing a reference, in particular upgrading the position of your own name, and
- the (rarely occurring) omission of references in order to avert hampering the author.

The last point refers to tacit omission of faulty papers, and only once I found a direct reference to such an act in Bottle (1973): *"Reference omitted to avoid embarrassing its author"*, an action that of course results in just the opposite effect. Deliberate misquoting is considered as scientific misconduct, and is discussed in Section 7.8 of Paper III.

Typos are a nuisance, of course, but confusing abbreviations for journal titles are also to be avoided. For example, "Ann. Phys." stands for *Annales de Physique*,[67] but also for *Annalen der Physik*.[68] The A&A instructions list three journals starting with A&A in Appendix D.[69] Another point of confusion, now becoming extinct, is the habit of

[66] *Félicitations pour vous préoccuper du problème des désignations. C'est une question trop souvent ignorée et qui provoque bien des difficultés par la suite. J'ai la même reaction que vous : si les étoiles ont déjà eté citées il est bien préférable d'utiliser les dénominations antérieures . . .*

[67] One of the EDP Sciences journals.

[68] Wiley InterScience.

[69] Simplified abbreviations of frequently used journals (`http://www.aanda.org/images/stories/doc/aaguide_v7.0_201001.pdf`).

replacing the name of the author or book by the Latin expression *ibid* or *op. cit.*[70] These abbreviations should only be used when a reference source is referred to several times in a short space – remember, memorizing several ibidems will contribute to overloading the reading parser.

A frequent error occurs as a consequence of downloading, when the correct URL is quoted, but when the reference to journal and page number is wrong, or is omitted. For example, Sterken (2006) is cited by Stein (2006) with a correct URL, but the paper is not attributed to its author, but to the two Editors of the Proceedings book (*i.e.* Sterken & Aerts 2006), and Stein (2006) fails to give the correct reference. The principal consequence is that the erroneous citation is copied into another paper without the author/copier ever having seen or read the original document, hence perpetuating the error in a cascade of papers. This, unavoidably, biases bibliometrics (see Sect. 18 of Paper III).

The most famous example of citing without having seen the original – over more than half a century – involves a paper in the medical literature entitled *O úplavici; Předběžné sdělení*[71] written by the Czechoslovak author Jaroslav Hlava in 1887. In the same year, the *Zentralblatt für Bakteriologie und Parasitenkunde* states[72] "Uplavici makes the interesting communication...", as such creating a fictitious author "Uplavici, O." and truncating the article title to its second part *Předběžné sdělení*. The mistake led to numerous citations of the work of "Dr. Dysentery", and after exactly 50 years the situation was remediated by Dobell (1939), who lists all varieties of this author's name: O. Hlava, Hlava Uplavici, and even the two "persons" Hlava and Uplavici.

Misquotations can survive for centuries, take for example *e pur si move*: these are the words which Galileo Galilei is said to have uttered on the 22nd of June 1633, after abjuring the heliocentric doctrine. The origin of this legend is hard to trace, and many variants of the expression are found in the literature: *e pur si muove*,[73] *eppure si muove, eppur si move, eppur si muove*, and so on. Berthold (1897) traces the first written version "eppur si move" to a publication in 1757 by Giuseppe Baretti (1719–1789), but there is no solid trace to the original statement, if it was uttered at all. As Galileo was Pisan, and the Italian language – in particular the literary language – was basically a form of Tuscan, it was not unlikely that the words reportedly uttered may actually have been *"eppur si muove"*. Galileo himself, in his essays, was a very good stylist and in any case the distance between his written language and his spoken was unlikely to be great (Lepschy 2006). Galileo's alleged uttering is most probably as much of a fiction as is the fable about the EUREKA of Archimedes.

Another source of miscites are the so-called textbook errors, which often are repeated with little or no opportunity for correction. One example, pointed out by Wróblewski (1985), is the incorrect description of Ole Rømer's work on the velocity of light in many physics texts and books dealing with the history of science. Figure 3 illustrates the extent of this misinformation: though Rømer never gave any numerical value for the speed of light, dozens of authors attribute a numerical value to him. About half of the values hover

[70] *ibidem*: in the same place; *opus citatum*: the work cited.

[71] *On dysentery; a preliminary communication.*

[72] "Uplavici macht nun jetzt die interessante Mittheilung..." 1887, Jena verlag, p. 538.

[73] And yet it moves.

around 220 000 km s^{-1}, the remaining ones are in the range 300 000–350 000 km s^{-1} (see Paper II for further discussions on this bimodal distribution), and this is due to the calculation being extremely sensitive to the value used for the Astronomical Unit, a quantity that was quite poorly known in Rømer's times.

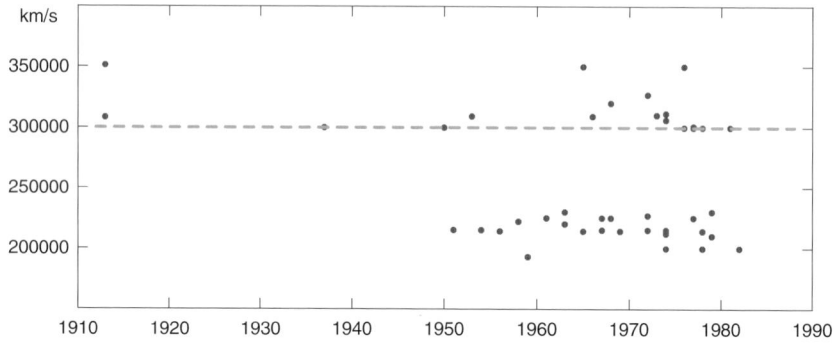

Fig. 3. Numerical value for the speed of light quoted in history of science and physics texts and books as "derived by Rømer". The dashed line represents the physical constant c. Source: Sterken (2007).

Misquoting by changing content occurs when the original content is modified, truncated or cited out of context. Note that even changing the font in part of a sentence can change the meaning. If you really wish to quote, and also emphasize part of the quotation, you must specify (for example in a footnote) "italics mine" or something similar.

A very special case of misquoting by changing content is when the original quote comprises an error or misspelling: in such a case one should not merely correct the error, but reproduce the passage *verbatim*, and insert the Latin word *sic* in square brackets right after the error to which it refers. *Sic* means "as such", and helps the reader's parser out of the confusion about who is wrong: the quoter or the quoted.

Mind, though, that the application domain is much wider than typographical errors. Take, for example, the following sentence from *Lady Chatterley's Lover* (D. H. Lawrence's controversial novel written in 1928): *"Why is the star Jupiter bigger than the star Neptune?"* From an astronomer's viewpoint, and out of context, this statement is wrong, since it refers to planets. But from the author's perspective it is not necessary inappropriate, since the sentence is uttered by Sir Clifford Chatterley, and obviously refers to the celestial lights Jupiter and Neptune, where "bigger" certainly denotes perceived brightness. In addition, the mindset of the present reader is totally different from the one of the writer nearly one century ago. Here, *sic* would be of good use, since quoting out of historical context also is a mode of misquoting. Overuse of *sic* should be avoided: for example, on page 25 the spelling of "Colours" is different from the way it appears in the rest of this book, yet this divergence needed not to be flagged with *sic*.

Inaccurate quotations and citations, whether intentional or not, simply increase the noise in the information transmission system (see Sect. 1 in Paper II), and cheat the citation rankings. This also occurs when citing obsolete papers: avoid citing or attacking positions that are no longer defended.

12 Instructions and guidelines

This Section deals with instructions, and focusses specifically on submitting a manuscript to an Editor of a conference proceedings book, but some of these instructions and guidelines are also useful to authors of journal research papers.

There are two sets of guidelines: Publisher instructions, and Editor instructions. The first set of rules is described in detail in the Publisher's *Instructions for Authors*[74] that define the "house style" of a journal. Respecting these rules is just mandatory, for journals as well as for Proceedings books.

12.1 Editorial instructions

Editorial instructions are normally provided by the Editor(s) of a book. One of the major mishaps when editing books or conference Proceedings occurs because of authors' poor understanding of the system of references, in particular the `thebibliography` environment in LaTeX. Though the use of this environment allows most versatile handling of citations and references, many authors do not realize that formatting errors in the reference list can propagate through a complete book and, consequently, destroy the coherence of the bibliographic citations and pagecites. Even a manuscript that correctly compiles as a stand-alone paper, may render another paper's citations chaotic and useless. A second nuisance is when an author does not respect margins, especially with overflowing running titles, URL addresses, Figures, Tables, and their captions.

Editors of a book or conference proceedings must practice very systematic and tight working procedures.

- Editors frequently deal with over a hundred LaTeX files, and perhaps several hundred graphics files, when compiling a single book, hence authors are advised to name their LaTeX and encapsulated postscript files in a systematic way, such as `surname.tex`, `surname-fig1.eps`, `surname-fig2.eps`, etc. For example, for Sterken & Aerts (2006), I received several `firstname.tex` files – an impossible task considering the duplication of half a dozen first names in the audience. It is also not uncommon that an Editor covering a meeting in, say, Brussels, receives multiple files `brussels.tex`, `fig1.eps`, `fig2.eps`, `etc.`: in such cases all these files have to be renamed in order to create a workable and efficient compilation environment, and this process takes time and leads to errors.

- The confusion between Encapsulated Postscript (`.eps`) and Postscript (`.ps`) formats, which are standard (ASCII) formats for importing and exporting PostScript language files in all environments (Section 5.4 in Paper II). The file must include a header comment starting with `%!PS-Adobe-3.0 EPSF...`, and a `%%BoundingBox` comment that describes the bounds of the illustration. The `.eps` file can contain any combination of text, graphs, and images. An `.eps` file is the same as any other PostScript language page description, with some restrictions, and is usually a single page. Some authors label their Figure files with the `.ps` extension. While not

[74] For A&A, see the author's guide on `http://www.aanda.org/content/view/136/184/lang,en/`

basically wrong, this habit creates confusion, as the `.ps` extension is generally used for multipage documents (a complete paper or book). Mixing `.eps` and `.ps` extensions for one-page documents to be inserted in a TEX manuscript certainly does not help journal and production Editors keeping a tidy job.

- If page limits are imposed, they should be respected. In practice, allowing even a one-line overshot unavoidably results in two excess pages, as papers quite often start on odd pages. Authors should be aware that also Editors have page limits, in the sense that extra pages must be paid for from their budgets.

- Some authors submit subsequent versions of their LATEX file after their first submission already underwent considerable editorial labor. It is just not possible to track down such changes, nor is sending back a source file to the author a workable option, since the Editor may lose all control over the changes made previously. Moreover, Editors often simultaneously modify multiple TEX files for adding index entries. It requires the editorial experience of just one book to know that not every author understands how to efficaciously contribute to the editorial process.

- Late (or no) submission is one of the most irritant problems, as it has a direct impact on the editorial planning. Some excuses are very common: disk crash, laptop stolen, etc. The worst justification for a really long delay I ever received was *"I am waiting for the permission to include one more illustration"*. The most pleasant statement I welcomed was *"I apologize for the delay in sending the files (for which there is no real excuse)"*. Proceedings book editing is always done along with the complete load of other work like observing, teaching, data analysis, meetings, vacation, family business, etc., and a delay of a couple of weeks can make a hell of a difference in an Editor's planning. And if the manuscript is never submitted, there is not much an Editor can do, except boldly publish the title of the talk with a statement in the way Schielicke (1998) did:

  > TO THE REGRET OF THE MEMBERS OF THE BOARD OF THE ASTRONOMISCHE GESELLSCHAFT, OUR 1997 AWARDEE OF THE KARL SCHWARZSCHILD MEDAL WAS NOT IN A POSITION TO PREPARE THE MANUSCRIPT OF HIS LECTURE, AS WAS AGREED BEFORE.

Let it be clear that Publisher instructions always override Editor's guidelines, and that these conventions are not just a Publisher craze to take the fun out of your work, but are given to ensure cost-effective streamlining of the publishing process towards a coherent and handsome publication in the house style of the Publisher. When I was editing a conference Proceedings book, one author wrote me that he refuses to use the `\section` command, because it spoils space, and he replaced the command by one of his own making, rendering his paper different from all others in the book. Another author just submitted an MSword file including mathematics, expecting that the Editor would do the homework. Needless to say, both approaches were totally unacceptable.

Publishing has changed dramatically in the past decade: simply compare a conference book of the 1980s with a modern one. Many of the older books have a mixture of typefaces and fonts,[75] variable line spacing and variable margins, and a most heterogeneous collection of reference formats. Modern Proceedings books are aesthetically more pleasing, and more efficiently organized, in the first place because of tight instructions.

After all, accurate rules also prevail in electronic banking (accuracy of amounts and accounts), observing proposals (coherence and deadline), grant applications (deadline, structure and logic), even placing an order for your favorite music at a webstore implies abiding to strict procedures.

12.2 Copyrights

Copyright deals with the exclusive legal right to reproduce in publications, and is usually transferred to the Publisher through a *Copyright Assignment*. Most scientific authors – young and old – just blindly sign such a copyright-transfer contract, and many a participant is rather surprised when confronted with the content of such a form, which is mostly quite cleverly worded. Take as an example the following paragraph in the Copyright Assignment from the *Astronomical Society of the Pacific* (ASP) addressed to you, the author:

> ... BECAUSE YOU DESIRE TO HAVE THIS PAPER SO PUBLISHED, YOU GRANT AND ASSIGN THE ENTIRE COPYRIGHT FOR THIS PAPER EXCLUSIVELY TO THE ASP. THE COPYRIGHT CONSISTS OF ALL RIGHTS PROTECTED BY THE ORIGINAL COPYRIGHT LAWS OF THE US AND OF ALL FOREIGN COUNTRIES, IN ALL LANGUAGES AND FORMS OF COMMUNICATION.

This means that you transfer your entire copyright to the Publisher. In this particular case, you retain the permission to reproduce your work elsewhere, provided you give appropriate credit. Note that some copyright assignments leave you with much less liberty than this one. Beware that you cannot legally sign two Copyright Assignments and submit the same manuscript to two copyright-demanding Publishers, see Section 7.12 in Paper III.

Almost all book Publishers take away your copyrights, and very few journals leave copyrights to you (the copyright holder for the papers published in A&A is not the journal, but the European Southern Observatory, ESO).

Evidently, the financial consequences augment with the commercial value of the material that you copy: just be prepared to fight lawyers if you breach copyrights with a well-established Publisher, museum, or library. Let me give you one practical example. In Sterken (2005), I reproduced three graphics: a color reproduction of two portrait paintings on the cover, and one black-and-white reproduction of a comic strip inside the book. Permission to reproduce the portrait of Ole Rømer was granted free of charge by the Rundetårn Museum in Copenhagen: hence I reproduced it inside the book (in black-and-white), and also on the cover of the book (in color). Permission to reproduce the portrait of Giovanni Domenico Cassini was granted by Observatoire de Paris for a fee of less than 50 Euro per reproduction, hence I reproduced it on the cover only. Permission for a single black-and-white reproduction of a *Tintin* comic strip was granted for an almost ten times higher fee, and several well-specified restrictions were imposed (no modifications

[75] Typeface and font are explained in Section 7.1 of Paper II.

of text balloons, for example). Likewise, the permission to reproduce the photograph of the tapestry on the front cover of this book, was purchased from the *Röhsska Museum of Fashion, Design and Decorative Arts* (Göteborg, Sweden) by Publisher EDP.

13 FAQs about the editorial process

13.1 Who should be coauthor?

Appropriate authorship assignment is of great significance and value to the scientist, because it contributes to the building of reputation, to the assignment of responsibility, and to recognition of achievement. In addition, correct authorship is crucial for the scientific grant administration, because authorship and citation are observables that contribute to the scientific rating of a project or of a person.

It is well-known that senior scientists do influence decisions on authorship, and that some senior scientists sign for authorship at the expense of junior researchers. And it is also known that senior authors have moved to first authorship at the cost of others, that authorship is granted to chairpersons of departments as a matter of convention, or that senior authors are listed just to boost the paper, see Drenth (1998) and references therein, and also Sections 7.3–7.5 in Paper III.

The *Vancouver guidelines*[76] attempt to define the contributions that allow coauthorship, and suggest that authorship should be based **only** on substantial contributions to conception and design – or analysis and interpretation – of data, **and** profound involvement in the publishing process. A very important element – for us, astronomers – is subsequently formulated in the so-called *Copenhagen compromise*,[77] which adds that **the acquisition of data merits authorship – provided that those involved in the acquisition of data also meet the other criteria for authorship**. So, the guideline is that authorship credit should be based solely on substantial contributions to:

1. **CONCEPTION AND DESIGN, OR ACQUISITION OF DATA, OR ANALYSIS AND INTERPRETATION OF DATA,**
2. **DRAFTING THE ARTICLE OR REVISING IT CRITICALLY FOR IMPORTANT INTELLECTUAL CONTENT, AND**
3. **FINAL APPROVAL OF THE VERSION TO BE PUBLISHED.**

All three conditions must be met.

These guidelines reject fund raising, and general supervision of the research group as justifications for authorship, and also the bare collection of raw data – though these contributions should be acknowledged. The guidelines do, however, make it clear that between them the authors must take responsibility for all aspects of the work: **any part of an article critical to its main conclusions must be the responsibility of at least one**

[76]The *Vancouver Group* refers to the *International Committee of Medical Journal Editors (ICMJE).*
[77]ICMJE, May 2000 meeting, `http://www.icmje.org/urm_main.html`

author. This brings up the question of how to correctly evaluate papers coauthored by literally dozens – if not hundreds – of authors, especially in the event of large projects that manage tight deadlines. In such cases it is, alas, quite common practice to enwrap a complete team in authorship, with barely enough time for each envelope member to critically respond to draft versions of the manuscript. Such practices not infrequently lead to complications, see Paper III.

The *Vancouver guidelines* are, in fact, already implied in service-observing contributions, as major observatories request an acknowledgment of the service-observing share of the work (for example, phrased as *"the authors thank N for doing a wonderful job in organizing the service observations"*). But in practice it does happen that service observers only hear *a posteriori* that a manuscript with their name on has been submitted, so they do not even get a chance to respond nor to contribute to the writing.

A question I am often asked is whether the use of a computer code written by someone else is a contribution that merits authorship. I see such a case as an analog to service observing: if the code was not specially written for a specific paper, and if the writer of the computer code does not participate in any of the other aspects of the analysis and the writing of the paper, then that person should not be a coauthor, but should be mentioned in the acknowledgments instead.

Both types of examples imply no involvement in analysis or interpretation, nor in writing the paper: they thus fail to meet any criterion of responsibility or accountability. Many a service observer or computer programmer, who involves as coauthor without participation in the data analysis or in the writing process, has later on been confronted with the results of inappropriate handling of the raw data, or with improper application of a computer program.

In summary, most scientific papers have several contributors. Some of these people are to be listed as authors, provided that some requirements are met. The remaining contributors (language editors, referees, supervisors, service observers, data archivists, librarians, ...) are to be mentioned in the acknowledgments paragraph at the end of the paper, see also Section 13.10.

13.2 What is honorary authorship?

Honorary authorship is authorship given to colleagues who did not contribute to the research or to the communication of the results, and is akin to *gift authorship*: granting authorship to people who did not make any intellectual effort. This kind of present is often mechanically assigned to senior colleagues since long retired, but the mechanism is also applied for profit by borrowing the name of a celebrity. Honorary and gift authorships not only are demotivating for those who did the real work, such arrangements can be quite risky for the recipients. Section 13.3 states that the first author, in general, bears the responsibility for the complete work. But if there are senior honorary coauthors in the list, and if the paper is rejected because it does not fit the scope of the journal, then the first-author responsibility rule leads to a paradox, since the seniors should have stopped the submission to an unsuitable journal in the first place.

Avoiding inclusion of gratuitous coauthors, evidently, also applies to yourself on other people's papers: exclude yourself if you have not contributed to a paper for which you are invited to sign as coauthor – although it must be recognized that the rising trend of papers with dozens of authors does not make such decision easy. Small teams do sometimes work with a kind of reciprocity, *i.e.*, mutual exchange of participation in each other's papers. This is a habit to avoid because you may end up disappointed (reciprocity is not always guaranteed and is often forgotten), and you may even be blamed for grave errors. For more details on related issues in the context of multisite observing campaigns, I refer to Sterken (1988).

13.3 How to settle the order of author names?

In some fields of science, authors are always listed in alphabetic order,[78] but in astronomy author names are mostly arranged in order of decreasing contribution to the work. A solution that I applied during quite some time in my early postdoc years, was to decide the author sequence at the very end of the writing phase, just before submission. But that ideal does not fit every team, and turned out to work only when no more than two or three authors were involved. Even two-author teams have shown to run in trouble when it comes to deciding who comes first.

Remember that the first author, in general, bears responsibility for the work. The perceived importance of the last author is more difficult to evaluate: it can be seen as "the last is the least", or that the last name is the supervisor (with possibly the same negative appreciation), or that the last author can be an important senior contributor who insists on that position.

The main point is that the code used by the authors in constructing the hierarchical order is only known to the authors, and not available to the reader or Editor (in fact, the code is relatively undefined). Therefore, more and more journals (especially in biomedical sciences, but also *Nature*) require or encourage the authors to describe their specific contributions to the paper they submit. The acknowledgments paragraph at the end of the paper is the ideal place to explain what each author has contributed.

Sheskin (2006) proposes the *analytic hierarchy process*, a method to enable coauthors to determine their fractional contributions to a multi-authored paper. The procedure consists of dividing the paper into primary and secondary elements – for example, "Methods" in the IMRAD scheme would be a primary element, comprising light curves, frequency analysis, spectra, ... as secondary elements. Each author's contribution to each part can then be readily assessed, and the relative merit can be calculated. The implementation of this method remains to be seen.

One should always discuss authorship and author ranking in an open way: if you think you should be the first author, then start a dialogue with the others, and give your arguments. **Never** change author sequence on your own initiative only: dialogue is the only solution if you do not wish to permanently damage your relationship with colleagues with whom you collaborate (see the case discussed in Sect. 8.7 of Paper III).

[78] In mathematics, for example.

13.4 What if a coauthor passes away?

It does happen that a coauthor – even the first author – dies before a manuscript is composed or submitted. If that person has contributed in a substantial way to the project, then one should consider including that person in the list of authors. The proper way to do this is to mention the complete affiliation of the place where the deceased was working while committed to the project, and add in a footnote a statement expressing that this coauthor has passed away on a specific date.

13.5 Can the list of authors change during the review process?

Yes, it happens, but it should be avoided. It can occur that a listed author asks to be removed, but such an action mostly points towards shortcomings during the preparation phase of the manuscript. There are few good reasons to add names after submission, for instance it can happen that a referee contributes so much to the content of a paper, that the author(s) may decide to offer coauthorship to the referee. Extending coauthorship to the Editor of a journal, however, is considered as inappropriate (Hames 2011).

13.6 Who submits the manuscript to the Editor?

The *Corresponding Author* is the person who submits the paper for publication, and who is responsible for the manuscript as it moves through the editorial line. This person, most often, is the first author, and manages all correspondence with the editorial office, and coordinates the communication with the referee(s) through the manuscript management system, using the author identification provided by the journal. In addition, this author releases the proofs, and deals with the administrative aspect of organizing the payment of page charges. In my opinion, it is crucial that a PhD student – if that person is first author – also becomes corresponding author.

The BMJ Group, who publishes *The British Medical Journal*, proposed the concept of *guarantor*: the author who is prepared to take responsibility for the whole article. This system is not used in any of the astronomy journals I know of.

My experience as an author and as a referee is that quite a number of journal Editors do not engage in direct contact with their authors and referees, and most communication is taken care of by an editorial assistant,[79] at least as long as there are no disputes or conflicts. I must say that, as a referee, I have always enjoyed the direct contacts with Editors of journals who still nurture the principle of direct communication with their referees and authors.

13.7 Must I wait for a coauthor's consent to submit?

Yes, you must wait till all authors have read and agreed on the last version before it can be submitted. Likewise, coauthors have the obligation to verify and validate not only the

[79] Editorial assistants, in most cases, are not scientists.

submitted manuscript, but also the revised version. I illustrate the hassle with three cases in which I got involved.

1. In the case of a two-author manuscript where I was second, one of my students submitted to a journal without telling me, nor even showing me the last version of the manuscript. The referee required that the paper be completely rewritten, which led to a major delay and even withdrawal in the end.

2. One paper, with six authors, spent many months in the reviewing process, with multiple passes through the referee's hands. The final version had undergone a major facelift, including corrections and removal of erroneous parts, so I requested the first author to contact every coauthor in order to obtain their consent before submitting the final version. Four of the coauthors had not followed the evolution of the manuscript during its revisions – which included fundamental changes – yet they merely consented without any discussion whatsoever. One coauthor never even replied.

3. As Editor of a Proceedings book, I once received a paper signed by three authors. Not only was the page limit transgressed, the syntax also revealed that the native-English speaking coauthors could not possibly have seen the submitted manuscript. So, I copied my reply to all coauthors, asking to shorten the paper and to language-edit the manuscript. The second author complied, and submitted a revised version, after which the first author requested that the changes be undone. This way of working is most inefficient, steals time from all parties, and spoils the professional relationship between all persons involved.

You may not realize how easy it is for the reader (and referee and Editor) to tell such state of things apart. Submitting a manuscript before all coauthors have approved is bound to lead to problems, creates a very embarrassing situation for some coauthors, and may even produce the vision that the supervisor does not even bother to read the students' manuscripts before they are submitted. **Journals require that the submitter implicitly or explicitly declares that all authors have seen and agree on the manuscript that is submitted**, so when one or more authors do not confirm their consent, the manuscript should simply not be submitted.

The *Vancouver guidelines* clearly state that drafting the article or revising it, is a basic **duty** of every coauthor. But the **right** to at least see, read and approve any submitted version is a fundamental right that should not be denied to anyone on the author list, see also Section 13.2. In my editorial practice I am still astounded to see, again and again, that so many senior researchers seem not to know this rule.

13.8 When to submit a manuscript to `astro-ph`?

The idea to submit your manuscript to a journal, and at the same time to deposit it at `astro-ph`, is not widely recommended for the following reasons:

1. the deposited material is not peer-reviewed;

2. the file is archived with reference to the journal to which it was submitted, so in case that the paper is refused and submitted to another journal, the reference will be incorrect;[80]
3. if the peer review reveals gross errors, then the flawed manuscript may remain available to the community at large. If the revision process takes considerable time, it is not impossible that someone reacts with a counter paper even before your paper is finalized.

Let me illustrate the problem with a real example. An author submits a manuscript to a journal, and at the same time deposits it at `astro-ph`. But the referee disagrees with some of the conclusions, and substantial parts of the paper have to be removed. The paper is finally accepted, but the total revision time (with two passes through the referee's hands) takes a full 11 months, and the acceptance year is one calendar year later. In this particular case, the `arXiv e-print` today still renders the originally submitted flawed version, whereas ADS does lead to the correct abstract and bibliographic code, but with the link pointing to the 1-year older version available at `astro-ph` (see also Sect. 3.2.1 in Uta Grothkopf's paper in Part 1 of these Proceedings). Bibliometric bias is not unlikely to develop when only the `astro-ph` server is consulted.

In fact, the situation described above also occurs when presenting results in poster format at a meeting, if the poster is based on a submitted but not yet accepted journal paper. If the referee finds the main paper unacceptable for publication, then the poster paper should be withdrawn of course. This, obviously, seldomly happens.

13.9 How to communicate with the referee?

Communication with the referee is always done via the Editorial Office, and not directly with the referee. This interaction is carried out by the corresponding author. The four main categories of responses by a referee are:

1. acceptable with minor revision;
2. acceptable after major revision to be verified by the referee;
3. outright rejection – mostly for scientific content, but also for poor language, inadequate graphics, and badly structured or incoherent narrative, and
4. rejection because the paper does not fit the scope of the journal.

The last outcome is always somewhat regrettable, because it is a result that could have been foreseen[81] by the senior coauthors. There are two ways of pursuit here: either submit to another journal, or oppose the Editor's decision (for example, by providing a precedent case of a similar article that was accepted for publication in the same journal). In the latter case, it is by and large better to let the senior coauthors do the fighting.

The third scenario is the most painful to undergo, although the second category can also be quite difficult to deal with, because soft-looking and polite statements by a referee

[80] This feature even is a source of information for Editors.

[81] For example, if you submitted a review paper to a journal that does not accept reviews.

may hide a lot of disagreement (in just the same way as a soft and positive letter of reference may not get you the job you are looking for).

Learn to correctly read the referee's message: sometimes the referee asks a question that looks stupid, at first sight. That does not necessarily mean that the referee is slow of mind: it is very probable that the referee, by playing devil's advocate, points to some confusing or vague elements in some passages. Few written statements are misunderstood because the reader lacks intelligence: most of the incorrect interpretations result from foggy writing (like using words and verbs from Table 1 in the wrong context). Referees – like lawyers – surely interpret ambiguous sentences in the way **they** think is just, not necessarily in the way the author thinks is correct.

Explain the changes that you have made to the manuscript (do not just silently expect the referee to find out what you have changed), and spell out why you prefer not to follow some of the referee's suggestions. If you plainly disagree with the referee, then open a dialogue and provide supporting evidence for your viewpoint – through the Editorial Office of course.

When receiving a very negative referee report, never answer immediately on line, but wait at least twelve hours before responding by email or letter. There is a fair chance that, reading the report over the next day, you may find it less shocking or to a lesser extent impossible to digest.

Be forgiving of other people's mistakes: also the referee makes typographical errors, especially because the reports are always due by a deadline, and the closer to the deadline, the less time is left for the referee to proofread his or her report. I once requested an author to add a reddening vector and error bars to their "Figure 5", but in fact I mistyped the Figure number, and meant instead to refer to their Figure 6. The author replied *"Figure 5 is a power spectrum and then how to add reddening vector and error bars?"* Such sarcasm does not boost a smooth and fast reviewing process, and is as inadequate as insulting your referee at a sports competition. On my suggestion that proofreading by a native English-speaking person would be very beneficial, the same authors replied *"Actually our manuscript was sent to a native English speaker for proofreading with charge. ... It would be appreciated if you would revise the manuscript."*

In fact, any form of ironic riposte is out of place in a response to the referee, not to speak of brutal replies which, regretfully, appear from time to time.

13.10 When to thank the referee?

Referees remain, very often, anonymous. Yet, sentences like *"The author thanks the anonymous referees for their valuable comments"* regularly occur in the literature. Kozak (2009) lists several considerations why authors are keen to acknowledge referees:

1. the authors think that the comments they receive are of such a high value that the referee's contribution must be acknowledged;
2. the authors think that it might help them to have the paper accepted because it makes the referee feel happy;
3. it is just that author's standard procedure;

4. they are asked to do this by the journal Editor.[82]

Kozak believes that the majority of referees are aware that they may be acknowledged with the second motive only, and that they may have become quite resistant to this kind of acknowledgment. Concerning the acknowledging-by-default of the referee, he mentions that there are even authors who are so eager to thank their referees that they do so in a first submission! Concerning thanking because the Editor of a journal asks to do so, he makes it clear that he does thank referees only if their comments are useful, a very correct attitude indeed.

In summary, it is required to thank the referee if that person merits it by having discovered major flaws, or by having made significant suggestions to improve the scientific quality of the paper. When the referee suggests only minor changes, an undeserved flattering acknowledgment may even be seen as an insult.

13.11 Can a student be referee?

Students may occasionally be asked to review a paper of a senior scientist. The first thing to consider is whether there is any conflict of interest – that is, when close friends or colleagues with whom you collaborate are on the author list. You should then normally refuse becoming the referee. Do not ruin given chances so early in your career, and never hope to hide behind anonymity: abuse of such powers will always backfire on its originator (see Sect. 8.7 in Paper III).

13.12 With whom can I share a referee report?

At the moment that you receive a referee report, there are only three individuals who know its content: the referee, the Editor, and yourself. The only other people that have the right (and the sole right) to read its content are the coauthors. You may decide to share the content with your supervisor, but a referee report is not intended to become public domain. Vice versa, if you become a referee, then your report should not be shared with anyone, not even for language editing. If you wish to consult other colleagues, you must first confer with the Editor in order to avoid involving other experts who may already have been excluded by the Editor or the authors.

A related question is if referees can identify themselves to authors: this can be considered – if the policy of the journal tolerates disclosure of the referee's identity – and it should only be done via the Editor. Vice versa, authors should not attempt to identify referees, and I refer to more discussions on this topic in Steve Shore's paper in Part 1 of these Proceedings.

13.13 May I reproduce complete sentences?

Several students asked me whether they must get permission to reproduce complete sentences. In fact no, because of the so-called *Principle of Fair Use*. This principle is a

[82] Or quite often by the thesis supervisor (point added by myself).

privilege that allows users to copy without permission for non-commercial purposes of criticism, research and teaching. Thus, **when citing one or a few sentences, provided you give due credit, no permission is needed**. But the principle is vague, and when in doubt, it is better to ask for permission, though you should realize that most journals are published in a commercial environment ("for profit"), and that you will likely have to pay a fee. The procedure is very simple: visit the website of the *Copyright Clearance Center*[83] (CCC), which offers easy-to-use licensing services, select the journal you wish to reproduce from, complete the order form, and the copyright permission will end up in your shopping cart. If you reproduce copyrighted content in a book (for example, the list starting on page 55 that somehow exceeds the notion of fair use), you must ask the Publisher of the book to purchase the reproduction rights. Note that the price for non-profit use ("academic") is substantially lower than for commercial use.

The ease of `copy/paste` text blocks into a collection of digital notes can lead to problems, especially when sentences and paragraphs are copied without the originating reference (full journal reference, or URL). Hence, authors should keep a tidy system with clear separation of own notations and copied phrases (for example, by systematically using boldface for the latter). But also oral communications (private as well as at meetings) can easily lead to "forgetting" the source of information.[84]

I return to the first case mentioned in Section 13.7, where I was struck by one (and the only one) complete paragraph without any typographical errors. Asked from where this passage was taken, the author replied that he did not know. One sentence was quite intriguing because of erroneous syntax with a needless double negation: *"We note that the choice of an incorrect annual alias value usually has no little or no effect on the subsequent search for other frequencies."* Passing this phrase through iThenticate[85], the system returns a 96% similarity index with a sentence in Breger *et al.* (2005). Correcting "has no little or no effect" to "has little or no effect" yields a 100% match. **The idea or principle expressed in this sentence is very explicit, and should thus not be copied without reference to its source text**. There are two ways to correctly handle this situation: either by quoting the sentence *verbatim* as it is done here, or by rephrasing it in the sense *"As Breger et al. (2005) emphasize, the choice of an incorrect annual alias value usually has little or no effect ... "*.

This example illustrates the maxim that even duplicators need careful proofreading, as is very well known from experience over the ages of copying by scribes. Perhaps the most famous example of a similar copying error is the so-called *Wicked Bible* published in 1631 by the royal printers in London, where for the seventh commandment the word "not" was omitted, resulting in *"Thou shalt commit adultery"*.[86] The books were consequently destroyed, and only a handful copies escaped, now making these books very expensive collector's items.

---

[83] `http://www.copyright.com/`

[84] Tacit knowledge and cryptomnesy are explained in Section 7.10 of Paper III.

[85] CrossCheck: `www.crossref.org/crosscheck.html` (see an example in Sect. 16.5 of Paper III).

[86] Exodus 20:14.

13.14 How to cite or quote a straying paper?

When using secondary sources, like review papers, it often happens that you cannot access the quoted primary information in that review. In such a case, you should be careful not to substitute the reviewer's opinion for the views expressed in the primary source. As an example, I refer to Section 2, where I cite Perkowitz (1989), although I have not been able to consult the original quotation – even by tracing and contacting the copyright holder, who took two months only to answer *"If you would kindly provide your Visa or MasterCard number and the expiration date... "*. Therefore, the reference at the end of the paper contains the note "quoted by Kuyper (1991)". This is quoting from second-hand information in the absence of primary sources: it is less secure, but it does not mislead the reader, and it is acceptable because it is fully documented. The same approach was followed when citing an unavailable paper, see page 35 referring to Giuseppe Baretti.

13.15 How to sort words that incorporate diacritics?

The alphabetic sort order depends on language, but also on culture – even within the same language family. Moreover, some languages use *diacritics*: marks below, above or through a character, or a combination of characters by which the sound of the letter is changed. Examples are č, ç, œ, ø, ö, ñ: these are different characters, not different fonts of a same character – see, for example, the short sentence in Czech language on page 35. The ASCII character set does not contain any diacritics, hence an ASCII sort will lead to problems, unless assisted by manual intervention. In English, a letter with diacritic is sorted after the letter without diacritic. But should we sort ae–af–æ or ae–æ–af? The *Chicago Manual of Style* (15th edition) says that for English-language sorting, words beginning with or including accented letters are alphabetized as though they were unaccented: see, for example, the position of Öpik in the Name Index at the end of this book.[87] Some letters with diacritics can be spelled without them, for example u and o with umlaut (ü, ö) can be written, respectively, as ue and oe, which is fine as long as the same spelling is applied throughout the text, and in the reference and index lists.

Sorting Greek-letter names is a special case: the LaTeX automatic sorting produces a list alphabetized according to the English spellings of the Greek letters, rather than in the order of the Greek alphabet. But when sorting star identifications in which a specific star is identified by a Greek letter, followed by the genitive form of its parent constellation's Latin name,[88] sorting should first be according to the constellation name, followed by an alphabetization following the Greek alphabet.

Sorting is also important when using acronyms, such as those of photometric systems. The Johnson *UBV* photometric system[89] has three passbands with very specific wavelengths progressing from blue to red, and the same is true for the Strömgren *uvby* system, where the central wavelengths run from 350 to 550 nm. But a GOOGLE search for *uvby* yields about 25 000 hits, while a search for the non-existing *ubvy* yields 8 000! Some

[87] This feature is mostly incorrectly rendered when sorting in LaTeX.

[88] The so-called Bayer designation established by Johann Bayer in 1603.

[89] These examples refer to systems designed by Harold Johnson, Bengt Strömgren and Theodore Walraven.

photometric systems list their bands in opposite sequence, for example the Walraven *VBLUW* thus follows a wavelength progression from red to blue in its acronym. Some designations have diacritic characters (subscript or superscript indices), or mix upper and lower case letters. It is most important not to change the pre-defined order or the case of these designations (see also the genealogy tree in Fig. 13 of Paper II).

13.16 To cite the URL or to cite the complete bibliometric reference?

The consequences of all too quickly pasting an URL were described in Section 11.9, and even if the URL link delivers the correct paper, the `astro-ph` reference referred to in Section 13.8 does not always lead to the latest and correct version of the accepted manuscript. So, even importing reference lists does not exempt the authors from verifying the entries. Not to speak of even more confusing statements like *"This article originally appeared in Icarus Vol. 187/1, March 2007, pp. 132–143. Those citing this article should use the original publication details."*[90]

13.17 Why publish your data?

There are several reasons for publishing or for publicly archiving primary data:

1. some papers use observational data that are defective in terms of standardization and calibration, and such flaws may lead to incorrect conclusions (see Sterken 2011 for an example). Hence, the publication of primary data may provide the basis for rectification eventually;
2. data sets – especially time series data, but also spectral energy distributions – are always incomplete, hence inclusion of data in a publication will allow subsequent studies to fill in missing phases, and even to finetune the derived frequencies;
3. archived data sets allow for meta-analysis, examples abound in Virtual Observatory projects (see, for example, Solano 2006);
4. data, especially historical archives, may later be used in ways that the originator of the dataset never could have imagined;
5. sharing detailed research data is associated with increased citation rate, see Piwowar *et al.* (2007), and also Section 18.4 in Paper III;
6. preservation of data provides the option of avoiding duplication of expensive experiments;
7. at the same time, the availability of original research data allows reconstruction of the theory of an instrument with which previous observations were collected;
8. calibration of astrophysical parameters (like effective temperature or metallicity as a function of color index) often rely on model-atmosphere calculations, and without the numerical model data, such calibrations can never be considered for upgrading;

[90]Footnote to the title of the paper by Belton *et al.* (2007) reproduced in Icarus Vol. 191, 2007, pp. 310–321, see also Section 7.12 in Paper III.

9. computer code should also be published in full (even though future machines will not be able to handle the code format) because a well-documented published code can be of great value in designing new data-reduction and analysis packages.

13.18 Why do journals engage in publishing data?

There is a substantial difference between a *data paper* (or data in a paper) and a *data archive*. The latter merely is a set of tabulated information, the former usually also elaborates why the data were obtained, explains how the data were acquired, includes a discussion of the accuracy, and possibly gives some illustrations of results already published elsewhere. Journals, through the organized structure of the editorial flow, and supported by verification through peer review, are gatekeepers of good scientific practice. They sustain agreements with archives (like the accord between A&A and CDS, see the papers by Agnès Henri and Laurent Cambrésy in Part 1 of these Proceedings) that assure that archived data are solidly linked with the originating journal article, that the data are rightly formatted with a correct description of the metadata,[91] and that they bear a robust time stamp – that is, that the content of the web archive is not modified *a posteriori* without proper documentation.

It is always good practice to check the journal's policy on enclosing data before you submit a paper.

13.19 How to cope with writer's block?

Writer's block or *writer's anxiety* is the inability to complete a manuscript due to many reasons. A major ground can be self doubt, fear of failure, or the fright of making significant errors. But also the seeking to produce a perfect prose at the first try, or being crushed by deadlines. A frequently occurring problem is information overload, effectively blocking the organization of thoughts and research notes. Writer's block occurs to all classes of text writers: fiction as well as non-fiction, song writers, poets, movie-script authors and journalists.

There are strategies or techniques to escape out of such dead-end streets, for example by first tracing a mind map or writing an outline, then by composing the first draft, and subsequently by reworking and polishing. A good start is by following the checklist presented on page 55. And if nothing helps, put the job aside for a while, and concentrate on something completely different. Or, look for a fine *Muse*[92] to help drawing out of you the first paragraphs of your opus.

Severe writer's block syndrome occasionally occurs when one author (usually the principal author) feels uneasy about the presence – or the rank in the authorlist – of one or more somewhat inert coauthors. The misgivings about such inactive associates can render any creative labor impossible, effectively creating a block that unavoidably will turn into a total stop – and the more pressure these inert people exert, the more massive

[91] See Laurent Cambrésy's paper in Part 1 of these Proceedings.

[92] In Greek mythology, the Muses are the nine goddesses who inspire artists and writers.

this block will develop. From my personal experience I know that none of the above-mentioned hints – not even a choir of Muses – could possibly solve such a mental block. You should always realize that you are not the first to live through such a situation, and you are definitely not the last one to undergo it, and that there is only one solution: just sit down and write, so that you can turn the page on that project as fast as possible, and look forward to working with other people. Or quit science.

13.20 Should I allow someone else to completely rewrite my manuscript?

Never! But there are always ways to make a manuscript more readable by rewriting some sentences or even complete paragraphs. If

1. the manuscript has a proper structure (as explained in Sect. 9);
2. the language is straightforward and direct, without winding digressions;
3. the principles explained in Adams' and Adams, Halliday, Peter & Usdin's contributions in Part 1 of these Proceedings, have been followed;
4. careful and multiple proofreading was done;
5. correct citation practice was implemented (including verification of references); and
6. you ran the text file through a good spelling checker,

then I see no reason whatsoever to have your manuscript completely rewritten by a post-doc or a supervisor.

Such forced rescripting may lead to damaging consequences. First of all, the rewriter may claim a higher rank in the list of authors. Second, if the senior rewriter was not involved in all aspects of the research, he or she may change your conclusions, which in turn leads to a modification of accountabilities. Third, the young scientist's reputation is harmed in the long run – even if only by the brunt of gossip. And, worst of all – as I experienced myself – the terrible feeling so well expressed by Melanie Safka:

> LOOK WHAT THEY DONE TO MY SONG MA ... IT IS THE ONLY THING THAT I COULD DO HALF RIGHT ....

PhD supervisors should manifestly avoid such scenarios. **It is the student's indisputable right to stand by their song – that is, if it is truly the product of their own creativity, and if it is at least half right in each of the 6 items listed in the previous paragraph.**

Rewriting does occur in the commercial press, for example TIME Magazine states that *"Letters may be published in Time Magazine and edited for purposes of clarity and space"*. And it happens, in a milder form, by Editors of conference proceedings, see Section 13.21.

13.21 What to do when disagreeing with editorial changes?

There are three levels of editorial changes: straight typographical errors in the body of the text (including formatting errors in the references), linguistic changes, and outright censorship.

Typographical and formatting errors, if detected, are normally corrected in the background by the Publisher's copy editor or by the language editor, *i.e.*, without asking the author's consent. Such changes – sometimes referred to as *language washing* – are almost always justified, and should not be disputed. But be careful: in the most recent book I coedited (Milone & Sterken 2011), the copy editor had replaced all mentions of CCD (Charge Coupled Device) by CD (Compact Disc?), probably because that editor's dictionary file suggested this replacement. This only illustrates that you should proofread again after receiving the page proofs.

Changes in style and syntax are more tricky, because the language editor may misunderstand what the author exactly wants to say. Moreover, the language corrector, in most of the cases, does not hold a degree in astronomy or in another science,[93] and this can lead to misconceptions. I have had only once in my career a disagreement with a language editor: in Sterken & Jaschek (1996) I described a light curve with a so-called *stillstand* phase.[94] The Cambridge University Press language corrector changed all occurrences of that term into the linguistically correct *standstill*. I then argued with the Editor that the stillstand phenomenon had been historically coined as such, and that it had become a jargon[95] word just like *bremsstrahlung* or *metallicity*[96]. The Editor accepted, and the phrasing appeared in the way I intended. For almost all other suggestions, I was quite happy with the language editing because it improved the readability of my paper, or because I could not find a better wording than the one suggested by the language editor. It is, nevertheless, important that the author has the opportunity to discuss, or even disagree with some of the proposed changes. Stronger even, such changes should never be made in the background, *i.e.*, they should be made **by** the author, and not **for** the author.

Outright censoring, without the author's consent or without the author's knowledge, should never occur. I experienced it twice in conference proceedings: once an unfavorable citation was removed, the second case even involved the removal of an entire Figure and the discussion connected to it. Since I only discovered this when the books were delivered almost one year later, protests were to no avail. Such behavior is rightly considered as editorial misbehavior, see Section 10.2 in Paper III.

So, how must such conflicts be handled by the author? The journal Editor should be the first person to contact in case of disagreement with the language corrector. Just like answering the referee, you must take care that your complaint embeds the first two storytelling elements referred to on page 2: the *content* of the message, and the *form* of the message. If the form is bad (rude language, many typos), you will never be able to convince anyone that you know better about syntax and style than the language editor does. Likewise, if the content falls short in terms of argument or proof, you cannot win over the Editor.

---

[93] Claire Halliday, one of the language editors at A&A, holds a PhD in astronomy.

[94] A plateau on the rising branch of the light curve of the pulsating star BW Vulpeculae.

[95] Paper II deals with similarly "wrong" jargon: the using of *font* for *typeface*.

[96] Metallicity refers to the proportion of matter in a star made of elements other than H and He.

13.22 Is it too much to expect authors to read the papers they cite?

This question is a *rhetorical question*[97] by Young (1993a), who pointed out that by misquoting, an erroneous formula had been attributed to him. A few months later, he reports to have found the same incorrect equation in another report published since. He thus concludes (Young 1993b) that it is evident that the latter authors have simply copied both the formula and the reference from the previously mentioned paper, without checking if either was correct. He subsequently states to have no objection when those who do not understand the physics use an incorrect formula, but that he does object to having the wrong formula attributed to him. Hence his concluding question

IS IT TOO MUCH TO EXPECT AUTHORS TO READ THE PAPERS THEY CITE?

This question also bears on Öpik's (1977) own story: a critic first completely misrepresents his work, and then – rightly – sets out to destroy this fantasy of his own imagination (which was just the very opposite of what Öpik was saying). So, in order to avoid the creation of a myth on the basis of second-hand information, Öpik (1976) published a letter of rectification in *Science*, which was immediately followed by a rather vague and irrelevant rebuttal by Ulrich (1976).

Young's example refers to copying mistakes from journal to journal, and such cases can be easily traced. Similar copying errors from journal to book, and in book translations, do frequently occur, and are much more difficult to spot. One such case – unfortunately again with a formula by A. T. Young – was spotted by Tuvikene (2008) while using a formula for atmospheric extinction[98] copied from Young (1974) by Hall & Genet (1982). Interestingly, the latter authors give the complete and correct bibliographic reference to Young (1974), though that reference cannot be found in ADS, which only gives the *bibcode* of the book in which Young's contribution is a chapter. As Hall & Genet (1982) is not any more readily available in libraries, and since they did not quote the primary source (Young & Irvine 1967), it was not possible to find and check the original formula, and so an erroneous extinction correction that was programmed in a computer code had to wait for detection and correction during almost half a century!

Note that erroneous results in both examples would not be visible at face value, but would become appreciable in the somewhat extreme limits of the parameter domain wherein these formulae were applied.

13.23 How to review a paper?

If you are asked to referee a paper, the first thing you must ask yourself is if you are really competent in the matter. The second point is whether you can comply with the time that the Editor offers for completing the review (typically three to four weeks for a normal research paper, much less for a Letter). Finally, you must check whether there is any *conflict of interest*, *i.e.*, if you have any stake that compromises the action of being a

[97] A figure of speech in question format, without expecting an answer.

[98] $X = \sec z[1 - 0.0012(\sec^2 z - 1)]$ but with the exponent 2 in the argument $\sec z$ omitted.

reviewer. Obstacles could be that you are in strong competition with the authoring team, or even that you are (or have been) part of the project (see the case discussed in Sect. 8.7 of Paper III), or that the other author is a member of your institute, or is one of your best – or worst – friends. If you have the competence, if there is no conflict of interest, and if you have the time, then you may accept the invitation.

I have nowhere seen explicit guidelines, but a most helpful guide for critical assessment of a research article is the *"checklist for critiquing a research article"* by Kuyper (1991): she gives a structured list – following the IMRaD scheme – of 20 cardinal points to be considered when reviewing a paper:[99]

- Introduction

  1. Read the statement of purpose at the end of the introduction. What was the objective of the study?
  2. Consider the title. Does it precisely state the subject of the paper?
  3. Read the statement of purpose in the abstract. Does it match that in the introduction?
  4. Check the sequence of statements in the introduction. Does all information lead directly to the purpose of the study?

- Methods

  5. Review all methods in relation to the objective of the study. Are the methods valid for studying this problem?
  6. Check the methods for essential information. Could the study be duplicated from the information given?
  7. Review the methods for possible fatal flaws. Is the sample selection adequate? Is the experimental design appropriate?
  8. Check the sequence of statements in the methods. Does all information belong in the methods? Can the methods be subdivided for greater clarity?

- Results

  9. Scrutinize the data, as presented in tables and illustrations. Does the title or legend accurately describe content? Are column headings and labels accurate? Are the data organized for ready comparison and interpretation?
  10. Review the results as presented in the text while referring to data in the tables and illustrations. Does the text complement, and not simply repeat, data? Are there discrepancies in results between text and tables?
  11. Check all calculations and presentation of data.

[99] Reproduced with permission after purchase by Publisher EDP of reproduction rights from the copyrights holder.

12. Review the results in the light of the stated objective. Does the study reveal what the researcher intended?

- DISCUSSION

13. Check the interpretation against the results. Does the discussion merely repeat the results? Does the interpretation arise logically from the data, or is it too far-fetched? Have shortcomings of the research been addressed?
14. Compare the interpretation to related studies cited in the article. Is the interpretation at odds or in line with other researchers' thinking?
15. Consider the published research on this topic. Have all key studies been considered?
16. Reflect on directions for future research. Has the author suggested further work?

- OVERVIEW

17. Consider the journal for which the article is intended. Are the topic and format appropriate for that journal?
18. Reread the abstract. Does it accurately summarize the article?
19. Check the structure of the article (first headings and then paragraphing). Is all material organized under the appropriate heading? Are sections subdivided logically into subsections or paragraphs?
20. Reflect on the author's thinking and writing style. Does the author present this research logically and clearly?

Not only can most of these 20 points be directly applied to astronomy, the list is also an excellent guide or blueprint in planning the writing of a paper.

13.24 How to revise a multi-authored paper?

This is a matter of optimizing discussion and communication, a prime condition for successful collaborative writing. For a paper with multiple authors, there are three ways to deal with the referee report, and the feedback on it:

1. THE HUB-AND-SPOKE NETWORK: a system where all communication traffic moves along spokes connected to the central hub, which is the corresponding author;

2. THE MERRY-GO-ROUND METHOD: sending the manuscript source file down the list of authors, so that every contributor makes changes and then sends the file to the next coauthor;

3. THE PARALLEL COMMUNICATION MODEL: the first author sends the referee report as a group mail to all coauthors, who (normally) return their feedback in parallel lines with copy to everyone involved.

The first method has the disadvantage that it is slow, that information is only partially shared, that the hub (you) is the bottleneck in the network, and that a short inactivity or failure of the hub will halt the complete system – just like a router failure in a communication network between computers will stop all communication. But the major drawback is that information exchange is not open, that there is no archiving, and that *clique* forming[100] is more likely to occur. Once such a situation is established, it becomes much more difficult to formulate a coherent response towards opposing views of a referee.

The second *carousel* course is the least creative, and results in papers of the lowest finishing quality. Moreover, it is utterly inefficient, and not seldomly leads to multiple people simultaneously modifying the same source files. Not to mention that not every one uses the same text editor or LaTeX settings, nor do we all have the same level of skill in typesetting and managing LaTeX.

The third approach, on the other hand, is more frank, since each coauthor can take note of everyone else's opinions. It also provides a straightforward view of who is really active in the team.

I recommend by far the parallel approach, and I have seen many a revised paper lagging because the *hub* inadequately filtered (*i.e.*, read, misread, translated or transferred) the opinions of one author to the other. The parallel route cuts down on possible disagreements between subsets of coauthors. The parallel communication model provides a strong edge in handling the problems referred to in Section 13.21, notably when dealing with language editing. Imagine, for example, that there is only one native English-speaking person among the half a dozen coauthors of a multi-authored paper, and that all editorial changes are taken care of by the first author whose mother language is not English. Even though language editing, as a rule, does ameliorate syntax issues, it may happen that slight changes in particular idiomatic expressions (such as *stillstand*, see page 53) are suggested, and that such nuances are not detected by the corresponding author, while they would be noticed by the native English-speaking coauthor if the parallel communication model would have been implied. Thus, without feedback from all authors – even at the language-editing level – the published paper may in the end contain unwanted expressions, and thus arouse feelings of frustration in all parties involved.

Figure 2 of Paper II illustrates noise sources in the model of information transmission, and each of the three communication scenarios described above contributes its own level of noise on the sender's side of the communication train. It is of utmost importance to keep all those noise levels as low as possible.

13.25 When can I ask for another referee?

Asking for another referee is a right, however, you must first make sure that your request is based on solid grounds, *e.g.*, if you think that the referee is plainly wrong, or if the report contains rude personal attacks. And really think twice before you act. For example, in the first case mentioned in Section 13.7, the first author suggested to ask for a new referee, a

[100] A subset of the coauthor population may form a common view of the other – or even just one – coauthor(s).

request in disagreement with the second author's opinion, resulting in withdrawal of the manuscript.

In general, it is not wise to start a head-on fight with a referee, unless you are sure that you are right and the reviewer is darn wrong: it is better to display some diplomatic modesty in the way Wright & Armstrong (2007) advise:

> ... IF I BELIEVE A REFEREE IS MISTAKEN IN HIS/HER CONCERN, AND I KNOW A WAY TO DEFUSE THAT MISTAKEN CONCERN WITHOUT TELLING THE REFEREE THAT HE/SHE IS MISTAKEN, THEN I WILL USE THAT WAY BECAUSE THE PROBABILITY OF SURVIVING THE REVIEW PROCESS DECREASES WHEN REFEREE CONCERNS ARE CHALLENGED RATHER THAN ACCEPTED.

But if you decide to oppose to the referee's views, remember from Section 13.9 to formulate your arguments in a non-offensive way, otherwise you will be judged on the form of the message, and not on its content, and you will lose the battle. As Wright & Armstrong (2007) rightly point out, it is a matter of surviving the revision process.

13.26 Why do Editors edit, and why do referees review?

Editors of non-commercial journals do not operate on a classic *pay-for-work* scheme, though they receive indemnities for the editorial work that they carry out besides their regular research or teaching functions. Nor do referees receive any remuneration. So why do they bother to do it? Well, Editors and referees receive very interesting papers to review in the first place, and this leads to professional enrichment, and helps them keeping up with the advancements in their fields.

I have once witnessed a senior referee presenting his "bill" at a meeting of Editors and Publishers: taking one work day for one paper to review, with about a dozen papers per year to referee, results at the end of a career in about two years of *work-without-pay*. Although I concur with the calculation of labor time,[101] my personal vision is that I still consider myself to benefit a major profit in terms of new facts learned from these in-depth readings. Moreover, I consider refereeing a form of service to the community, and even an effective mode of parental investment towards the students. **Contributing to the editorial process through refereeing should never be the subject of a cost-benefit analysis**, and I am convinced that when money becomes a prime motive for refereeing, the system will then really break down. Editors need independent referees because, without reviewers it is simply not possible to run a high-quality journal, see also Section 8 (on referee misconduct) in Paper III.

Reporting a new scientific result always implies imaginative thoughts (hypothesis, speculation, ...) and criticism (error budget, validation ...). It is very often difficult to produce yourself an exhaustive and complete critical analysis of all aspects of your own scientific reasoning. So, criticism or judgment by an external person (or by a competitor) – just like proofreading by alien eyes – may bring out deficiencies that escaped the authors' attention (or that were poorly explained). The external referee is just part of the

[101] Klahr (1985) estimates the total cost of one panel review to be equal to 3 person-days.

cooperative and open enterprise called science, and merely adds to the dialogue between imagination and critical judgment, with the referee's feedback not seldomly constituting a true bonus to the information content and credibility of the reviewed paper.

14 Summary

Communication of science is as important as the research results themselves. Top results just get higher exposure when they are transmitted in a *state of the art* fashion. Sandage (1989) underlined this most properly for Edwin Hubble's writing style:

> It was Hubble's mastery of the language that gave some of his papers such dominance over prior work by others. Often the problem had in fact been solved, but without the same elegance of style, power of presentation, and excellence of summary possessed by Hubble when he was at his best. Clearly, the lesson for students is learn to write at the same time that you learn to do great science.

But Hubble was not just anyone: he had received a formal training as a lawyer, and as such had learned how to clearly explain evidence, hypothesis and conclusion to his opponents. None of us can possibly dream of having such a background, but we can take a lesson from lawyers: **know your facts, understand your data, explain your points, and learn from your adversaries**. And, if you receive comments on your manuscripts, then

1. learn to properly handle invaluable feedback, and try not to get upset or feel offended: many remarks are fair, and most are made to help you;
2. understand that errors are normal for beginners (even for old hands), that failure is a necessary aspect of research, and that the best way to learn is to learn from your own and from other people's errors;
3. know that most Editors and referees try to help you, and that most papers get published after all.

Above all, exercise self-examination, with respect to the scientific content and significance of your paper, as well with regard to the presence and rank of your name in the author list. And never forget that it is quite normal that writing an innovative paper always takes much more time than scribbling a copy-and-paste article.

Postscript

In Sterken (2006) I explain that the comma serves to indicate a pause, and that omission or transposition of a comma may completely change the meaning and the direction of a phrase. I demonstrated the alteration of the meaning by using the following sentences in which only the number and place of the commas differ:

> Chris said, Conny let's meet tomorrow, and
> Chris, said Conny, let's meet tomorrow.

I used the same example in a lecture in Paris,[102] and after my talk a colleague came to me, and said

> YOUR TALK SADDENS ME: IF STUDENTS NEED TO LEARN EVERYTHING ABOUT COMMAS AND PUNCTUATION MARKS, WHERE WILL THEY EVER FIND THE TIME TO DO SCIENCE?

I answered that the education in scientific writing is like the grooming in *haute cuisine*, where excellence is not only achieved in elaborate preparations of exquisite food, but also in avoiding artificiality and pretension when the food is served.

Molière expressed this aspect of punctuation and syntax so well in his comédie-ballet *Le Bourgeois Gentilhomme*,[103] in which Monsieur Jourdain asks his teacher for help rendering the sentence *"Beautiful marquise,*[104] *your beautiful eyes make me die from love"* in a more elegant way. The *Maître de Philosophie* then offers him five sentences that differ only in the emplacement of the words, and in the punctuation marks, see Figure 4. Each of these sentences has exactly the same meaning – that is, as long as the punctuation marks are correctly placed. In the end, Monsieur Jourdain asks his teacher which formulation is the most elegant, and the reply simply was: **the very first one you proposed**.

That story underlines that you are not expected to write like Shakespeare, nor like Molière, but that your writings should honor the basic characteristic of scientific style as explained on page 12: your style should be direct, plain, assertive, context-independent, and brief.

Belle Marquise, vos beaux yeux me font mourir d'amour.
D'amour mourir me font, belle Marquise, vos beaux yeux.
Vos yeux beaux d'amour me font, belle Marquise, mourir.
Mourir vos beaux yeux, belle Marquise, d'amour me font.
Me font vos yeux beaux mourir, belle Marquise, d'amour.

Fig. 4. Five phrases with the same words rearranged, though with different punctuation as indicated by vertical arrow symbols. Source: Molière, *Le Bourgeois Gentilhomme* (1670). Typeface Arial, font Arial 18 pt, reduction factor 50% (see Paper II).

I am grateful to PhD students Wu Yue and Zhang Li Yun for kindly providing the Pinyin transliterations reproduced in Figure 2, and to Veerle J. Sterken for proofreading a first version of this manuscript. I thank the Publisher EDP for purchasing the reproduction right to list the 20 points in Section 13.23.

[102] *Comité pour les Pays en Développement*, Institut de France, Académie des Sciences, 3 December 2009.
[103] *The Bourgeois Gentleman*, by Jean-Baptiste Poquelin (1622–1673).
[104] A *marquise*, or marchioness, is a noblewoman ranking below a duchess and above a countess. But this word has also a second meaning: a *marquise diamond cut* is one of the more elegant (oval) diamond shapes.

References

Allen, R.J., 1970, Stephan's quartet?, A&A, 7, 330

Belton, M.J.S., Thomas, P., Veverka, J., *et al.*, 2007, The internal structure of Jupiter family cometary nuclei from Deep Impact observations: The "talps" or "layered pile" model, Icarus, 187, 332

Berthold, G., 1897, Ueber den angeblichen Ausspruch Galilei's, Bibl. Math. p. 57

Bertout, C., & Schneider, P., 2004, Introducing structured abstracts for A&A articles, A&A, 420, E1

Bertout, C., & Schneider, P., 2005, Editorship and peer-review at A&A, A&A, 441, E3

Bottle, R.T., 1973, Information Obtainable from Analyses of Scientific Bibliographies, Library Trends, 22, 60

Breger, M., Lenz, P., Antoci, V., *et al.*, 2005, Detection of 75+ pulsation frequencies in the d Scuti star FG Virginis, A&A, 435, 955

Dobell, C., 1939, Dr. O. Uplavici (1887–1938), Isis, 30, 268

Drenth, J.P.H., 1998, Multiple Authorship: The Contribution of Senior Authors, J. Amer. Med. Assoc., 280, 219

Dubois, P., 1995, personal communication, 16 November 1995

Duerbeck, H.W., 2005, Book Review: Astronomy in and around Prague, J. Astron. Data, 11, 5

Feynman, R.P., 1985, Surely You're Joking, Mr. Feynman! (New York: W.W. Norton & Co)

Hall, D.S., & Genet, R.M., 1982, Photoelectric Photometry of Variable Stars (Fairborn, Ohio), p. 9-4

Hames, I.,, 2011, EASE forum, `http://www.ease.org.uk/forum/index.shtml`

Klahr, D., 1985, Amer. Psychol., 40, 148

Kozak, M., 2009, Eur. Sci. Edit., 35, 102

Kuyper, B.J., 1991, BioScience, 41, 248

Lauer, T.R., Faber, S.M., Ajhar, E.A., *et al.*, 1998, AJ, 116, 2263

Lemaître, G., 1927, Un Univers homogène de masse constante et de rayon croissant, rendant compte de la vitesse radiale des nébuleuses extra-galactiques, Annales de la Société scientifique de Bruxelles, XLVII, 49

Lepschy, L., 2006, personal communication

Li, C., Fu, J.-N., Zhang, Y.-P. & Jiang, S.-Y., 2007, A new study of the long-term variability of the SX Phoenicis star CY Aquarii, J. Beijing Normal Univ., 43, 621

Malthus, T., 1798, An Essay on the Principle of Population, London

Matzinger, P., & Mirkwood, G., 1978, J. Exper. Med., 148, 84

Maunder, A.S.D., 1902, MNRAS, 62, 57

Maunder, W., Mrs., 1904, MNRAS, 64, 224

Maunder, A.S.D., & Maunder, E.W., 1908, The heavens and their story (Robert Culley, London)

Merton, R.K., 1995, Social Forces, 74, 379

Milone, E. F. & Sterken, C. (eds.), 2011, Astronomical Photometry: Past, Present, and Future (New York: Springer Verlag)

Oort, J.H., 1970, A&A, 7, 405

Öpik, E.J., 1976, Solar Models, Science, 191, 1292

Öpik, E.J., 1977, About Dogma in Science, and Other Recollections of an Astronomer, Ann. Rev. Astron. Astrophys., 15, 1

Perkowitz, S., 1989, Commentary: can scientists learn to write?, J. Techn. Writing Comm., 19, 353 (quoted by Kuyper 1991)

Pinker, S., 1994, The Language Instinct (Penguin Books)

Piwowar, H.A., Day, R.S., & Fridsma, D.B., 2007, PLoS ONE `www.plosone.org`, doi: 10.1371/journal.pone.0000308

Pottasch S.R., & Steinberg, J.L., 1969, A&A, 3, 381

Qiu, J., 2008, Identity crisis, Nature, 451, 766

Salager-Meyer, F., 2009, Academic equality and cooperative justice, TESOL Quarterly, 43, 703–709

Sandage, A., 1989, Edwin Hubble 1889–1953, JRASC, 83, 351

Shermer, M., 2007, Scientific American, October 2007, 23–24

Sheskin T.J., 2006, Sci. Eng. Ethics, 12, 555

Schielicke, R.E. (ed.), 1998, In Reviews in Modern Astronomy 11: Stars and Galaxies (Hamburg, Germany: Astronomische Gesellschaft), 1

Shore, S.N., 2009, A&A, 500, 31

Sokal, A., 1996, Transgressing the Boundaries: Towards a Transformative Hermeneutics of Quantum Gravity, Social Text (Duke University Press), 46/47, 217

Solano, E., 2006, Virtual Observatory: What Is It and How Can It Help Me?, in Astrophysics of Variable Stars, ed. Sterken, C. and Aerts, C. (San Francisco: Astronomical Society of the Pacific), ASP Conf. Ser., 349, 29

Stein, K., 2006, J. Amer. Diet. Assoc., 106, 1947

Sterken, C., 1977, A&A, 57, 361

Sterken, C., 1988, The Coordination of Astronomical Observations in Coordination of Observational Projects in Astronomy, ed. C. Jaschek & C. Sterken (Cambridge University Press), p. 3

Sterken, C., 2000, Mode Identification through Photometry: Full Colours?, in Delta Scuti and Related Stars, Reference Handbook and Proceedings of the 6th Vienna Workshop in Astrophysics, ed. M. Breger and M. Montgomery (San Francisco: ASP), ASP Conf. Ser., 210, 99

Sterken, C., 2005, The Light-Time Effect in Astrophysics: Causes and Cures of the $O-C$ Diagram, ed. Sterken, C. (San Francisco: Astronomical Society of the Pacific), ASP Conf. Ser., 335, 178

Sterken, C., 2006, Advice on Writing a Scientific Paper, in Astrophysics of Variable Stars, ed. Sterken, C. and Aerts, C. (San Francisco: Astronomical Society of the Pacific), ASP Conf. Ser., 349, 445, `http://www.aspbooks.org/publications/349-0445.pdf`

Sterken, C., 2007, On the Provability of Heliocentrism. I. Ole Roemer and the Finite Speed of Light, Sartoniana, 20, 19

Sterken, C., 2011, Just one new measurement of the B[e] supergiant Hen-S22, IBVS, 6000

Sterken, C., & Aerts, C. (eds.), 2006, Astrophysics of Variable Stars, ASP Conference Series, Vol. 349

Sterken, C., & Manfroid, J., 1992, Astronomical Photometry: a Guide (Kluwer Academic, Dordrecht)

Sterken, C., & Jaschek, C., 1996, Light Curves of Variable Stars, A Pictorial Atlas (Cambridge: Cambridge University Press)

Sterken, C., Samus, N., & Szabados, L. (eds.), 2010, Variable Stars, the Galactic Halo and Galaxy Formation (Moscow: Sternberg Astronomical Institute of Moscow University), p. 125

Sweetland, J.H., 1989, Errors in Bibliographic Citations: A Continuing Problem, The Library Quarterly, 59, 291

Thomas, K., 1991, in A Cultural History of Gesture: from Antiquity to the Present Day, ed. J. Bremmer & H. Roodenburg (Polity Press, Cambridge), p. 1

Tukey, J.W., 1977, Exploratory Data Analysis (Addison-Wesley Publishing Company)

Tuvikene, T., 2008, personal communication

Ulrich, R.K., 1976, Science, 191, 1293

Watson, J.D., 2007, Avoid Boring People (Oxford University Press)

Wright, M., & Armstrong, J.S., 2007, Verification of Citations: Fawlty Towers of Knowledge?, Munich Personal RePEc Archive, 4149, `http://mpra.ub.uni-muenchen.de/`

Wróblewski, A., 1985, de Mora Luminis: A spectacle in two acts with a prologue and an epilogue, Am. J. Phys., 53, 620

Young, A.T., 1974, Observatyional technique and data reduction, in Methods of Experimental Physics, Vol. 12, Astrophysics, Part A: Optical and Infrared. ed. Carleton, N.P. (New York, Academic Press), p. 152

Young, A.T., 1993a, Scintillation noise in CCD photometry, Observatory, 113, 41

Young, A.T., 1993b, Propagation of errors, Observatory, 113, 266

Young, A.T., 1994, in The Impact of Long-term Monitoring on Variable Star Research: Astrophysics, Instrumentation, Data Handling, Archiving, ed. C. Sterken & M. de Groot, NATO Advanced Science Institutes (ASI) Series C: Mathematical and Physical Sciences (Dordrecht: Kluwer)

Young, A.T., & Irvine, W.M., 1967, Multicolor photoelectric photometry of the brighter planets. I. Program and Procedure, AJ, 72, 945

Some additional readings:

Anderson, L.K., 2006, McGraw-Hill's Proofreading Handbook (London: McGraw-Hill)

Butcher, J., 2004, Copy-editing: The Cambridge Handbook for Editors, Copy-editors and Proofreaders third edition (Cambridge: Cambridge University Press)

The Chicago Manual of Style 2003, The Chicago Manual of Style: The Essential Guide for Writers, Editors, and Publishers, 15th edition (Chicago: The University of Chicago Press)

Sullivan, K.D., & Eggleston, M., 2006, The McGraw-Hill Desk Reference for Editors, Writers, and Proofreaders (London: McGraw-Hill)

Scientific Writing for Young Astronomers
C. Sterken (ed)
EAS Publications Series, **50** *(2011) 65–170*
www.eas-journal.org

WRITING A SCIENTIFIC PAPER II. COMMUNICATION BY GRAPHICS

C. Sterken[1]

Abstract. This paper discusses facets of visual communication by way of images, graphs, diagrams and tabular material. Design types and elements of graphical images are presented, along with advice on how to create graphs, and on how to read graphical illustrations. This is done in astronomical context, using case studies and historical examples of good and bad graphics.

Design types of graphs (scatter and vector plots, histograms, pie charts, ternary diagrams and three-dimensional surface graphs) are explicated, as well as the major components of graphical images (axes, legends, textual parts, etc.).

The basic features of computer graphics (image resolution, vector images, bitmaps, graphical file formats and file conversions) are explained, as well as concepts of color models and of color spaces (with emphasis on aspects of readability of color graphics by viewers suffering from color-vision deficiencies).

Special attention is given to the verity of graphical content, and to misrepresentations and errors in graphics and associated basic statistics. Dangers of dot joining and curve fitting are discussed, with emphasis on the perception of linearity, the issue of nonsense correlations, and the handling of outliers. Finally, the distinction between data, fits and models is illustrated.

1 Introduction

This paper discusses elements of visual communication by way of images, graphs, diagrams and tabular material. This chapter is not intended as a manual, but rather as an elaborated field guide to scientific graphics, offering case studies and historical examples of good and bad graphics. As in Paper I, some redundancy and overlap of topics was built in on purpose, so the reader is not forced to study this paper by reading all Sections in sequence, but can consult any Section and follow the cross-references to previous and following parts in this paper and in the two associated papers.

[1] Vrije Universiteit Brussel, Pleinlaan 2, 1050 Brussels, Belgium

©The authors, published by EAS, EDP Sciences, 2011
DOI: 10.1051/eas/1150002

The scope of visual communication with graphics embraces all aspects of graphic design: the creative side of making graphs, the technical aspects related to the use of computer-controlled input and output devices, and the interpretive aspects when reading graphs made by other persons.[1] Section 2 explains the process of visual communication in relation to textual communication. Then follows an example of a communication without graphical content, and an example of graphics unaccompanied by text. This is followed by a detailed description of design types, and of the structural components of graphs. Subsequent Sections deal with basic aspects of graphical resolution, and with some technical features of color printing. Finally, we deal with verity of graphical content.

This paper contains several graphs copied from books and journals, and some of these graphs are reproduced on a smaller scale because they are only meant to illustrate a principle. Readers who wish to look at the scientific content of these graphs are referred to the original publication given in the list of references. This paper, as well as Paper III, also includes a selection of purpose-made graphics that accentuate the design principles explained in the text.

Why do we need such an extensive chapter on making graphs? There are several reasons.

1. An inspection of contemporary scientific graphics reveals an abundant amount of blunders: some scientific papers have even been rejected because of bad graphics (see item 11 in Sect. 5.3 of Paper III).
2. Just as authors nowadays became typesetters, they also turned into draftspersons – though with the assistance of computer graphics software. When I started my scientific career as a young postdoc, most of our graphics were made by a professional or semi-professional draftsperson who was skilled in document design, and to whom we would give a more or less tidy draft on graph paper. That person, using pen and ink, would then turn the draft design into a final graph at a scale fit for photographic reproduction. The first consequence of this division of labor was homogeneity: the person making the final product used more or less standard parameters (symbol size, line thickness, label fonts, ...). In addition, that person would tell you if you had forgotten to add tick marks, or units, or labels, and also would give you guidelines for improving your next graph. This era has definitely come to an end: young scientists are now entirely on their own.
3. A wide selection of commercial software offers integrated facilities for producing statistics from spreadsheets, and doing regression analysis and graphing in semi-automatic ways – nevertheless, many such graphs need subsequent optimization.
4. You want to persuade the reader with the best visual arguments. So, while keeping the presentations truthful and guileless, you also wish to present your product in the most pleasing and aesthetic format.
5. Although numerical quantities (and their statistical errors) from computer-aided analysis nowadays prevail, many conclusions remain based on graphical forms of

[1] Interpretation refers to the construction, by the reader, of the meaning of a message. That reading often differs from the designer's or the writer's intended meaning.

data presentation. Such as the position of a set of stars in some diagram: for example, the many varieties of Hertzsprung–Russell diagrams (see also Sect. 11.8) from which spectral types and evolutionary ages are derived.

6. Astrophysicists rely on calibrations and models, such as the period-luminosity relation of Cepheids, or color index versus effective temperature calibrations, and such relationships can be easily applied by means of a given formula. The example on page 162 illustrates this for spectral classification through photometry, a technique that relies on the position of a star in some diagram. The discussion shows how the graphical representation of the data tells you about the relevance or about the ambiguity of the conclusions. If conclusions are directly derived from graphical representations, it is of utmost importance to make sure that the graph correctly represents **all** the data.

7. Last but not least, just like the writing process contributes to the analysis and to the conclusions, the graphical design process sometimes reveals something that you had not yet seen in the data, equations and tabular material – some kind of *advancing insight*, as explained in Section 2 of Paper I.

2 Data graphics and visual communication

2.1 Written communication

The process of composing a scientific paper or a book often is a graphical procedure, because most scientific works consist of a body of textual elements, with embedded graphics. Paper I lists some crucial differences between oral and literate modes of communication. Communication by way of visual information, however, differs in many ways from oral and written modes of information transfer.

The mathematician Claude Elwood Shannon (1916–2001) founded information theory with his paper *A Mathematical Theory of Communication* (Shannon 1948), in which he derived the logarithmic formula for information content of a message. This technically complicated paper was followed by a much more readable book, with almost the same title[2], by Shannon & Weaver (1949). Shannon defines communication-related quantities that are *"... measures of information, choice and uncertainty"* – much akin to entropy[3] as described in thermodynamics. His simple model of a communication system is shown in Figure 1, and I use this widely copied picture as a metaphor:[4] the *information source* conceives a message, the *transmitter* transforms the message to a *signal* that is sent over a *channel* to the *receiver*, which subsequently passes it on to the *destination*. The processing of information (mainly through language) is essentially stochastic[5] in nature, and is complicated by a vexing perturbation called *noise*.

[2] *The Mathematical Theory of Communication*, *i.e.*, only the indefinite article was changed.

[3] Entropy is a measure of how disorganized a system is.

[4] A metaphor is a figure of speech in which a word or phrase literally denoting one kind of object or idea is used in place of another to suggest an analogy between them (from the *Merriam–Webster Online Dictionary*).

[5] A stochastic process is a process whose subsequent state is determined by a random element.

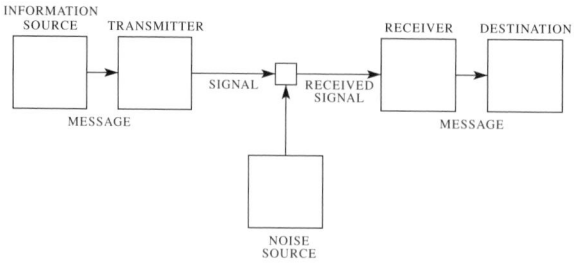

Fig. 1. Schematic diagram of a general information transmission system. Source: Shannon (1948).

In written communication, the information source and the destination are, respectively, the writer's brain and the reader's mind. In oral communication, the transmitter and the receiver are the voice and the ear. In scientific writing, these functions are taken care of by the medium that carries the message (paper or screen), and by the sensing organ (visual or touch sensation). But, whereas oral communication is person to person, scientific writing mostly emanates from multi-author composite brains. Unfortunately, the transmitting medium does not carry the original idea, but a message transformed – across time and culture – by the faculty of language, and by the inherent limitations of graphical information. Likewise, the receiving medium imprints a transformed message onto the receiver. Noise sources interfere at both ends of the chain, and contain all forms of pollution: typographical mistakes, syntax errors, translation flaws, graphics errors, chart junk, miscommunication between coauthors, etc. And the transmission channel – the journal – unavoidably adds its very own noise components: via the Editor and the referees, through language editing, via the Publisher, and through the printing process, as illustrated in Figure 2. There is, though, a fundamental difference between Shannon's communication and our transfer of knowledge: Shannon reduces communication to pure transmission of information. In scientific writing, we also deal with the **meaning** of the information, a meaning that is not only *extracted* from the transmitted message, but is subsequently also *reconstructed* by the receiver. In addition, there are several instances of **feedback**, *viz.*, from the coauthors, the Editor, and the referees.

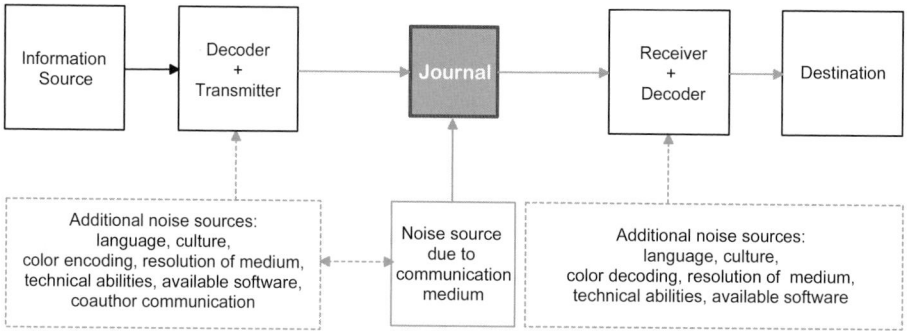

Fig. 2. Schematic diagram of an "extended Shannon" transmission model of communication.

2.2 Processing visual information

The foremost characteristic of the mode of written communication is its outspoken sequential and discrete character: words and sentences are sent and received as discrete elements in a kind of time series. The way how visual information is received and processed differs in yet another manner: the receiver (the eye) gets the message in one blink, and transfers that image to the destination (the brain) for analysis and interpretation. That assimilation of a graphical message by the brain, however, is not instantaneous, but embraces a complicated mental process that consists of breaking down the graphics in different elements. This process, in turn, leads to another time sequence of message elements – a series that differs from person to person – and that, to a considerable degree, depends on the ambience of the moment. So, just as the writing and reading processes are linked, as described in Paper I, the assimilation of graphics also engenders crucial elements of "reading" and "writing".

The opening sentence of Young's (1845) "lecture 38" on vision says:[6]

> THE MEDIUM OF COMMUNICATION, BY WHICH WE BECOME ACQUAINTED WITH ALL THE OBJECTS THAT WE HAVE BEEN LATELY CONSIDERING, IS THE EYE; ...

Yet, a most interesting aspect of the eye as a medium of communication is that the viewer of graphical information is continuously caught in a trap of illusions caused by the very process of vision. At the same time, it is exactly through the untrue mental representation by the receiver's mind that grayscale and color printing with discrete dots is possible at all.[7]

Even more so than when producing text, the creation of graphics demands careful attention by the first element in Figure 2: the information source, or the mind. The other components in the communication chain – the transmitters (camera, draftsperson, artist, plot program), receivers and sources of noise (contrast, resolution, color, ink, ...) – can be controlled to some extent. As photographer Patrick Demarchelier says:[8] *"Your eye is the first camera. The camera is just an accessory between you and the subject"*. In other words: **the design phase of your graphical presentation is the very first phase in the whole process**. The next step is the transfer of your technical and artistic ideas into your artwork[9]. Tufte (1983) states:

> ALLOWING ARTIST-ILLUSTRATORS TO CONTROL THE DESIGN AND CONTENT OF STATISTICAL GRAPHICS IS ALMOST LIKE ALLOWING TYPOGRAPHERS TO CONTROL THE CONTENT, STYLE, AND EDITING OF PROSE.

I cannot underline enough the utmost importance of this first step: its meaning will become fully clear in Section 11, notably by the quote of Ronald Fisher (1925) on page 148.

---

[6] Thomas Young (1773–1829).

[7] Notably by continuous assimilation of high-resolution information, see Section 6.1.

[8] Time Magazine November 2008.

[9] Artwork is the general term used for all kinds of graphics.

2.3 Why use graphs at all?

Graphs and graphical illustrations are made for several reasons, *viz.*:

1. to provide a first contact with the data;

2. to facilitate communication (see, for example, Fig. 21 that visually explains different time standards);

3. to save space;

4. to detect trends and patterns (trends and geometric forms demand less visual memory than tabular information), and

5. to assist you in making decisions and interpretations.

Above all, graphical presentations should help the reader understand the facts and arguments presented in the paper.

2.4 Images, graphs and tabular structures

Artwork is just much more than a collection of pictures: **graphics comprises any illustrative material received by Publisher and Printer**, thus also non-typeset characters such as mathematical equations in graphical format, foreign language characters (*e.g.*, the Pinyin characters shown in Fig. 2 of Paper I), movies, animations, and data Tables.

Graphics come in one of three types: images, graphs and tabular structures. An image is a picture or photograph, a graph (often referred to as a plot) presents interesting data and complex ideas with precision and efficiency, and a Table gives an overview expressed by a condensed and orderly structure, with elements formatted and sorted, and very often of machine-readable conception. Technically spoken, artwork comes in two flavors:

1. *linework*: graphs of all kinds, such as diagrams, drawings, cartoons, and

2. *halftones*: "photographs" rendered in print. Movies and animations simply are time-series compositions of above-mentioned types.

3 Text without graphics and graphics without text

There is one important step that precedes Demarchelier's "first step": the decision whether to support your argument with graphics foregoes the phase of conception and design. This is an important decision, because omitting convincing graphical arguments may very well result in a paper that totally escapes attention. To demonstrate that graphics and text are intimately related, I now proceed with an example of a paper without any supporting graphics, and then I present an illustration of the opposite situation: interpreting an unfamiliar image in the absence of any textual support.

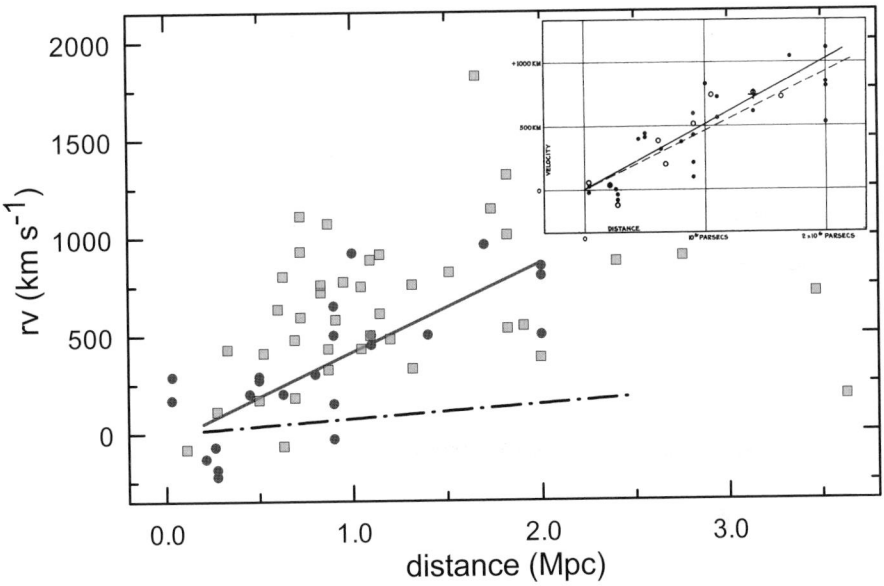

Fig. 3. Relation between observed distance and radial velocity of Hubble's "nebulae" (•: data from Hubble 1929; ■: Lemaître's available data, compiled by Duerbeck & Seitter 2001). The full line is the linear fit to Hubble's data, the dash-dotted line indicates the modern relation. The inset shows Hubble's original diagram (the dashed line in Hubble's drawing is just another way Hubble fitted his data). Source: Sterken (2009).

3.1 Consequences of non-graphical communication: an example

A painful case of non-graphical communication is the paper by the Belgian Georges Lemaître (1894–1966), one of the first relativistic cosmologists. Edwin Hubble (1889–1953) published in 1929 a list of distances[10] and radial velocities[11] of 42 "nebulae" (galaxies), and so derived the empirical velocity–distance relation as shown by the inset in Figure 3. The constant of proportionality, $H_0 = 465 \pm 50$ km s^{-1}Mpc^{-1}, was later called the *Hubble constant.*[12] Hubble proposed two solutions: one using the 42 points, and a second one based on the same data binned in groups (therefore there are two linear fits in his graph). Lemaître's theoretical deductions, based on Einstein's theory of General Relativity, and combined with the available data – in fact the same data used by Hubble – led to his vision of an expanding universe with a *linear* relation between recession velocity and redshift, and an expansion constant of 625 km s^{-1}Mpc^{-1}. Lemaître was the first who referred to a "beginning" universe.

As already explained in Paper I, Lemaître (1927) formulates his hypothesis most explicitly already in the title of his work: *Un Univers homogène de masse constante et de*

[10]Distances derived from the period-luminosity relation of Cepheids.

[11]Radial velocities from Strömberg (1925).

[12]The subscript in H_0 refers to the value of the Hubble constant at this moment.

rayon croissant rendant compte de la vitesse radiale des nébuleuses extragalactiques, in which he concludes that the radial-velocity data lead to an apparent Doppler effect that is caused by the change in the radius of the universe.[13] Lemaître's publication, unfortunately, appeared in the *Annales de la Société Scientifique de Bruxelles*, and thus the paper was not widely read. It is though remarkable that four years later a translation appeared "communicated by A. S. Eddington" (Eddington 1931), and that this translation does not even mention Lemaître's 625 km s^{-1}Mpc^{-1}numerical result. It thus seems that **omitting graphical representation of crucial new findings in a paper reduces its visibility and diffusion**, especially when published in a not so widely available journal.

3.2 Non-textual communication: an example

The front cover of this book reproduces, with permission of the *Röhsska Museum of Fashion, Design and Decorative Arts*,[14] the Tapestry *Astronomie* depicting the Muse *Astronomia* together with one or two astronomers, and several shepherds gazing at the heavens. It is a Flemish tapestry, probably made in Paris,[15] with linen, wool and silk. Its dimensions are 240 × 340 cm, and it was photographed by Swedish photographer Håkan Berg, who included a *Kodak Control Color Patch* strip and a *Kodak Gray Scale* calibration exposure (reproduced on the back cover of this book). A 600 dpi[16] resolution scan of an 18 × 24-cm photoprint was provided by the Museum.

Before discussing some technical aspects of this image, let us first have a look at its content, and at the message we think it conveys. The object is a typical Renaissance tapestry of the type that has been described as *"the mobile frescoes of the North"* by Delmarcel (1999). The central figure is a crowned lady who points at the sky, and holds a banderole with the inscription *Astronomie*, together with a couple of symbols of $*$ and $\circ$ shape. The second person, perhaps an astronomer, holds an armillary sphere[17] while looking up. The folded lower edge of his robe reveals some letters P, T, O, M,[18] and the sleeve of his robe also carries characters.[19] The *sphaera* displays some zodiac signs – the central one, no doubt, is Aries. A second male person is seated and takes notes, he may be a scribe or a chronicler. That person's robe is also adorned with text characters. In front of him is what looks like a positional observing instrument of the astrolabium type. Four other people (one on a ship) contemplate the heavens with great awe. The landscape clearly is not the Low Countries; the sky shows a waxing gibbous Moon, and more than three dozen stars. Half a dozen stars with tails are present, among which a very conspicuous "wheel of fire" with six fiery outlets.

---

[13]Lemaître (1927): *"v/c mesure donc l'effet Doppler apparent dû à la variation du rayon de l'univers."*

[14]*Röhsska Museet för Konsthantverk och Design*, Göteborg, Sweden.

[15]*circa* 1500–1510, according to Delmarcel & Brosens (2009).

[16]dpi: dots per inch.

[17]Johannes de Sacrobosco (or John of Holywood, 1195–1256) published a standard work on astronomy entitled *Libellus de sphaera*, featuring a similar handheld sphere on the cover.

[18]Perhaps referring to Ptolemy?

[19]Renaissance tapestries often display labels, with a mix of French and Latin words, and Greek symbols, in mixed fonts.

What is the meaning of this scene? Without a caption it is hard to understand all the iconography, but it is very likely that the tapestry is one of a tapestry cycle dedicated to the seven *artes liberales*: music, astronomy, geometry, grammar, dialectic, rhetoric, and arithmetic (Brosens 2000) – although no trace of the other tapestries of such a set was ever found. The celestial phenomenon probably refers to the passage of a comet, or/and to an exceptional meteor shower, or storm. It cannot possibly be the Leonid meteor storm of October 24–26, 1532 because the tapestry was made long before that date. There was a Leonid shower recorded in China on October 24–25, 1498 (see the Table on p. 274 of Littmann 1998), and Rada & Stephenson (1992) mention that in Damascus, on AH 902[20] *"... star shooting developed and the sky began to clear of clouds. There were around 20 shooting stars in an hour"*. This was about the last exactly-dated shower[21] recorded in both medieval Arabic and European records for about 330 years. The image of the Moon, with an illuminated part of 84%, could perhaps be an accurate time indicator.[22] Ephemeris calculations based on the NASA Jet Propulsion Laboratory Horizons Systems[23] reveal that on 27 July 1497 the Moon was a waning crescent (9% illuminated), whereas on 25 October 1498 (Gregorian Calendar date 3 November 1498) the Moon was waxing gibbous with an illumination fraction of 77%, which corresponds quite well to the ratio measured on the image.

What, then, does the fiery wheel depicted in Figure 27 represent: a comet? But according to the extensive compilation of Chinese astronomical records *Zhongguo Gudai Tianxiang Jilu Zongji*, compiled under the auspices of Beijing Observatory in 1988, there are no cometary records in Chinese history in AD 1497 or 1498 (Stephenson 2010). Reference to Ho Peng Yoke's (1962) cometary catalogue shows that the comet of AD 1499 was seen[24] in Korea, as well as in China, but not in Europe, and that it is not considered as one of the "great comets".

Or is the celestial iconography only a metaphorization, with the celestial background to be seen as one very special *millefleur* design so typical for Renaissance tapestries? I looked at millefleur compositions at the Paris Cluny museum[25] on many wall hangings from the same period. They are very thematic, with several tapestries displaying cartoon-like sequences, depicting lords and their entourages against an almost abstract background. I perceive one very striking difference between those tapestries and the tapestry discussed here: all other millefleur patterns (various kinds of flowers and birds) are distributed all over the scene, whereas in the case under discussion, the celestial objects are only where they belong: in the sky. And the former never seem to receive any attention from the personages in a tableau, whereas in the latter case (except for the Muse) it is just the opposite. Since all millefleur details at the Cluny museum look very realistic, we might thus be tempted to conclude that the celestial scene is not a millefleur, but an

[20] *Anno Hegirae*, or year 902 of the Islamic calendar, corresponding to 27 July 1497 in the Julian Calendar.

[21] Most likely a Perseid shower, see Figure 2 of Rada & Stephenson (1992).

[22] The image on the tapestry is not perfectly symmetric, and shows a quite wide terminator, hence the accuracy of my measurement is not better than 3–5%.

[23] `http://ssd.jpl.nasa.gov/horizons.cgi`

[24] The comet was visible till 6 September 1499.

[25] *Musée national du Moyen Âge*, `http://www.musee-moyenage.fr/ang/index.html`

artistic recording of a very special celestial event. In addition, the fiery stars quite well resemble the starlets in Albrecht Dürer's *Apocalypse* woodcuts,[26] made in 1498.

Rests the question whether this rendering of the Moon is to be taken literally or symbolically: the *man-in-the-Moon* image, preferably in crescent phase shown in three-quarter or full profile, was a quite common icon from the 12–13th century on (Montgomery 1999). There is thus no way to know whether the waxing Moon in the tapestry is a symbolic or a high-fidelity image.

What is the use of bringing up this image in the context of this book? There are three facets of publishing related to this discussion, *viz.* the graphical characteristics, the descriptive aspects, and the production features.

1. **The graphics**. The technical aspects of the reproduction in terms of image resolution and color are discussed in Sections 8 and 9. The tapestry itself is, in fact, a very low-resolution object (the physical pixel[27] elements are about 0.46 lines mm^{-1} in the vertical direction with ~ 0.73 pixels mm^{-1} horizontally), photographed and reproduced at very high resolution. Thus, a good-quality reproduction in print will reveal the original pixels, whereas a poor-quality reproduction will blend the original image elements, see the examples in Figure 27. The textual elements inside the art object are somewhat ambiguous: the characters are of different fonts, and some of the symbols even overlap.

2. **The documentation of the original message**. There are many uncertainties in the reconstruction of the artist's undocumented message when only looking at the scene, without any clue of who designed the tapestry, nor when the design was made, nor what was the geographical origin of the scene. Nor do we know whether the phenomena were rendered in an objective way. Any reconstruction based on historical sources remains ambiguous, since the place and time of the observation is not exactly known, and because it is not always possible to convert a given date in the Islamic calendar to an unambiguous modern date format with absolute accuracy. Not to forget that such an analysis is prone to many potential mistakes: authors of historical texts do not always clearly indicate which time system they use,[28] and modern computer applications do not always allow the processing of time in systems other than the Gregorian calendar.

3. **The production**. Renaissance tapestries were created by following a cartoon drawing of the same size (made by an artist-illustrator for a customer or sponsor). Assuming that the client was the person who witnessed the celestial phenomenon, this first step is already beset by some amount of communication noise. Especially since the weaver only could see the back side of the weave, thus the drawing of the cartoon must necessarily have been a mirror image of the finished picture. Delmarcel (1999) mentions that, occasionally, mistakes were made in drawing and coloring

[26] Dürer (1471–1528) made a series of 15 woodcuts of scenes from the *Book of Revelation*.
[27] The pixel (picture element) is the smallest discrete component of a (digital) image.
[28] Ho (1962) explicitly flags Gregorian-calendar dates, but does not do so for the Julian-calendar timings.

those cartoons, so that left-handed figures appear, and inscriptions are displayed in reverse.[29] What means that a precise dating based on the Moon's phase may become completely inaccurate if such a reversal would unfortunately have happened. Concerning the intrinsic colors of the artwork, it is most fascinating that medieval weavers already knew how to place threads of different colors so that the eye of the spectator would experience intermediate colors,[30] see Section 9.4.

The foregoing discussion underlines the need for accuracy in any graphical rendering, and in all textual and tabular commentary related to it, as well as in all communications with persons involved in the publishing and production process.

4 Design types of graphs

The first point I wish to clarify is the difference between a plot and a drawing: the former refers to a graphical rendering of tabulated data points, or of data calculated by means of a mathematical function. The latter forms of design are graphs literally drawn using special mouse-driven or algorithm-driven drawing software. A special case of drawings are the schematic diagrams shown in Figures 1 and 2. Text boxes, for example in Section 10 (pages 143 and 150), are also graphical elements (though they are strictly typeset).

More than 50% of the graphs encountered in scientific journals are relational, *i.e.*, the graphs are about data where there is a link between the variables – not just between two quantities, but quite often of multivariate[31] character. Many of these graphical representations are, in one way or another, ordered in series: for example time series (light curves), wavelength series (spectra) or cyclic series (phase diagrams).

In what follows I make a distinction between points/lines and surfaces, a difference that is not related to the true spatial dimension of the graphical content, but to the fact that rendering one-dimensional quantitative information by symbols and lines often yields a different visual impression of the data, than when viewing the same quantities as surfaces.

4.1 Points and lines

4.1.1 Scatter plots

The scatter plot is widely used to show data of two or more related variables. It is particularly useful when the variable of the y-axis[32] is (thought to be) dependent upon the values of the (usually independent) x variable. The data points in a scatter plot are represented

[29] The situation is very similar to modern printing, even more so to reproducing photographs from negatives.

[30] Mortimer (1731), on James Le Blon's method of weaving tapestries, states "The Observation of the *compounded Colours* reflected from two Pieces of Silk, of different Colours, placed near together, first gave him the Thought of what Effect of weaving *Threads* of different Colours would be, when all the *Threads* were so fine, as not to be distinguished at a small Distance one from another."

[31] A multivariate system is one whose properties are determined by many independent variables that are individually subject to sampling and measurement errors.

[32] Here, the x and y-axis denote, respectively, the horizontal and vertical axis in the graph.

by symbols that are sometimes joined by straight lines or by curves, see also Section 11.5. This paper includes several scatter plot graphs.

4.1.2 Vector plots

A vector plot is a scatter plot that does not only represent x, y data, but also carries additional information, see for example Figure 4, where the symbols are vectors with directional and magnitude (length of vector) information. Vector plots can get overcrowded when the vector grid is too dense. Note that in Figure 4 there is some overlap between legend[33] and data. This is a pseudo-color image, resulting from converting the gray levels of a monochromatic image to a color image using a lookup table. The shaded band on the right is the color-code scale, which is the key for quantifying the graphical data.

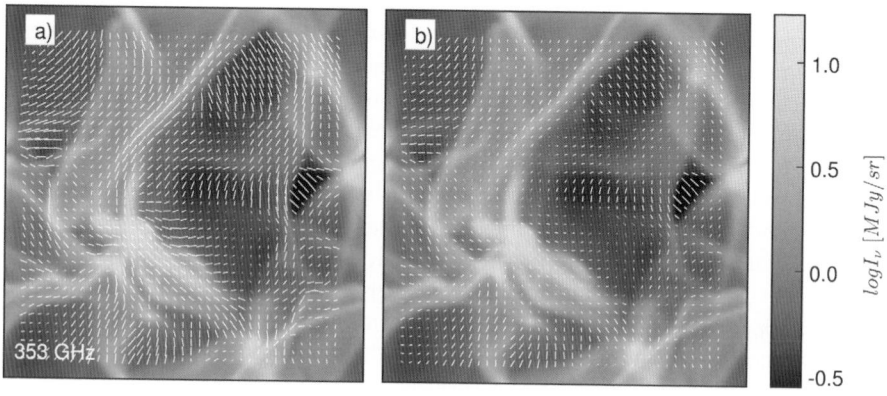

Fig. 4. Simulated polarized dust emission: two models. The scatter plot symbols represent polarization vectors, the colors are log intensity. Source: Pelkonen *et al.* (2009).

4.1.3 Line graphs

Line diagrams are one of the most common tools used to present information. A line graph reveals trends and relationships between data, and illustrates visually how two variables vary with each other.

4.1.4 Time series and phase diagrams

Time series are plots of a quantity (color index, magnitude, velocity, chemical composition, ...) as a function of time. A phase diagram[34] is a time-series curve folded with the

[33] The legend of a graph is explained in Section 7.5.1.

[34] Not to confuse with phase diagram in physics, which is a graph showing the limiting conditions for different phases of a substance while undergoing changes in parameters such as pressure and temperature, see also Section 4.3.7.

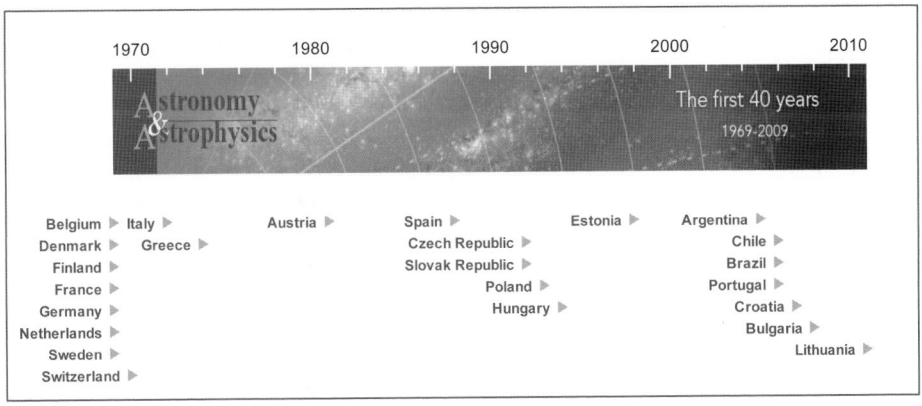

Fig. 5. A&A country membership timeline. The light- and dark-blue tints of labels and symbols are the colors used on the journal's cover (as listed in Table 7).

period of variability, so that the measured quantity is plotted as a function of phase. Phase diagrams are one special way of visualizing cyclic data. Attaching an extra cycle (or part of a cycle) to the time (phase) axis helps the eye to see the cyclic pattern, and avoids the reader to have to jump back and forth between phases 1 and 0. This redundancy is very useful, see the phase axis in Figure 14.

4.1.5 Timelines

A very special time-series graph is the so-called *timeline* – a graphics that conveys a sense of change over time, for example a geologic time scale, or the timelines of evolution or nucleosynthesis. Figure 5 shows some timeline events related to the history of A&A.

Timelines are indispensable for describing large projects and experiments, and for documenting the construction of instruments and telescopes. Time, expressed in Julian Day or civil year (AD or BC)[35] is mostly increasing from left to right along a horizontal x-axis, and can be negative (Fig. 25), but needs not necessarily be linear in scale, nor continuous (Fig. 41). Figure 21 is another example – literally a time line related to the definition of time – including many elements of graphical design discussed in this paper, such as partial grids and omission of the y axis (see also the caption of that Figure). Figure 42 illustrates the use of the cycle number on the x-axis, together with the duplicate time axis in units of year.

4.2 Box-whisker plots

Box-whisker plots display data by showing the minimum of a sample, the lower quartile (Q_1, see also page 153), the median (Q_2), the upper quartile (Q_3), and the highest

[35] AD stands for *Anno Domini* and BC for *Before Christ*. BCE can also be used: *Before the Common Era*.

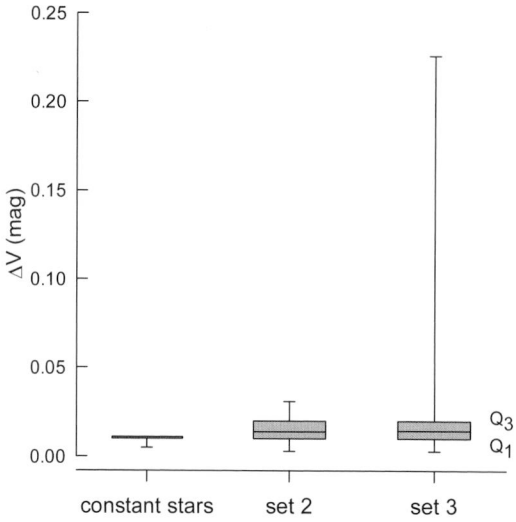

Fig. 6. Box-whisker plot for amplitudes of photometric variability of high-luminosity O and B stars observed by Appenzeller (1972). The sample minimum and maximum are indicated with short horizontal bars, the quartiles are labeled Q_1 and Q_3, the unlabeled horizontal line (Q_2) is the median or second quartile. Set 3 comprises the complete set of amplitudes, set 2 is set 3 *minus* "outlier" R 81, and the first box represents stars with amplitudes ≤ 0.011 (2× the mean error of one observation). See also Sterken (2011) for an application of box-whisker plots to large datasets.

data point, **without any assumption of the underlying statistical distribution of the data**. Figure 6, for example, is a box-whisker plot that displays the range of photometric amplitudes for measurements obtained by Appenzeller (1972) – *i.e.*, $\Delta_{\max} V$ from his Table 1. A further discussion of the use of box-whisker plots to detect outliers is given in Section 11.4, see also Figure 45.

4.3 Surface plots

As stated in the second paragraph of this Section, "surface" here refers to surface aspects applied to the representation of data of any kind, thus not necessarily to three-dimensional data surfaces only.

4.3.1 Histograms

A histogram is a graphical display that shows the frequency distribution of a single variable by dividing the dataset into bins. For each group, a rectangle of base length equal to the range of values in that specific bin is constructed, with the height of each bar corresponding to the count or frequency (or percentage) for the class that it represents. A histogram has an appearance similar to a vertical bar graph (see next Section), but when the variable is continuous, there usually are no gaps between the bars. When the variables are discrete, however, gaps should be left between the bars. One can draw the histogram with adjacent bars touching, or with bars that do not touch – as long as the bin width (for the statistics) does not change. Such histograms are equivalent, and the non-touching version has the look of a bar chart. It is often useful to detach the baseline from the bars, as is done in Figure 28, provided the reader is told how to quantify the vertical scale.

The histogram deals very well with large datasets, and easily reveals the presence of any unusual points or gaps in the data. For good examples, see Figures 5 and 6 of Laurent

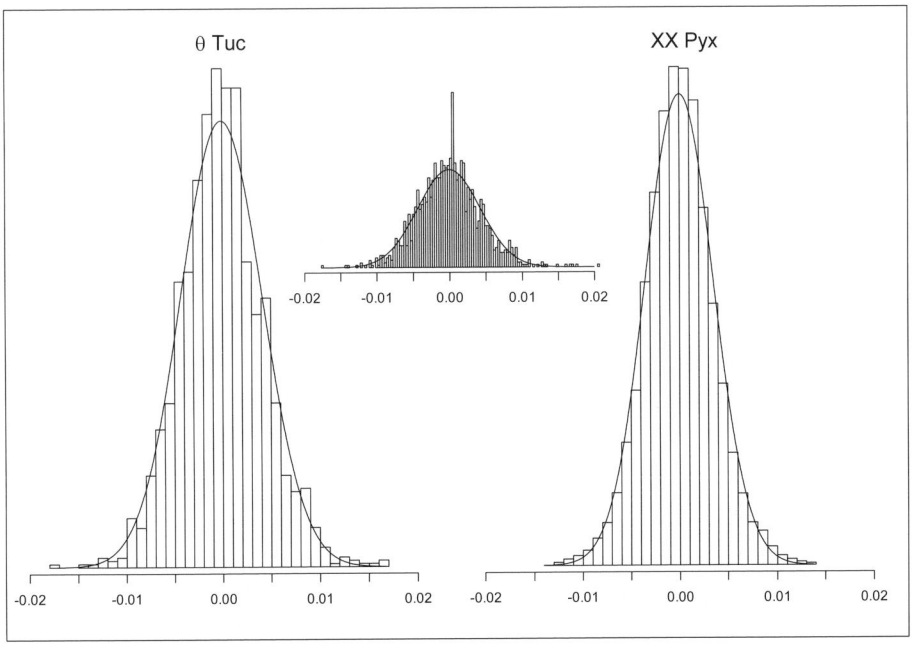

Fig. 7. Histograms of magnitude measurements of θ Tucanae and XX Pyxidis from Sterken (2000). The inset (scale factor 0.75) was made with the same θ Tuc dataset and three times smaller bins.

Cambrésy's paper in Part 1 of these Proceedings, and also Figure 28 below. One glance at a histogram reveals a lot about the statistical properties of the variable: standard deviation, symmetries, outliers (see Sect. 11.4), etc. The histograms in Figure 7 illustrate this very well: they show the distribution of magnitude residuals[36] after removal of all pulsation frequencies from the magnitude time series of two δ Scuti stars. While the XX Pyx residuals support a rather normal distribution with $\sigma = 0.0035$, the θ Tuc residuals are distributed differently, with a standard deviation that exceeds the former by 25%. It is not possible to disentangle the contributions due to intrinsic stellar variability of θ Tuc, and the deviations caused by incomplete frequency solutions or insufficient data content. This example illustrates the assertion of Hoaglin *et al.* (1983) that

> AN ANALYSIS OF A DATASET IS NOT COMPLETE WITHOUT A CAREFUL EXAMINATION OF THE RESIDUALS.

The middle graph shows how the visual impression is modified by merely making the bin widths smaller by a factor of three.

Stacked charts (*i.e.*, each bar is divided into relative proportions of more than one variable) are quite popular but, honestly, I always have a problem to get their message at

[36]Residuals are what remains after a model has been subtracted from the data: residual = data *minus* fit.

a first glance; I find out that I really have to visually "measure" the relative differences before I can make up my mind. Adjacent-bar histograms do not pose this problem.

Histograms are very often used for inspecting and adjusting the tonal range of an image: they show the distribution of pixels within various intensity ranges, and a properly exposed photograph with full tonality will deliver smooth histogram statistics. Figure 4 covers a complete tonal range with good coverage of all densities from dark to bright, whereas in the reproduction on the cover of this book the dark tones predominate, with a total absence of bright shades, thus more shadows than highlights – as is quite normal for this kind of art object. Figure 43 shows information about each of the color channels in the cover illustration: the left side of each diagram displays the darkest shadows, the right side shows the brightest tones, and the height shows how much of that tone is present. Similar – yet even more extreme – histograms result from astronomical imaging where the numerous dark-sky pixels produce a histogram with typical "underexposed" rendering.

4.3.2 Bar graphs

Bar graphs, also called step plots, usually present data grouped in class intervals, and as such they easily reveal patterns and trends. Bar graphs consist of a horizontal or vertical axis, and a series of labeled bars perpendicular to that axis. The width of each bar is arbitrary, and adjacent bars are often separated by a blank space. The vertical bar graph displays data better than the horizontal bar graph, and is preferred when possible, though horizontal bars perpendicular to the ordinate axis offer the advantage that a horizontal label can easily be added to each bar, see Figure 31 in Paper III.

Bar graphs, like pie charts (Sect. 4.3.3), leave you a free choice in the ordering along the x axis: items may be ordered alphabetically, or ranked according to the magnitude of the depicted quantity, or following any other suitable criterion.

A bar graph and a histogram differ in the following: in a bar graph, frequency is measured by the height of the bar, whereas in a histogram, frequency is measured by the area of the column. Figure 31 in Paper III illustrates the combination of both types of graphs on the same dataset.

4.3.3 Pie charts

A pie chart displays the values of a given variable (*e.g.*, percentage distribution of subsamples) using a circle divided into segments. The area of each segment is the same proportion of a circle as the subset is of the total dataset and, consequently, the sum of the segments amounts to 100% of the pie – see Figure 29 for a historic example (Newton's color wheel of 1704). The pie segments are often labeled with their actual values, and one or more segments are sometimes separated from the rest of the pie for emphasis. Pie charts with less than a handful of slices should not be constructed, and charts with an excessive number of very thin slices should equally be avoided (see Fig. 8).

Pie charts are very good tools for showing quantitative information when emphasizing ratios of components in a system (see also Sect. 4.3.7), but the viewer of such graphs should be aware that, when comparing quantities visualized by surfaces, the perception of surface will lead to overestimation of the real ratio. Figure 8 shows the relation between the percentages and the corresponding surface areas (linked by a parabolic curve): whereas the ratio of the percentage levels of the two major segments is about 1.3, the ratio of surface areas is 1.6, which leads to a perceptual misrepresentation that worsens with increasing ratios. The leftmost inset replicates Figure 2 of Bertout's paper in Part 1 of these Proceedings, with area color codes "picked" from the original postscript file (see also page 139 on color picking).

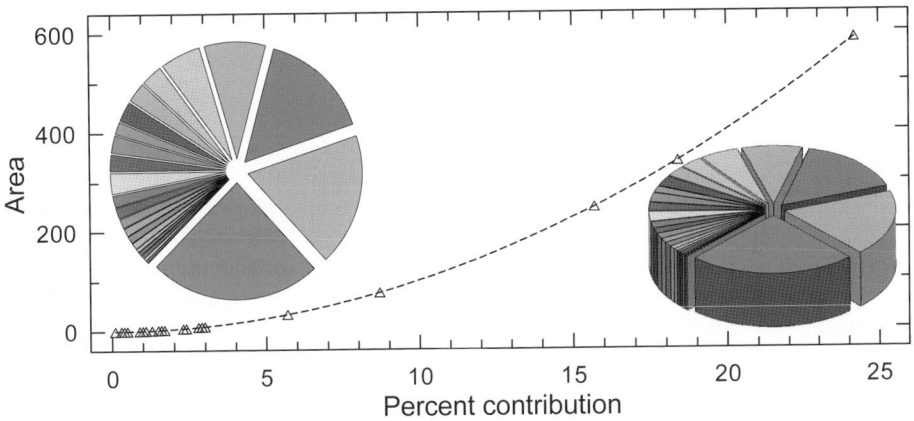

Fig. 8. Submissions from A&A sponsoring countries in 2010 in two pie chart graphs. The line is the parabolic fit of percentage contributions *versus* area for the left pie chart.

Pie charts (and also bar charts) are often presented in pseudo 3-D by adding "depth" through a *lip* of substantial thickness. Adding such an extension (thus using volume for representing a quantity, see the rightmost inset in Fig. 8) will enhance the deceptive perception even more, especially when the pie with the largest component is placed in front. Note that both insets have the same linear dimension along the x-axis. The explosion factor provides a different viewing aspect in 2-D and in 3-D: the three-dimensional concept takes less space, but produces higher distortion, *i.e.*, it increases the *lie factor* (see Sect. 11.3).

The only real difference between the original graph and Figure 8 is that in the former the variables are ordered alphabetically, whereas in the latter they are aligned according to rank order of increasing contribution. This pushes all small slices together, and illustrates the point made already that inclusion of many thin slices should be avoided. Small slices are notoriously difficult to label without overlap or without distracting the eye with an overdosis of labels. The original graph is a good example of proper labeling: labels and the element that is labeled (symbol, area or line) are linked, and should thus be close in space. Note also that the smaller the slices, the larger must be the *explosion factor*, *i.e.*, the amount of eccentricity added at the center of each slice so that their central parts do not stick together.

4.3.4 Bubble graphs and pictographs

Bubble graphs are similar to scatter plots, but they display numerical quantities with plot symbols that have a surface area proportional to the value of the variable. They are also two-dimensional representations of one-dimensional data. Bubble graphs are very popular for making starmaps, but – like pictographs – they are known to misrepresent the data, because the visual sensation through surface rendering is not a linear function of the original quantity.

Pictographs use picture symbols to display statistical information, and are used most often by professional graphic designers. Pictographs should be used with care, because the graphs may, either accidentally or deliberately, misrepresent the data quantities.

4.3.5 Contour maps

Contour maps (also called isolines or isograms) of a function of two variables are a set of lines, along which the depicted function is constant. Such graphs allow visualization of extremely high data densities, provided that an appropriate contour interval is selected: too close contours may occasionally blend. The gradient of the function is perpendicular to the contour line, and lines are often numerically labeled. Labeling is mandatory when the spacing of the lines is uneven. Contour maps are ideal additions to pseudocolor maps, but only when an explanation of the coloring conventions is supplied (in the caption, or as a legend).

4.3.6 Sky maps

Sky maps are projections, and the most common type in astronomy is where declination or latitude (galactic, heliocentric, ...) goes along the y-axis and right ascension or longitude runs along the x-axis. All projections distort, in one way or another, hence the map designer should be vigilant about aspects of conformity, *i.e.*, distance, azimuth angle, scale, and area conservation.

An example of a very handsome and versatile sky map is the annual *Skygazer's Almanac* in *Sky & Telescope*, which is a handy two-page foldout chart that summarizes the rising and setting times of the Sun, Moon and planets. Each chart depicts multiple timelines with celestial events plotted on an hourglass shape, showing in well-chosen colors when celestial cyclic phenomena are observable.

4.3.7 Ternary diagrams

Two-dimensional ternary[37] diagrams (also known as triangular diagrams) show relative proportions (in percentages) of compositional data[38] in a three-component system. Each component is normalized on a scale of 100%, and the coordinates are called barycentric coordinates. Points outside the triangle have at least one negative coordinate (see Fig. 30). Ternary diagrams are thus used for displaying 3-D data in a 2-D graph with a triangular

[37]The adjective ternary means "having three components or elements".

[38]Not to confuse with *composite data types*, which are data types constructed in a computer program.

coordinate system. These graphs are, in some sense, comparable to pie charts, where the components also carry only relative information. There are three reasons to include ternary diagrams in this discussion:

1. they are used to illustrate the relative percentages of three-color primaries in colorimetry (see Sect. 9.5.1);
2. they are (for astronomers) relatively unfamiliar, thus need special attention when applied or read, and
3. they are very useful for supporting non-quantitative arguments, see Figure 36, and also Figure 13 in Paper III.

Each corner of the composition holds 100% of the corresponding component, and all mixtures of two components lie on the edge of the triangle that opposes the apex of the missing component. The closer a point lies to an apex, the more of that component is in the mixture. It is quite common to label the three apexes, but the percentages of the whole, or the normalized scales (0–1), are mostly not given. Users of ternary graphing software should verify that autoscaling and normalization occurs on the full data range, and not only on the range of the data in the plot.

Figure 30, for example, displays two versions of Maxwell's color space,[39] also called *Maxwell's triangle* – one of the very early ternary diagrams. Figure 33 also illustrates the principle of ternary plots, though not with an equilateral triangle, but via transformation to cartesian-type x, y axes where each x, y contains only relative information through the implicit condition $x + y + z = 1$.

Ternary diagrams are not easy to read and to interpret, moreover they are not invariant to some coordinate transformations – for example, adding a constant in order to render all coordinate values positive transforms the data content of the graph. Some basic statistical quantities (variance and correlation) are not invariant, since most statistical techniques are designed for free-ranging data, and not for bounded variables in ternary visualizations. Patterns (contour lines) in such diagrams may thus have a different meaning, and spurious correlations may result, Euclidean rulers will shrink when approaching the edges,[40] grid lines can change, and this all has an impact on outlier metrics (see Sect. 11.4). The triangular inset in Figure 33 accentuates not only distortions of the shape of embedded curves, but also illustrates the mental switch needed to read off coordinates along the triangle's sides.

4.4 Genuine 3-D surfaces

Genuine three-dimensional surfaces (for example, point-spread functions and 3-D models) need real 3-D visualization. Figure 9 shows an example of a surface plot of the distribution in solar latitude of the total sunspot area as a function of time (for the time interval 1877–1902). Three-dimensional graphs have very specific elements that need attention, *viz.*:

[39] Color models and color spaces are explained in Section 9.4.

[40] Distances between pairs near the center and along edges may seem equal, but are not, see Figure 36.

- contrary to what is said in Section 7.5.4, grid lines can be most useful when applied as a surface mesh, since mesh plots are ideal for visualizing, though they are much less suitable for precise quantization – grid density is thus an important parameter: if the density is too coarse, the rendering of the surface becomes inadequate;
- it may happen that some sides of a 3-D surface are hidden by what is in front, or that a data aspect is obscured by an axis, see Figure 9, where the highest peak almost coincides with the vertical axis at the back;
- an important visual facet is the *lighting angle* (45° in this case) that produces the shadows: a change in this angle can bring forth a totally different view;
- *depth*, the third dimension shown at a selected angle (solar latitude in Fig. 9), also strongly influences perception, and affects legibility of axis labels;
- opaque background walls may lead to uncontrolled clipping of data and outliers, although this is not the case in Figure 9, since there are no data beyond the axes limits;
- comparison of both graphs in Figure 9 shows that replacing missing values by zero-values produces a totally different perception, and even adds a spurious high-frequency signal.[41]

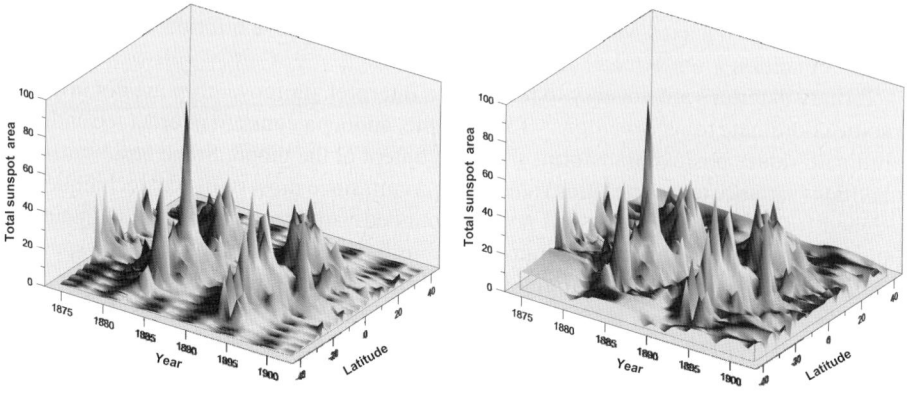

Fig. 9. Sunspot data from Maunder (1904). *Left*: the missing values from Maunder's Table were included as zeros. *Right*: based on the original Table without replacing the missing values with 0.

The dataset shown in both panels of Figure 9 is a very famous one, which led to the well-known *Maunder butterfly diagram* that demonstrates that the latitude of sunspot occurrence varies cyclically on a time scale of about 11 years, with sunspots forming at high latitudes during the minimum of the cycle, and getting closer to the solar equator towards the cycle maximum. And not only the data are crucial, also the initial Maunder

[41] Replacing missing values by zero-values in the wings of a histogram may yield a totally spurious best-fit theoretical distribution function.

graph was fascinating, even the story of the publication of the original data, and the subsequent graphical renderings by two different persons leading to contradictory conclusions, is most intriguing, and deserves some explanation.

Maunder's (1904) paper commences with the following statement: *"Rather more than a year ago the Astronomer Royal communicated to this Society a paper which he had desired me to prepare"*.[42] That paper summarized, in graphical and in tabular form, the results of more than 30 years of work in the measurement and reduction of about 9 000 photographs of the Sun, involving more than 250 000 measurements made in duplicate. The data comprise the total area of whole sunspots divided by the number of days of observation, expressed in millionths of the Sun's visible hemisphere, for each degree of solar latitude. That paper discussed the data with graphics consisting only of bar charts of the annual spot area as a function of heliographic latitude. These graphs showed gross trends, however failed to reveal any striking patterns.

Surprisingly, one year later Maunder supplements this very paper with "other diagrams",[43] together with new diagrams that he deemed necessary for refuting some conclusions – entirely based on Maunder's own data – since published by William J. S. Lockyer. He refers to his own "authorless" paper[44] as *"the Greenwich paper of a year ago"*. Two aspects are of interest in this discussion:

1. the somewhat strange way of publishing such a large and original dataset without author, and

2. why Lockyer (1903) and Maunder (1904) come to **different graphical interpretations of the same data**.

Not to speak of Lockyer – who very well knew who had composed the 1903 data paper on sunspots – who refers to it in terms of *"the fact that the Astronomer Royal has quite recently published the values of the amount of sunspot area for each degree of latitude for each hemisphere ...*", without even mentioning the composer[45] of that paper.

Maunder (1904) presents curves of distributions of spotted areas in zones of 5° latitude, and states that his Figures are "practically" the same as those of Plate 15 in the anonymous 1903 paper, but that the new presentation in 5° zones describes the facts better than the more detailed (bar-)graphs given previously. Indeed, the 1°-resolution graphs are dominated by isolated outbursts and their features are, as he describes it *"a feature of solar book-keeping, not of solar physics."* In other words, **he describes the positive effect of binning in bringing out salient features – and thus emphasizes the impact of binning on the perception of the graphical content.**

Maunder (1904) questions whether Lockyer's year-to-year joining of spot-centers to form "spot-activity tracks" (Fig. 10) is justified, and he regards the linking together of *"certain selected"* spot-centers as *"a matter of personal predilection, not as one of solar physics"* (see also the discussion on dot joining in Sect. 11.5). He then illustrates the

[42]This paper is MNRAS 63, 452. The Astronomer Royal in 1904 was William H. M. Christie (1845–1922).

[43]Diagrams, he says, he composed at the time when he prepared the 1903 paper for the Astronomer Royal.

[44]ADS refers to this paper as authored by N/A (Bibcode 1903MNRAS..63..452), see Paper I, Section 9.2.

[45]Here I use *composer* because the paper was not properly authored.

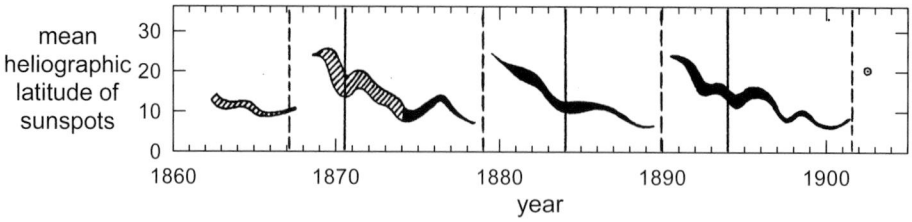

Fig. 10. Mean daily area of sunspots, Lockyer (1903) Plate 1. The caption of the original Figure does not specify the meaning of the different patterns of shading.

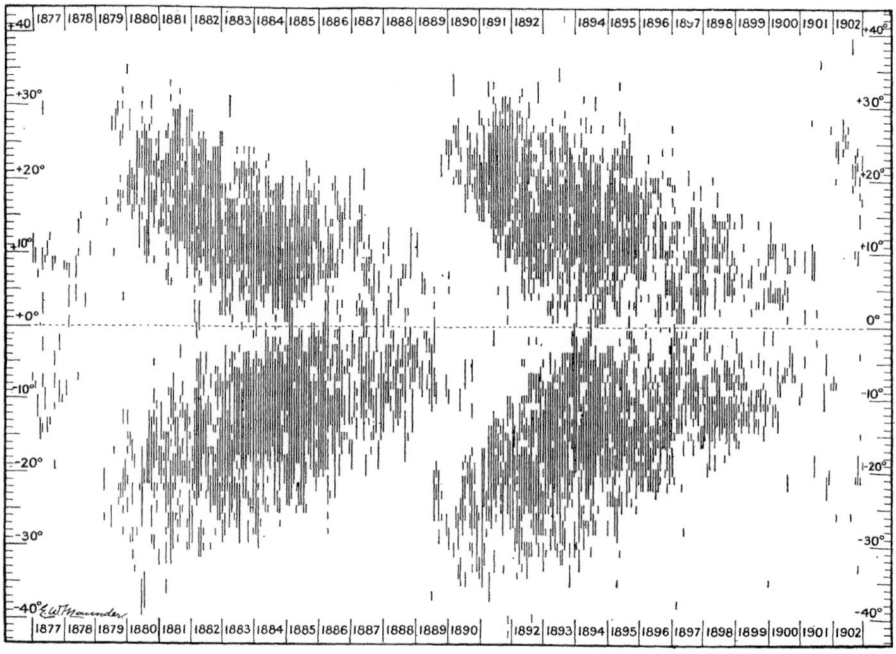

Fig. 11. Plate16 of Maunder (1904): distribution of spot-centers in latitude, rotation by rotation, of the Sun (not on the same scales as Fig. 10). Compare with Figure 52. Note that the time axis labels for 1891 and 1893 were omitted in order to avoid overplotting with data.

arbitrary character of Lockyer's sunspot-group associations by showing the distribution in latitude of spot-centers in the northern hemisphere, and finally presents his ultimate diagram in "Plate 16": the distribution of spot-centers in latitude, irrespective of their surface area, rotation by rotation of the Sun, covering a quarter of a century, *i.e.*, two solar cycles (Fig. 11). This visually convincing diagram is now known as his famous *butterfly diagram*. Figure 12 shows a modern version of this diagram, together with a graph that basically comprises the same data, with the spatial component removed by averaging (lower panel in Fig. 12). The Table below lists the principal differences between the modern and the vintage versions of these diagrams.[46]

Item	Maunder (1904)	NASA butterfly diagram
Ordinate scale	linear in solar latitude φ	linear in $\sin\varphi$
Data extent	restricted to $-45° \leq \varphi \leq +45°$	full range $-90° \leq \varphi \leq +90°$, including dataless regions
Dimension	2-D	3-D by means of red and yellow color coding
Database	contains no 0-values	contains zeros, which are rendered as white pixels

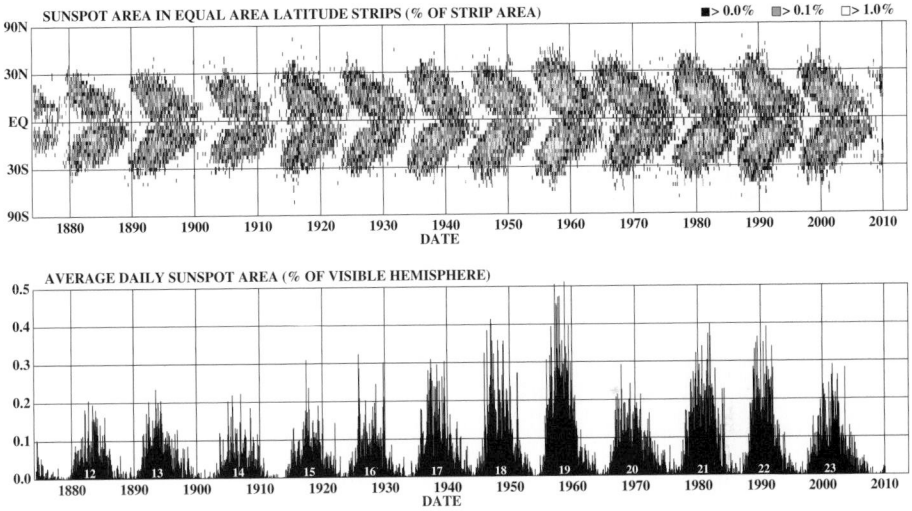

Fig. 12. Daily sunspot area averaged over individual solar rotations. *Top*: sunspot area in equal-area latitude strips; *bottom*: average daily sunspot area (in units of 0.001 visible solar hemisphere). The white labels 12–23 in the lower diagram indicate the solar-cycle number. Source: David Hathaway, NASA/MSFC.

4.5 Drawings and cartoons

Drawings and cartoons are often utilized to demonstrate a scheme or plan, or a model. Figure 13, for example, shows the family tree of traditional photometric systems published by Sterken (1992). This graphical metaphor clearly suggests that photometric systems result from pruning and grafting the tree – in other words, it illustrates that

[46]Note that this tabular structure is included, intentionally, without Table number.

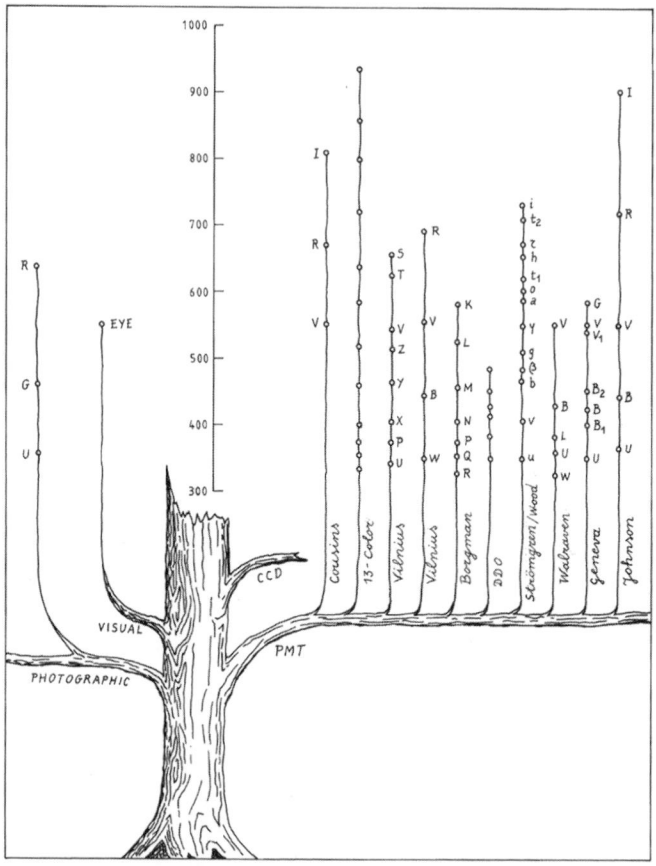

Fig. 13. Genealogy tree of classic photometric systems. Each main branch symbolizes a photometric detector, and the buds represent photometric bands arranged by wavelength along the vertical axis. Reprinted from *Vistas in Astronomy*, Vol. 35, *"On the future of existing photometric systems"*, C. Sterken (1992), with permission from ELSEVIER.

maintenance of a standard system is a prerequisite. This chart displays no strict quantitative information, although the vertical wavelength scale is correct. There is no need for an x-axis in this graph, since the layout of the horizontal axis is arbitrary, and the ranking of the items along the x-axis has no meaning, except for a purely aesthetic arrangement.

Note that the caption of Figure 13 gives the full bibliographic identification of the source, as requested by copyrightholder ELSEVIER. This reuse was free of charge, although it was required to obtain a license through RIGHTSLINK,[47] and sign a clause of compliance with the license terms and conditions.

4.6 Which design type of graph?

The Maunder–Lockyer story, explained in the previous Section, clearly illustrates that different graphical representations of the same dataset can very well change the viewer's

[47] `http://www.copyright.com/search.do?operation=show&page=ppu`

perception, and hence also the viewer's analysis and conclusions. Different kinds of graphs may thus have different powers of convincing the reader. So, before embarking on any graphical design, it is good practice to spend a moment of reflection about which type of graph will be used. This question is thus:

the data being equal, which graphical design conveys my message best?

The first and foremost answer is that you must try to avoid complexity by selecting the proper type of graph. Take, as an example, the phase diagram in Figure 14, the left part of which is reproduced from van Genderen & Sterken (2007). The manuscript submitted for publication, however, originally contained the right diagram of Figure 14 to illustrate a small modulation ($\sim 3\%$) of the optical brightness around phase zero of the very eccentric ($e \sim 0.5$) binary BP Cru, but the Editor of the journal could not be convinced. Replacing the bar graph with the scatter plot finally made our statement more acceptable. The only reason why we had opted for a bar-graph approach in the first place, was to enable direct comparison with similar diagrams in the literature (see van Genderen & Sterken 2007 and references therein). This example also shows that designs of graphs are copied or imitated from one source to the other, and as such get consolidated, unless "alien eyes" force the designer to reconsider the concept. The same applies to tri-variate data and ternary diagrams: they do not always provide the best mental impression to the eye untrained for their interpretation.

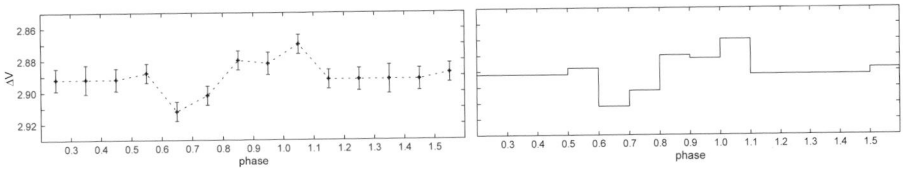

Fig. 14. The same data presented in different graphical forms. Source: van Genderen & Sterken (2007). Note how the labels and line widths suffer from the reduction factor.

Another element of design to consider is the aspect of symmetry: whenever part of a graph is a mirror image of another part, you may opt to omit the symmetric portion, because erasing redundant parts increases the information content rather than reducing it. That is, provided that you tell the reader that there is symmetry involved. A typical example of symmetric graphs is found in spectral-window frequency curves in Fourier analysis. Beware, though, that a symmetrical distribution on a log scale implies an asymmetric distribution of the argument variable.

5 Computer graphics

There are two basic types of computer images: vector images and raster (bitmap) images. The former are best suited for drawings (like scientific graphs), the latter for photographs, paintings, and image processing. Graphics files can contain raster information, vector information, or both.

5.1 Vector images

Vector graphics are images that are described by vector objects, *i.e.*, they are completely defined by points and curves (vectors), and by instructions (line type, line thickness, and color) for the vector-based drawing software to produce lines and curves by connecting the individual points (either with straight lines or with *Bézier curves*).[48] Well-designed vector drawings are resolution-independent and can be scaled to the maximum resolution of the printing device without any loss in sharpness. In other words: **vector graphics are scale (*i.e.*, output device resolution) independent**. Vector images are ideal for designing logos and other objects that have to be frequently resized, and because vector images only contain the data needed to describe the vectors, their file sizes can be quite modest, as long as the graphs are not too complicated.

5.2 Raster or bitmap images

A raster or bitmap image is a matrix of pixels with, *for every cell*, the instruction that describes its pixel content: examples of bitmaps are scanned photographic images and digital pictures. Bitmaps can take up a lot of disk space: an uncompressed good-resolution full-page color picture can easily amount to 50–100 Mb. Such images are resolution-dependent, therefore embedded text cannot be correctly scaled. There are basically four main types of bitmap images:

1. *line art*: contains only 2 colors, mostly black and white (these images are named bitmaps because only 1 bit [0 or 1] is used to describe each pixel);
2. *duotones*: contain shades of only two colors, often black and a second color;
3. *grayscale images*: contain shades of gray, but also pure black and white;
4. *multitone and full color images*: either tritone (three colors) or full-color (quadri).

An important term is *bit depth*, which is the number of bits used to store information about each pixel in the image. The higher the bit depth, the more levels of gray or color shading that can be rendered. Monochrome images are 1-bit images (each bit contains two levels of gray), grayscale images are 8-bit images. A four-color image usually is a 32-bit image with eight bits of data (2^8 or 256 levels of shade) per color channel,[49] yielding 16.7 million colors.

The term bit depth mostly refers to the number of bits per pixel in the image file, but can as well refer to the largest output number that a digital camera can produce. That number depends on the capabilities of the analog-to-digital signal converter. A 16-bit converter provides 65,536 levels of information, a 12-bit converter divides the signal into

[48] Bézier curves are segments resulting from cubic interpolations between control points, and have the property to preserve their smooth appearances, even at high magnification.

[49] Color channels are explained in Section 9.4.

4096 levels (although the 12-bit information is often stored in a 16-bit image format). Most CCD cameras use a 16-bit format for capturing and saving (for example in FITS format, see Sect. 5.4), and many digital cameras have settings that allow images to be shot and stored in the very flexible 16-bit RAW format (or in 16-bit and 8-bit JPEG as well). Note that there does not exist a common RAW format among camera types and brands, since each camera manufacturer has established its own proprietary RAW standard (NEF by Nikon, CRW by Canon, etc.).

5.3 Converting bitmap to vector data and vice versa

Transformation of images from raster to vector data is sometimes needed when scanning a photograph of a logo or a drawing (like Fig. 13), because such graphics have awkward magnification and file size properties. The transformation procedure to vector curves is also called *tracing*, a process comparable to *Optical Character Recognition* (OCR) of text, and results in lines and characters to be recognized and saved in vector format.

Vice versa, vector drawings must be converted to bitmaps for use on web pages, and many drawing-software packages allow saving vector data as bitmap files via the `Export` menu. Image-handling software can import some vector file formats through *rasterization*, and let you save the result as a bitmap.

In both kinds of applications, the user should not forget to specify the resolution and mode (grayscale or color) of the output bitmap, as well as the data-compressing method.

5.4 File formats

The reader (human or machine) needs to know how the data are stored in an image file. This information can be quite complex, and is called *metadata*, *i.e.*, "special data about data". Appropriate software is needed to reconstruct vector graphs, though in principle visualization of bitmaps is simpler – except for the data compression and all forms of included image previews.

Since image files are written as output of a computer application, and then read again by another application, this process very much follows Shannon's model of communication (see Sect. 2). Some elements related to file formats thus contribute to confusion and communication noise:

1. there is the *native format* designed by the developer of the software;
2. formats, by design, are made to deal with specific information, and are not necessarily transformable;
3. formats change with time (via higher software versions);
4. formats have been shown to contain programming bugs and hidden information – even malicious code containing safety hazards;
5. format is not the only aspect to consider: not all image-writing applications allow the user to specify the desired resolution, a problem when outputting information;
6. the perpetual problem of transporting data between operating-system platforms;
7. some formats are open standards, others are not, and

8. the data compression: some data-compression methods destroy some of the data during compression (so-called *lossy* compression, where the amount of compression is highly dependent on the image content), and thus modify the information content of the image.

So, if transforming from one format into another is such a problem, why then do we do it in the first place? There are several reasons, the first of which relates to making the graphs, and the second concerns getting the graphics published:

1. file conversion is needed if you collaborate with colleagues who use other software – even a previous version of the same software, and

2. bitmap and vector graphs can be saved in a wide variety of file formats, but very few of these are suitable for *prepress*[50] operations: Publishers are not prepared to deal with a wide variety of formats, and quite often limit your choice to TIFF or EPS.

I have a recent example of the first point: in one project, a participant sent us CCD frames in the *Pictor 416* FITS format, which could not be read with standard FITS readers (IDL[51] or SAOIMAGE[52]). But the frames could be handled by MAXIM[53], which in turn produces output in a FITS standard that is compatible with IDL and SAOIMAGE.

Table 1. Principal formats used in the scientific community, with file extension and bit depth per channel.

Ext	bit	Description
BMP	8	format that is not suitable for prepress
EPS	8	format that can contain bitmap as well as vector data
FITS	16	format for data capture, data reduction and analysis, not suitable for prepress
GIF	4–8	mainly used for internet graphics (so-called indexed color)
JPEG	8	applies a lossy encoding method, and is mainly used for internet applications
PDF	8	versatile file format that can contain any type of data
PICT	8	format that can contain both bitmap and vector data, mainly used on Macintosh computers
PNG	8	image format that employs lossless data compression, also used for web applications
RAW	16	image processing, not suitable for prepress (every manufacturer has own RAW standard)
TIFF	8–16	the best bitmap file format for prepress and other non-web applications

Table 1 is a concise overview of the principal formats in use in the scientific community, and the subsequent list gives some more detailed information about each type.

- BMP (*Windows bitmap format*): a Windows file format, ideal for Windows adepts as long as no transfers to other platforms are required.

[50]The *prepress* phase includes all processes and procedures that occur between the creation of an image and the final printing.

[51]Interactive Data Language.

[52]`http://fits.gsfc.nasa.gov/fits_viewer.html`

[53]`http://www.cyanogen.com/`

- EPS (*Encapsulated Postscript*): developed by ADOBE, and tuned for including graphics in typeset documents, see also Paper I.
- FITS (*Flexible Image Transport System*): the file header is in ASCII, the body contains graphics information in binary format. The FITS format was developed for exchanging scientific datasets, and it is commonly used to store astronomical photographs. It supports both 8-bit and 16-bit black and white images, but there is currently no standard for encoding color images.
- GIF (*Graphics Interchange Format*): used to place images on the Web. GIF uses *indexed color*, *i.e.*, each pixel's data only consists of an index number that identifies one of 16 (4 bit) or 256 (8 bit) selected colors of a user-defined *palette*[54] that is embedded in the file.
- JPEG (*Joint Photographic Experts Group*): a standard image file format for multiple computing platforms, allowing for variable levels of (always lossy) compression. JPEG always discards information that cannot later be recovered: saving and reloading a JPEG color image will always degrade its color information. JPEG is excellent for archiving images, *i.e.*, images that need no additional processing. JPEG2000 is a format similar to JPEG, though better because it uses a wavelet compression method and has a lossless option.
- PDF (*Portable Document Format*): an open standard for document exchange created by ADOBE Systems, used for documents, and characterized by portability (reliable viewing and printing across platforms).[55]
- PICT (*Apple Picture Format*): 8-bit format that can contain both bitmap and vector data, mainly used on Macintosh computers.
- PNG (*Portable Network Graphics*): was developed as a replacement for GIF. It supports all image types, and works across different platforms, and features lossless compression.
- RAW: contains the original pixel information directly from the detector, untouched by lossy compression and demosaicing[56] – in fact, the FITS format is a RAW format.
- TIFF (*Tagged Image File Format*): a very versatile and transportable format that is supported across many different platforms. It is a good choice, both for image file storage and for exchanging images with Macintosh, Windows or Linux-operated computers, but TIFF files are considerably larger than JPEG files. TIFF has many variants – some of which utilize lossy or lossless compression – and the user can occasionally suffer from incompatibilities between files generated by different applications. TIFF supports both 8-bit and 16-bit per channel images.

Submission to a Publisher in any format other than TIFF or EPS is not recommended, because these are the standard graphics file formats compatible with the composition software in the printing industry. JPEG is acceptable only if saved with the least compression settings (*i.e.*, highest quality).

[54]The concept *color palette* is described in Section 9.4.

[55]Invented by ADOBE, but not owned by ADOBE: the format is not proprietary – it belongs to the community.

[56]The processing pipeline that builds a color image from the outputs of the color-filtered detector pixels.

Note that all formats in Table 1 are ≤8-bit formats, except FITS, TIFF and RAW, which contain 16-bit information. The transition from 16-bit to 8-bit – if necessary – should be the last step in the image processing, *i.e.*, all modifications (color, brightness, cropping, contrast) of 16-bit images should be done in the 16-bit environment before saving in 8-bit. Converting 8-bit to 16-bit makes no sense because adding the extra bit depth does not add any information: ***image storage depth* is not to be confused with *image information depth*** (the format cannot add, but for sure it can destroy information content). Once a file is damaged by saving as JPEG, the harm can never be undone by converting it back to TIFF or PNG!

6 Photographs

Journals reproduce photographic images, but what you see in print are not photographs, but so-called *halftones*. In printer's jargon, the term halftone can also refer to the image that is produced by this reprographic process, so if a Publisher asks you how many halftones are in your paper, then the number of photographic images is meant. Another confusing misconception is the reference to color versus black-and-white photographs. We almost never deal with photographs that only contain black and white: most photographs are grayscale renderings. A real black-and-white image is Figure 16, which is a 2-bit composition: each pixel is either black, or white (0 or 1), without graytones (that have bit values 0–255).

6.1 Halftones

Photographs (but also images on computer, TV and movie screens) are continuous-tone images including a wide range of colors or graytones. But XEROX copiers and laser or inkjet printers only have black ink or toner – no shades of gray (thus 2-bit). For printing such an image, it must be rendered into a halftone by a reprographic technique – *the halftoning process* – that simulates the photograph by rendering a binary image of equally-spaced dots of one color (also called halftone cells) and of varying size and density, which create the illusion of variations of gray or continuous color.[57] The rendition fools the eye-brain detector into the perception of shades by using the white of the paper and the color (or black) of the toner.

In desktop publishing, halftones are automatically generated by the printing device when a file is sent for printing, and the halftoning parameters are by default contained in the postscript printer setup file.[58] The most important parameter is the *halftone screen frequency*, which is measured in lines per inch (lpi is the printing industry's term for resolution) and denotes the number of dots that fit in one inch. Note that the quality of the printing paper also plays a role: very fine halftone cells are of no use on low-quality porous "newspaper" paper, since the small cells will "bleed" and produce clumps

[57] In the past, photographs were reshot through a halftone screen, nowadays the process is entirely digital.
[58] File with a `.ppd` or `.PPD` extension.

of ink. Newspapers use 85 lpi, but journals and magazines use 133–175 lpi.[59] The line screens used by the printer of A&A are 170 lpi for the cover, 133 lpi for articles printed in black (on a rotary offset press), and 150 lpi for articles and pages printed in color (on a sheet offset press). For a four-color image (also specified as *quadri process*), four halftone screens at different screen angle are applied – one for each ink used in the printing process. A second parameter is the *halftone screen angle*: a typical sequence can be 45°, 75°, 90° and 105° for each of the colors (see Fig. 38).

As the lpi parameter comes into action only at the moment of printing, the result will primarily depend on the image resolution that was delivered (see also Sect. 8). The rule of thumb is to ensure that the delivered image should have a resolution of at least the resolution at the output side of the printing process. In other words:

GARBAGE GRAPHICS IN = GARBAGE GRAPHICS OUT = ALWAYS!

QUALITY GRAPHICS IN = QUALITY GRAPHICS OUT = SOMETIMES!

We are used to saying that a photograph is *taken*, but the correct technical term is *captured*. Any artwork image based on the photographic process is first captured, and then *rendered*. There are two types of rendering: let me call the first the *realistic* one, the second the *artistic* variant:

1. the former simply is the standard process of cropping and trimming an image, with unsophisticated optimization of contrast and color;
2. the latter is any kind of rendering that changes the straightforward interpretation of the graphical content through trimming of unfavorable picture elements, drastic changes of contrast, hue and saturation – in short any form of interpretive rendering that willingly or unwillingly changes the basic data content (see also Sect. 11, and Paper III).

6.2 Halftones and Moiré patterns

Moiré patterns are seen when a repetitive structure is overlaid with another such structure, with the line elements nearly superimposed. When scanning a photograph (*i.e.*, halftone) from a book or journal, Moiré patterns are produced by interference between the scanner sampling grid and the halftone screen of the original image. Such patterns may also appear when scanning a painting, because the line- and column pattern of the canvas will interfere with the scanner's sampling grid. All scanners will do this, for each of the color screens involved. For example, a 150 lpi halftone scanned at 600 dpi will result in a periodic 200 dpi Moiré pattern. To avoid any interference caused by the grids, line screens must be about 15° angles apart. Figure 15 illustrates the effect in two images of a polar-coordinate grid (with a line pattern of about 25 lpi) acquired with the same flatbed

[59] You obviously need not worry about paper quality when writing for a journal or for a proceedings Publisher, but you have to consider this aspect when producing a PhD-thesis manuscript at a local printshop!

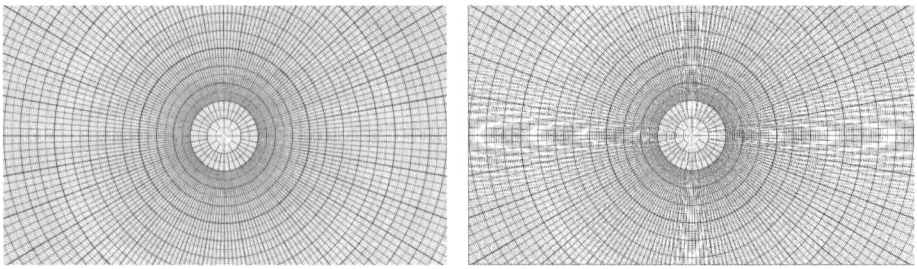

Fig. 15. *Left*: direct print of scanned polar-coordinate grid; *right*: image viewed on monitor screen.

scanner. The left image is the direct print of the scan, and the right image is a screenshot of the scan as it is rendered on the monitor screen[60] showing, respectively, mild and strong Moiré interferences. If you scan halftones, you must consequently process the result with a *descreen filter*. Moiré patterns also appear in printing when the halftone screen angles of two colors are less than 30° apart.

7 Elements of a graphical image

Graphs must be effective, simple, clear, and aesthetic. That means that each of the composing elements of a graph must be worked out with proper care: neglecting one of the elements may very well destroy the good look of an entire composition. Moreover, elements in a graph strongly interact with each other, and this interaction may change the aesthetic appearance, even the scientific message. Pure text as graphics also occurs, for example in Figure 4 in Paper I, but we have seen that graphics without any words are meaningless, hence an important issue is the typesetting of textual components, in the first place because the typeface in a graph is often different from the typeface in the body of the text.

All good graphics contain textual characters, some of which are added by you, the designer of the graph: axis labels, graph titles, data labels and legends. Many of the components of graphical images discussed below are automatically generated by the drawing software. Even if you master the settings and options of your graphics software, it is still useful to verify and adjust these elements, so they can optimally serve the reader. It is, therefore, important to explain a few basic facts about typography.

[60]MLCD 20″ monitor with 32-bit "highest" video setting, *i.e.*, 1600 × 1200.

7.1 Aspects of textual components in graphics

7.1.1 Typefaces and fonts

Typefaces and fonts also belong to non-data parts of a graph. *Typeface* refers to the visual design characteristics of a specific character type. A *font*[61] is a set of letters and symbols of a specific design. A font simply is a variant design of a typeface: a text (for example a letter) that comprises only one typeface can still contain many fonts. One example is the classical and popular typeface HELVETICA (a modified version of HELVETICA became a Windows font called ARIAL[62]). So if the typeface is HELVETICA, the font could be HELVETICA 12pt, containing all the letters, punctuation marks and symbols needed for typesetting in typeface HELVETICA at 12 pt size.[63] In the past, fonts were sets of metal shapes of a particular typeface; nowadays fonts are files on a computer that describe the typeface.[64] Although font and typeface have different meanings, the terms are often interchanged in the context of digital printing. The more so because digital fonts are resizable, hence size-specific font nomenclature referring to point size has become irrelevant. The correct phrasing is to speak of *choosing a typeface*, but *installing a font*.

Fixed-resolution bitmap fonts do not print or display very well on some devices, hence authors are advised to always use scalable fonts (also called *outline fonts*). Scalable fonts, when sent to a screen or to a printer, are transformed to dots at the output device, a transformation that is called *rasterization*.

There were originally many different formats for digital fonts, fortunately some standardization has developed, see Table 2. One of the first standards was ADOBE's Type 3 bitmapped fonts, which were *open source*[65] from the beginning, and which were destined for the publishing and graphic arts industry. ADOBE then developed outline Type 1 fonts – which included *font hinting*[66] – but the company did not disclose the source code. Apple developed its own TrueType scalable font technology in the 1980s, and TrueType fonts were built into the operating systems of Apple and Windows computers. Type 1 or TrueType are scalable vector-based fonts (see Table 2).

For documents using Type 1 fonts, the ADOBE TYPE MANAGER program will automatically create bitmapped fonts of any size for display without embedding the fonts in the PDF document. Type 1 Fonts are implied by specifying ADOBE TIMES Roman, COURIER, HELVETICA, and SYMBOL POSTSCRIPT fonts for the entire paper, *including all graphics*.[67] Problems can arise when using PostScript Type 3 fonts – for example the COMPUTER MODERN ROMAN font family used by LaTeX – that are written to a less strict syntax than the Type 1 fonts. The same applies when annotating photographs. It is the task of the

[61] The word *font* comes from French *fonte* (casting) or Latin *fundere* (to cast).

[62] A non-standard postscript font that may cause problems with specific `dvi` viewers.

[63] Section 7.1.2 explains the *point*, a typographic unit.

[64] Metal fonts were supplied in a case for each font (like HELVETICA 8pt Roman, COURIER 12pt Italics). Digital fonts are contained in directories – sometimes called *font suitcases* – and when typesetting in LaTeX, the directory structure should follow the TeX Directory Structure, see `www.tug.org/fonts/fontinstall.html`

[65] *Open source* describes practices in production and development that promote access to the end product's source materials (Wikipedia).

[66] Hinting is mathematically adjusting the display of an outline font so that it lines up with a rasterized grid.

[67] ADOBE Acrobat Reader's `Document Properties/Fonts` will display a list of fonts used in the PDF file.

Publisher's copy or production editor to look out for such complications (see also the paper by Agnès Henri in Part 1 of these Proceedings).

Table 2. Principal "open" and multi-platform font standards used in typesetting.

Type	Description	Device
PostScript Type 3	Bitmap or outline fonts, open source	Print only
PostScript Type 1	Outline font description of characters by means of points and curves. Resolution independent and scalable.	Print only
TrueType (TT)	Outline font designed as a competitor to Type 1 fonts used in PostScript format	Print and Screen
OpenType (OT)	A cross between Type1 and Truetype.	Print and Screen

7.1.2 Typesetting units of length: pixels, points and picas

In typesetting, and in graphics, there are several typical units of length. One class comprises the absolute units of length: centimeters, millimeters, inches, points, and picas. The other group includes relative units such as ems,[68] pixels, and percentages. Absolute units of length are used for printed media, and relative units are used for on-screen reproduction. The *point* is the smallest unit, see Figure 16 for a real-size representation of these units. The small bars on the upper right in Figure 16 are of 1-cm length, subdivided in intervals of 1 pt, plotted with minimal and with recommended line weights – only to illustrate the resolution capacity of the human eye and the limitation of this printed medium. Note that there is a difference between printer's and computer units. For desk-top publishing, the following relations are applied:

$$1 \text{ printer's pica} = \frac{1}{6} \text{ inch} = 4.22 \text{ mm} \qquad 1 \text{ printer's point} = \frac{1}{12} \text{ pica} = 0.35 \text{ mm}$$

7.1.3 Which typeface and which font to use in graphics?

There are basically three classes of typefaces:

1. the Roman type, characterized by variable line thickness, with the ends of each stroke ornamented by short line segments called *serifs*;

2. non-Roman type, with lines of equal width or weight, without serifs, referred to as *sans serifs* – a type that is quite popular in graphs;

[68] An *em* is equal to the point size of the current font, and is the same for all fonts at a given point size: 1 em in a 12 point typeface is 12 points.

3. the default typeface for mathematical symbols and equations used in LaTeX, is known as *Computer Modern*[69] (including symbols like •). CMMI is not the same as TIMES Roman Italic: variables and indices (like $l, 1, l_1$) differ – moreover, sans-serifs are not suitable for math and physics because I and l are too similar.

The principal guideline is to avoid overuse of typeface: there should be no more than 3 to 4 different typefaces in a single research paper.[70] There are lots of other types that are a mix of both basic design shapes. Roman is mostly used for the text body, and sans serif for titles and headers of Sections (see Table 3). Typeface is also used to differentiate between variables, for example in statistics there is a convention to distinguish between *parameter* and *estimate*. The former concept represents the *true* value for the population, the latter is the *estimated* value from the sample: hence σ stands for the standard deviation of a population, and s for the estimate of σ.

When using the Publisher's macros, you should not worry about which typeface to use, and where, because selection of appropriate typeface and font is taken care of in the background. Still, some attention is needed to make the type in the graphics conform with the type used in the text body.

Table 3. Typefaces and fonts used in regular A&A papers. Note that there are only two typefaces, though many fonts.

Element	Typefaces	Font	
Title	HELVETICA	HELVETICA bold	16 pt
Author name	TIMES	TIMES Roman	11 pt
Affiliation	TIMES	TIMES Roman	9 pt
Abstract	TIMES	TIMES Roman	9 pt
Body of text	TIMES	TIMES Roman	10 pt
Footnotes	TIMES	TIMES Roman	9 pt
Sections	TIMES	HELVETICA bold	11 pt
Subsections	TIMES	HELVETICA italic	10 pt
Subsubsections	TIMES	HELVETICA	10 pt
References	TIMES	TIMES Roman	8 pt
Figure and Table captions	TIMES	TIMES Roman	9 pt

7.1.4 Font size

Graphics software provides a font size indicator that allows to select the font size. The font size is the total distance between the highest and lowest point of a set (between top

[69] More correctly, CMMI: Computer Modern Math Italic.
[70] The same applies to powerpoint presentations.

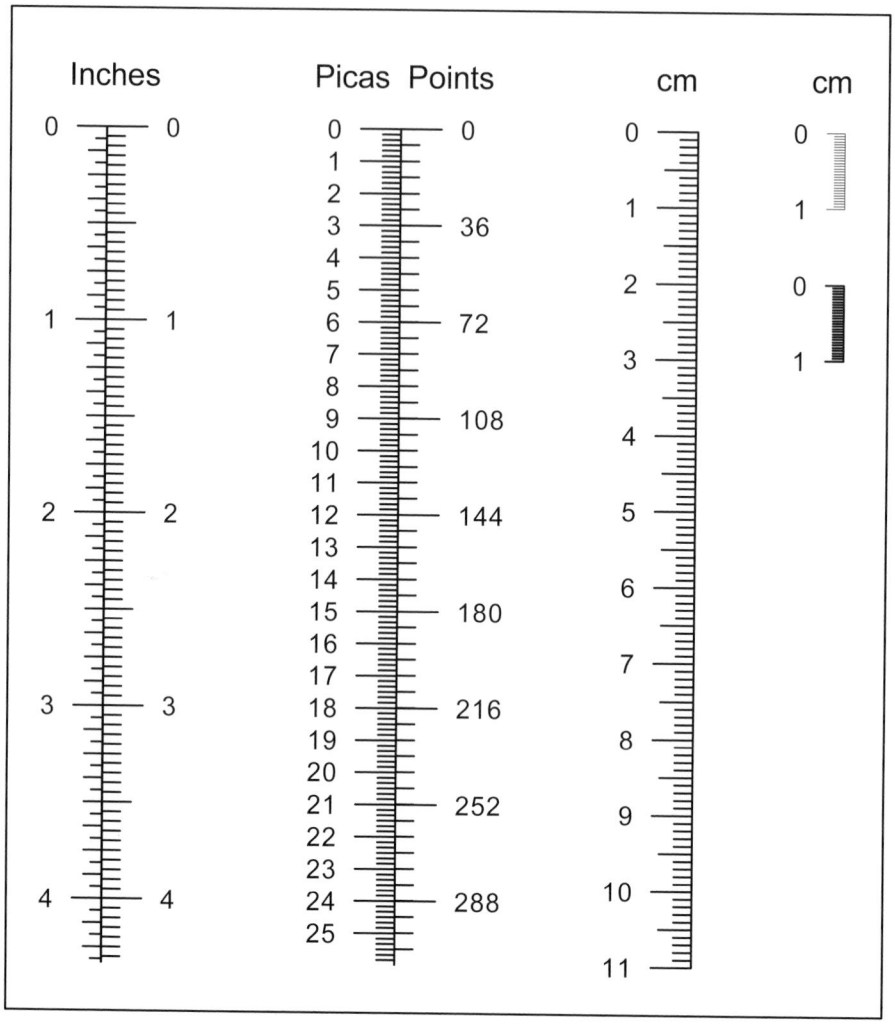

Fig. 16. Typographical units: inches (smallest tick marks $\frac{1}{16}$ [ticks on left side] and $\frac{1}{10}$ [ticks on right side], picas, points and cm. The bars on the upper right, subdivided in intervals of 1 pt (28.34 cm^{-1}), are meant to demonstrate the resolving capacity of the human eye and of this printed medium: the upper sample is drawn at the minimal line width, the lower sample has recommended line width (also called line weight) of 0.3 mm.

of character t and bottom of letter y), and is expressed in points. The width of the letter P in "Points" in Figure 16 is set in 12 pt ARIAL, but is only .75 the height.

Though most books and reports are set in 10–12 pt, you must be extra careful with text inside graphics, because graphs are quite often reproduced at a reduced scale. The smallest type to read in comfort is 8–9 pt, hence text in graphics should be set in 9 pt divided by the reduction factor (the ratio of original linear dimension to printed dimension). This guideline was not followed in Figures 9 and 14, simply because these graphs are included as examples only. The effect of different reduction factors is also noticeable in Figure 9 of Paper III.

A&A 500, 491–492 (2009)
DOI: 10.1051/0004-6361/200912200
© ESO 2009

Astronomy & Astrophysics
Special issue

COMMENTARY ON: BERTIN E. AND ARNOUTS S., 1996, A&AS, 117, 393

Modern astronomical surveys and the development of informatics

S. N. Shore

Dipartimento di Fisica "Enrico Fermi", Università di Pisa and INFN-Sezione di Pisa, largo B. Pontecorvo 3, Pisa 56127, Italy
e-mail: shore@df.unipi.it

Fig. 17. Example of A&A typefaces and fonts.

As mentioned before, the Publisher's macro takes care of typeface and font, and prevents you from inadvertent changes of typeface in the text. But font size can be overruled (for example, by decreasing the font size in Figures and in Tables, using `footnotesize` or `scriptsize`. It is good to always remain vigilant to the reduction ratio, so that your textual elements are not reduced to a size that is illegible.

Figure 17 illustrates some typefaces and fonts used on the titlepage of the paper by Shore (2009) in the A&A anniversary issue.[71] The journal name, the title of the paper, and the Special issue stamp, are in the sans serif typeface. All other typefaces are serifs, in different sizes and styles (also small capitals). The same applies to regular papers, where also Abstract and all Section headers are in sans serif, whereas the body of the paper is in serif, see Table 3 that lists the typefaces and fonts used in regular A&A papers.

7.2 Graph title

Graphs may or may not have a title, and quite often space can be saved by using the caption in place of a title. Avoid using the graph name (or the filename together with its extension) that you use internally (*i.e.*, the computer-generated codenames that come with the graph when it is printed to a file). The graph title should never mix with the data and the labels. A larger font may be selected for the graph title, even a different typeface. Figure 12, for example, originally had a title *"Daily sunspot area averaged over individual solar rotations"*, but to avoid visual clutter, the title was removed, and inserted in the caption instead.

[71] The pagewidth reproduction factor for this book is $\frac{\text{A\&A}}{\text{this book}} \sim 1.5$.

7.3 Axes

An axis is a base line with respect to which data are plotted or measured. The axis line should stand out, without dominating the graph with excessive line width or improper color. The most frequent mistake is to draw the axis line too thin, *i.e.*, using the minimum line width provided by the graphics computer package (typically 0.01 cm or 0.3 pt). Considering that the human eye can resolve 10 dots mm^{-1} (see also Sect. 8), a 0.01-cm line width is just on the limit – especially when considering that most graphs are scaled down in the printing process, and that print resolution also depends on the quality of the paper medium. To illustrate this point, I refer again to Figure 16 where the three vertical axes have line widths of 0.03 cm (~1 pt). The short segments to the right of the cm-axis have different line widths, and illustrate the problem of using very thin lines (*hairlines* in printers' jargon).

The point where the axes intersect (the origin) often is the point ($x = 0, y = 0$), but it could be any appropriate coordinate value. There is not even a need for the axes to intersect at all.

7.3.1 Axis orientation, and origin

There are no rigid rules about the orientation of axes, but it is quite customary to refer the x-axis to the independent variable, with the y-axis representing the dependent variable, and to orient the former horizontally. But there may be reasons to deviate from this custom. The x-axis is also referred to as *abscissa*, and the y-axis is often called the *ordinate*.

In astronomy papers, reverse axes are frequently chosen, in particular when dealing with magnitudes and color indices. Avoid that data points with the lowest ordinate fall right on the x-axis: this is not aesthetic, moreover there is a fair chance that symbols placed as such will interfere with the tickmarks (see the inset in Fig. 20 of Paper III). The solution is either to start tickmarking one unit below the datapoint with lowest y, or move the axis up or down. An exception is made in Figures 36 and 37, where data points are plotted on the axes, simply because they belong there, and also for the sake of showing the effect.

7.3.2 Axis tickmarks

Tickmarks are short dashes on an axis line to mark the scale along the axis. Together with the units, they help to estimate the numerical values of the plotted quantities, though the main purpose of most graphs is not to read off coordinates, but rather to show relationships between variables. Whether ticks should be on the up or down side of the horizontal axis (or left/right on the vertical) is a preference of taste, see for example Figures 24 and 25. The prime rule is that **tick marks should never interfere with the data** (for an example of such interference, see the leftmost data point in Fig. 42, and also in Fig. 20 of Paper III).

Tickmarks are a problem when they are too close, especially when too many small ticks occur inside a larger tickmark interval. Duplicate axes (*e.g.* a second x or y-axis at the top or at the opposite side of a graph) mostly have inverted tickmarks at the same locations as the marks on the axis that is duplicated. Except perhaps when showing time-series data, where it can be most useful to duplicate the Julian Date time axis with parallel units of years or decades. Choosing the tick spacing is important: a time axis, for example, should have tick marks at suitable intervals, *i.e.*, 1, 2, 5 or 10 years or longer, whereas tick intervals of 2.5 or 3 years are not so comfortable to handle. Similar attention is needed when tickmarking hours: you can either use decimal fractions of an hour, or sexagesimal multiples (10, 15, 20) of a minute.

Axes without tickmarks or without units should be avoided, except when the omission is done on purpose, as in Figure 18, which is a schematic diagram illustrating the difference between accuracy and precision discussed in Paper I. The data used for the histogram are the early estimates of the velocity of light[72] shown below the dashed line in Figure 3 of Paper I, thus limited to only one lobe of the bimodal distribution. Note that the bin size has been chosen so as to give a "good-looking" bell-shaped curve, but that the histogram (for whatever bin size that is applied) never approximates a true normal distribution.[73] The dashed line in Figure 18 corresponds to $215\,000 \pm 1\,800$ km s^{-1} resulting from a Gaussian fit to the data. The horizontal arrow bar labeled "precision" indicates the information used for inferring the precision, the other arrow bar illustrates how far accuracy and precision wander apart. Blind statistics thus brings forth an amazing precision of less than 1%, whereas the accuracy of this result is not much better than 30%. Using the remaining data (*i.e.*, same experimental results but with a different value for the astronomical unit), however, yields $314\,000 \pm 4\,000$ km s^{-1}, thus a seemingly higher accuracy but a much lower precision. Figure 30 in Paper III is another example of axes without tickmarks and units.

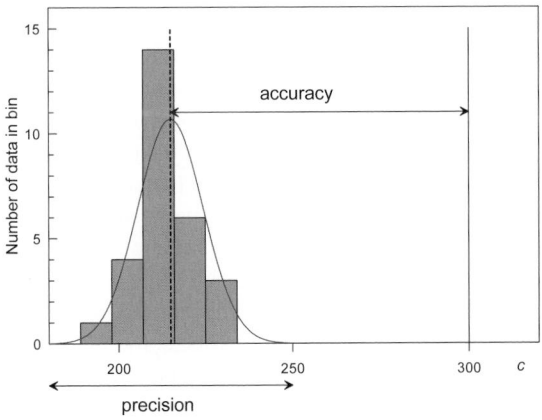

Fig. 18. Difference between accuracy and precision. The bell curve is a Gaussian fit (to a non-normally distributed sample). The solid vertical line corresponds to the value of the natural constant c, the dashed vertical line corresponds to 215 000 km s^{-1} resulting from the Gaussian fit. The unit on the x-axis is 1000 km s^{-1}, tick marks and labels on duplicate axes were omitted on purpose.

[72] Allegedly attributed to Ole Rømer, see Paper I.

[73] That is, with roughly 95% of the sample contained in the interval *mean* $\pm$ 1.96 s, where s is the sample standard deviation.

7.3.3 Axis title and units

The axis title must be in label size or larger, and must be accurate and correct. I have recently coauthored a paper that shows a color-magnitude diagram of an open cluster, featuring isochrones[74] and observational $B - V$ color indices.[75] Evolution tracks and isochrones are theoretical concepts which are, by definition, free of interstellar reddening, whereas color indices of stars are mostly reddened, so when overplotting an observational Hertzsprung–Russell diagram with theoretical curves, the reader must be clearly informed what is exactly plotted, so in this case it can only be the so-called dereddened color index $(B - V)_0$. Unfortunately, I could not convince the leading authors. Surprisingly, the literature on stellar evolution is packed with comparisons of isochrones and evolutionary paths with observational data, and this error is repeated all over: the argument for not correcting is that the meaning is clear from the context. As mentioned in Paper I, **scientific communication should be context-independent**, especially when it is so easily done right.

The axis label should indicate the plotted variable and the units in which it is expressed. The horizontal axis has the label horizontally, for the vertical axis there is the freedom to choose between parallel or perpendicular to the axis. The decision depends on the global aesthetics, and the economy of the page: long labels are preferably aligned parallel to the vertical axis[76], short labels can be put parallel to the horizontal axis.

Data units are sometimes not given at all (as along the abscissa in Fig. 18), or are plainly wrong, especially when units are truncated (*e.g.* JD 4606 for 245 4606) without further explanation. Errors are often made with *multipliers*, for example the x-axis unit in Figure 18 is 1000 km s^{-1}, and since it is awkward to use "kilo-km", the appropriate unit is 10^6 m s^{-1}, which would render the right side of the x-axis somewhat cluttered. One way out is to use the multiplier $\times$ 1000 km s^{-1}, but it often happens that in such a case /1000 km s^{-1} is written instead. Note that some variables are dimensionless, for example phase (ratio of two variables with the same unit) in a cyclic phenomenon.

Breaking axes should be done only when no other solution is possible, because the viewer cannot mentally evaluate the amount of the variable that is jumped, and will often subconsciously perceive the break as a small gap in quantity, see also the discussion of Figure 41. Figure 22 in Paper III is an example where the upper x axis includes a change in scale, whereas the lower x axis keeps one single uniform scale. Such a situation can lead to confusion, hence such difficulties should be mentioned in the caption.

Beware the abbreviations for units, and only use the standard SI nomenclature.[77] Table 4 lists the SI base units (that are dimensionally independent) and Table 5 gives the most common derived units used in astronomy. Note that prefix (hecto, mega,...) and unit form a single symbol that may be raised to a power without adding parentheses. The following list is a set of items related to units:

---

[74] In the context of stellar evolution, isochrones are lines of constant age, across a range of masses, in the Hertzsprung–Russell diagram.

[75] $B - V$ is a color index of the Johnson UBV system.

[76] That is, at an angle of 90° counterclockwise from the x-axis.

[77] International standard ISO 80000 is the successor of ISO 31, a style guide for the use of physical quantities, units of measurement and formulae in scientific papers. SI: International System of Units.

Table 4. The 7 SI base units.

Quantity	Unit Name	Symbol
length	meter	m
mass	kilogram	kg
time	second	s
thermodynamic temperature	kelvin	K
luminous intensity	candela	cd
electric current	ampere	A
amount of substance	mole	mol

- wavelength is no longer expressed in Ångstrom (Å) or micron, but in nano- and micrometer (nm, respectively, μm);
- wavelength λ in meter[78] and frequency ν in Hz are linked: $\lambda = 2.99792458\ 10^8/\nu$;
- when expressing magnitudes in a specific wavelength band or photometric system, it is not necessary to add "mag" or superscript m, since the expression $V = 9.0$ or $B - V = 0.25$, implies that the unit is magnitude;
- some symbols for non-SI units may lead to confusion: for example, straight quotes are used for inch and foot, but also for arcminutes and arcseconds (as well as for beginning and end quotes in a text);
- when using Julian Date, specify if the heliocentric correction has been applied (see the comment on page 153);
- when truncating Julian Date JD 245 5898.5 to 5898.5, the column header should list JD–245 0000 and not JD+245 0000;
- avoid Modified Julian Date (MJD), which can lead to 0.5-day confusion;
- when using Universal Time (UT), specify if it is UT1 or UTC[79] (see Fig. 21);
- special attention is needed for time units when using time format yyy/mm/dd and yyy/dd/mm: 2010/01/10 and 2010/10/01 are different dates;
- the same for decimal hours: hhmmss and hh.xxxx can lead to confusion;
- other formats for time and date should be carefully examined, see for example the case given in the footnote on page 135;
- another confusion with time is that some time signals are recorded in seconds, and others in decimals of a day, with a multiplier of 100 000: since a day has 86 400 sidereal seconds, one second is approximately $0\overset{d}{.}00001$, and the close match with $\frac{1}{100\,000}$ can be easily overlooked, and can thus easily lead to almost undetectable errors in frequency spectra of time series;

[78]The US spelling of *metre* is meter.

[79]UTC is a time standard based on International Atomic Time (TAI) with leap seconds added at irregular intervals to compensate for the Earth's slowing rotation. UT1 is defined as mean solar time.

- confusion between e, the base of natural logarithms, and e, the elementary charge;
- a space must be inserted between the numeric value and its following unit symbol,[80] and also between adjacent unit symbols (*e.g.* N m) – except when used in combinations like 1-m telescope, 8-pt characters, etc.;
- avoid mixing decimal separator and decimal comma (*e.g.*, 9.5 and 9,5) in text and graphics;
- when a number is less than unity, include 0 in front of the decimal separator;
- trailing zeroes are only allowed if they are significant, see Figure 52 that reproduces a Table with one and two decimals for the same variable, and where trailing zeroes were omitted;
- verify the correct use of upper case and lower case units: K is Kelvin, k is kilo;[81]
- symbols of units are not abbreviations, and therefore must never be pluralized (*i.e.*, write km, not kms), nor ended with a period;
- only one *solidus* (/) is allowed in compound units, unless parentheses are used, for example N/(m/s^2);
- the prefix and the unit form a whole (like μm), as such an exponent refers to all elements of this combination, see the (inverse) wavelength unit on the x-axis in Figure 19.

It is not forbidden to use different units for the same variable, for example in the case of multiple axes: multiple abscissae or ordinates (like time units, magnitude versus flux, or wavelength versus frequency) are very useful – duplicate axes are even very functional for the purpose of framing the graph. Figure 25 shows an x-axis with a somewhat confusing unit: the label indicates "Years (–BC/AD)", the time unit being year running from 4 000 BC to 2 000 AD, though the negative numbers are self-explanatory and could do without further specification.

7.3.4 Axis type

Not only the design type of a graph, as described in previous paragraphs, is to be considered, but also the most suitable axis format should be chosen. The most common axes are Cartesian, but some data – especially those of a cyclic character – deserve polar coordinates instead. In such graphs, the radial coordinate is defined in $R \in [0, \infty]$, and the angular coordinate in $R \in [0°, 360°]$ or $[0, 2\pi]$. Figure 40 shows a vintage polar coordinate graph with grid.

Phase diagrams, in principle, have x-coordinates in $R \in [0, 1]$, where the reader must mentally link the right and left edges. As mentioned already, a redundancy is quite often constructed by showing, for example, $x \in [0, 1.4]$ or $[-0.2, 1.2]$, a redundancy created to help the eye.

[80] A space character always stands for multiplication; preferably use unbreakable space in LaTeX.
[81] That is, the prefix, not the symbol for kilogram.

Table 5. Some supplementary and derived SI units used in astronomy.

Quantity	Unit Name	Symbol	Equivalent
velocity			$m\ s^{-1}$
acceleration			$m\ s^{-2}$
angular velocity			$rad\ s^{-1}$
angular acceleration			$rad\ s^{-2}$
frequency	hertz	Hz	s^{-1}
force	newton	N	$kg\ m\ s^{-2}$
pressure	pascal	Pa	$N\ m^{-2}$
work, energy	joule	J	N m or $kg\ m^2\ s^{-2}$
momentum			N s or $kg\ m\ s^{-1}$
power	watt	W	$J\ s^{-1}$
electric charge	coulomb	C	A s
luminous flux	lumen	lm	cd sr
illuminance	lux	lx	$lm\ m^{-2}$
time	Julian year	y[1]	365.25 d
	second	s	
	minute	m	
	hour	h	
	day	d	86 400 s
arc	degree	°	
	arcminute	′	
	arcsecond	″	
	milliarcsecond	mas	
plane angle	radian	rad	
solid angle	steradian	sr	
wavelength	nanometer	nm	10^{-9} m
	micrometer	μm	10^{-6} m
energy	electron volt	eV	$1.60\,2177\ 10^{-18}$ J
distance	astronomical unit	AU	$149\,598\ 10^6$ m
	parsec	pc	$30.857\ 10^{15}$ m
mass	solar mass	$M_\odot$	$1.9891\ 10^{30}$ kg
flux density	jansky	Jy	$10^{-26}\ W\ m^{-2}\ Hz^{-1}$

[1] Unfortunately there is no official symbol for the year, though "y" can be used.

Not all parameters or observables can be ordered along a numerical axis: some follow rank ordering – a so-called *ordinal scale* – that assigns values to objects based on their ranking. MK spectral types, for example, are mostly represented with a decimal subdivision, but especially in the "wings" of the spectral sequence (O 5–B 0 and K 5–M 0), spectral types are to be seen as labels, and not necessarily as numerals (Wing 2011).

7.3.5 Axis labels

A common problem with Figures is that the axis labels are often too small, illegible, inconsistent, or ambiguous. Figure 25 is an example with appropriate (but different) axis label and axis title font sizes. Incomplete labels are not uncommon, see Figure 26 where labels have been partially clipped in the reproduction process. Together with the fact that the overall image resolution is quite often on the low side, small labels – at reproduction ratios of 0.7 or 0.8 – will complicate the understanding of the graph.[82] A rule of thumb is to aim at 12 points for label size, as was applied in Figure 16. Another problem is missing labels: the *Astronomical Society of the Pacific* (ASP), in its *Overview of Publishing Process*[83] explicitly asks authors to check for *"labels missing around figures"*.

Another common error is forgetting axis labels altogether – though omission can also be done on purpose. Typical examples are the histograms that are inspected during image processing: the image-analysis software will mostly produce histograms with unlabeled axes without tickmarks, because these diagrams are constructed for qualitative inspection only, and are not meant for reproduction. One such example is Figure 28, where the y axis is unlabeled and without scale (assuming that the reader knows that the ordinate is normalized). The same with the x axis, except that it is often indicated that the pixel values run from 0 to 255 from left (dark) to right (bright). An additional scaled and labeled x axis was added for clarity.

7.3.6 Axis scale

The extent of a graph (along all axes) should cover a slightly larger range than the data. The scale can be linear, or non-linear, or the scale can appear linear when the plotted variable is not (for example, stellar magnitude is an implicit non-linear – that is, logarithmic – quantity). A non-linear scale means that data are plotted as a non-linear function of the independent variable. It is important to consider – at a very early stage – whether the natural scale of the data really is the optimal scale for presenting the graph, or whether the argument should be a function of the independent variable.

A logarithmic scale can be helpful when the data cover a large range of values. The base of the logarithm has to be specified, for example ln, $\log_{10}$ or $\log_2$. But the data range is not the only determinant for the choice of a particular non-linear scale: the shape of the function or relation to be plotted is crucial, as the subsequent discussion shows.

Stebbins & Whitford (1945) determined the mean colors of unreddened main-sequence stars grouped by spectral type, computed black-body colors for six wavelengths from Planck's formula,[84] and plotted the color indices as a function of $1/\lambda$. Figure 19 shows a reconstruction for B0, A0, G0 and M1 main-sequence stars.[85] The graph demonstrates at one glance that

---

[82] The effect of reproduction ratio on label size is clearly visible in Figure 9 of Paper III.

[83] `http://www.aspbooks.org/publishing_overview/`

[84] $E(\lambda, T) = \frac{2hc^2}{\lambda^5} \frac{1}{e^{\frac{hc}{\lambda kT}} - 1}$, where h is the Planck constant, c the speed of light, T the temperature of the black body and λ the wavelength.

[85] The calculated indices are not shown, but they agree fairly well with the observations (most of the points fall within the area occupied by the plot symbol, except for the leftmost point for spectral type A0).

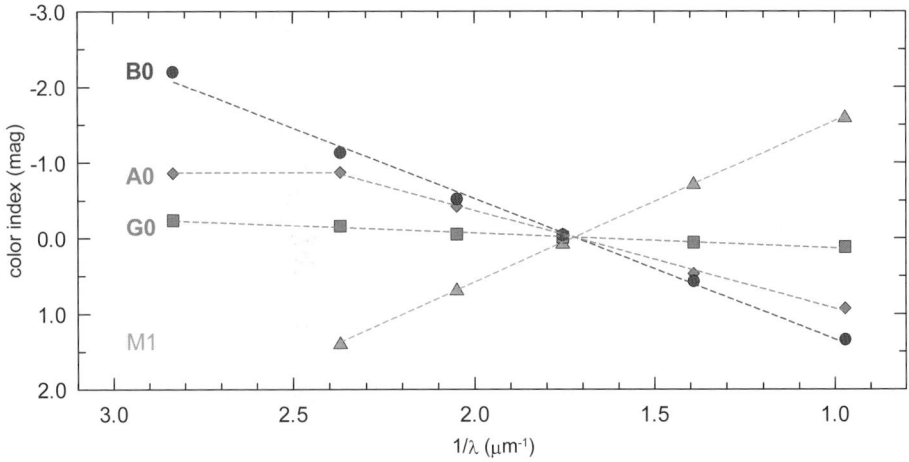

Fig. 19. Example of non-linear axis scales. Observed color index (logarithm of the ratio of two stellar irradiances) for main-sequence stars of different spectral type, as a function of $1/\lambda$. The dashed lines are linear fits that visualize black-body spectral energy distributions (for spectral type A0 the fit is constrained to $1/\lambda < 2.37$). Stebbins & Whitford (1945) list no data for M1 beyond 422 nm. Source: Sterken *et al.* (2011).

1. color index (logarithm of the ratio of stellar flux in two adjacent wavelength bands) is almost linear in $\frac{1}{\lambda}$;

2. some spectra show strong deviations, notably in the photometric band that contains the Balmer jump, and

3. the color indices have a common zero point at $\frac{1}{\lambda} = 1.75$ (570 nm).

This graph shows the power of selecting proper axis scales to render a complicated physical law as linear segments. The graph also displays the zero point of the color index scale.

There are no rigid rules about which type of axis scale to use, but if the data seem to follow a specific function, using a scale of the inverse function usually renders linearity. This approach is called *linearization*: for example, if the data follow an exponential law $y = ae^{bx}$, plotting along the inverse function $y^{-1} = \ln(a) + bx$ renders the plot linear. Logarithmic scales are the most popular non-linear scales because they are perfect for handling exponential and power-law information. Take, for example, Figures 7 and 8 of Paper III, which show the same bibliometric data on a linear and on a logarithmic scale. One of the datasets (case F) has a citation accretion curve that is strictly linear (on a logarithmic scale), which means that there is an *exponential growth* of this scientist's body of citations. Non-linear scales have a disadvantage, though: they are not as easily readable as linear scales (the logarithmic scale is notable for perceptive exaggeration of small numbers), and tickmarking is not always obvious.[86]

Figure 31 illustrates the use of linear and non-linear scales in the same graph. The x axis is expressed in linear units (arbitrary laboratory scale), the duplicate axis shows the corresponding physical units (nm) obtained through a third-degree polynomial fit of the data given in the original paper. Because the transformation is approximate, the tick marks are shown as $\diamond$ symbols. To avoid visual clutter, not all wavelength marks have been labeled; this example moreover illustrates the mixing of numerical and symbolic axis labels.

One aspect that relates to units is autoscaling, which can lead to a distorted image of the data. Autoscaling accommodates all data in the plot database, including outliers (errors of measurement, mistyped entries, flares, eclipses, etc.) and, as such, compresses the regular data to a scale that is not always appropriate for visual analysis. This leads to variable and inconsistent scales on the axes of different graphs belonging to one piece of work, rendering any comparison impossible.

7.3.7 Axes and framing

Two axes will form a so-called *half frame*, see Figure 43. Sometimes a full frame is formed by four functional axes, for example in Figures 39 and 42, but a frame is sometimes drawn for purely decorative reasons, see Figures 5 and 17.

7.3.8 Background colors and lighting conditions

The background color is an important part of the graph because color perception is affected by surrounding hues. Quite a large number of graphs are made with black or colored foreground symbols on a white background, simply because we print on white paper. The background color is an element to be decided at the design phase. Figure 20 illustrates the disturbing influence of some background colors: color perception changes when colors are viewed on differently-colored backgrounds, not to speak of the problems that color-vision deficient viewers may have, see Section 9.4.5.

Background colors and gray shading are always tricky, especially when printing in black and white. Figure 31, for example, shows a graph with intentionally applied gray shading for the mere purpose of making the yellow spectral line visible. This solution is not optimal for printing on white paper (yellow on white is almost invisible), but the yellow contrasts well on the gray background. Figure 21 is a beautiful example of a creative composition, here slightly adapted to avoid interference with the colored background.[87]

Another important element is the illumination conditions when viewing color combinations. In subdued light, the Purkinje effect (the tendency for the human eye to detect blue objects more readily than red under low light conditions) effectively modifies color content of an image, whereas a bright source of illumination may render colors very differently, as is illustrated in Figure 48.

---

[86] Standard tick values between major ticks on a $\log_{10}$ axis are 2, 3, ..., 9, and divide the interval in 9 parts.

[87] The original graph is available at `http://jjy.nict.go.jp/index-e.html`

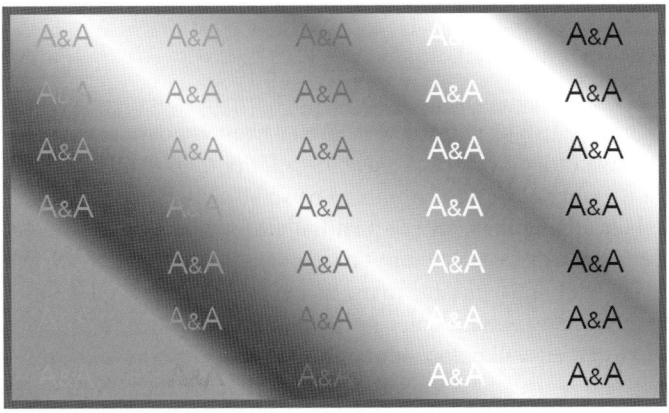

Fig. 20. Example of background contrast problems using, from left to right, *RGB*, white, and black labels. The typeface is HELVETICA, the font size 14 pt. The reduction factor is 0.365, what results in 8.6-pt printed characters.

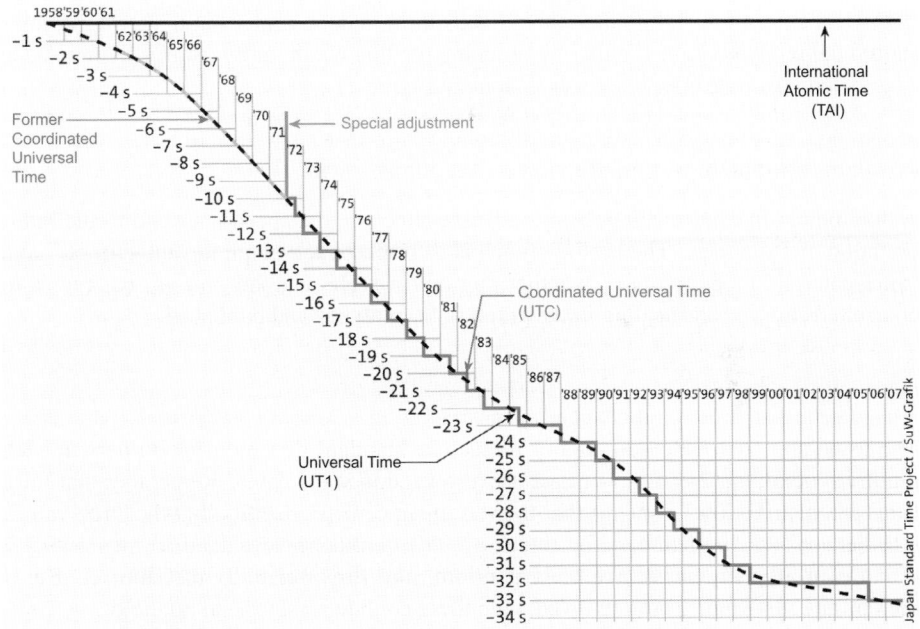

Fig. 21. Difference between uniform International Atomic Time (TAI) and Coordinated Universal Time (UTC), respectively UT1. The dashed curve 1958–1972 is the former Coordinated Universal Time. On January 1, 1972 the difference between TAI and UTC was set to 10 seconds. This graph is a timeline. Adapted from *Sterne und Weltraum*, October 2009, with permission (the only adaptation is a slightly different color rendering for enhancing contrast).

7.4 Plotting

7.4.1 Plot symbols and lines

Plot symbols should be neutral, and signs with an offensive or explicit political connotation should be avoided. The size of a plot symbol must allow for differentiation between several plotted variables, at the same time the size should be small enough to avoid overlapping symbols (see the overlapping data for the small slices in Fig. 8). Differently shaped symbols can be used to distinguish different variables in the same graph. When colors are used for coding, it is still indispensable to use shape as a redundant element of information to identify the variables, because color information is often lost in printing, and because not everyone can perceive all colors optimally (see Sect. 9.4.5).

Symbols can be omitted, and replaced by connecting lines (see also Sect. 11.5) when the symbol density is too high (see Fig. 41). Lines can either be full lines, or dashed, or dash-dotted. The most important point about lines is their width, especially when graphs are reduced in reproduction. Many graphs in this chapter have lines with native widths of 0.03–0.05 cm (or 0.8–1.4 points), though axis lines may be somewhat lighter. Grid lines should definitely be lighter, either by reducing the line width, or by using a lighter shade of gray (for example 60% black), as is done in Figure 21.

7.4.2 Autoclipping

Autoclipping, like autoscaling, brings adverse effects to the appearance and to the interpretation of a diagram. The best approach is to inspect each graph in "autoclip off" and "autoscale on" mode to check whether any out-of-scale data are present, but return to controlled clipping and scaling in the graph to be analyzed and published.

7.4.3 Error bars

The usual approach is to add plus or *minus* one standard deviation (or σ) bars to each plotted symbol, but 2-σ and 3-σ bars are also used in some cases. Remember that for a Gaussian distribution 1σ means that there is about 68% probability that the "true value" falls between the 1σ error bars, 2σ refers to a 95% confidence level, and 3σ means 99.7% – that is, **if all data errors are truly random, and thus normally distributed**. But if there is any systematic error involved, then these confidence levels do not hold.

It may happen that the error bars are of about the size of the symbol. In such a case the solution is to omit error bars altogether, and somehow quantify the data errors in the text or in the caption. As an alternative, you may display one representative error bar (without data point) at a location that is sufficiently off the locus of the data.

7.5 Graphics and texts as image components

Textual components in graphs, such as legends or keys, are often placed inside a legend box (see, for example, Fig. 39). When adding text, you must decide on the font type, on the font size, and also on the font style. It is very important to make text inside graphs consistent with text inside the body of the manuscript, especially with relation to mathematical symbols and units.

7.5.1 Legend or key, and labels

The legend (or key) explains the meaning of the different symbols or colors in a graph. If the graph contains lines, they may also be referred to in the legend. In Figure 19, I used direct labeling of points, so the legend was omitted on purpose, because it interfered with the data symbols. The problem of low contrast of yellow (typically selected for representing stars of class G) was circumvented by selecting brown.

It is best to avoid "ALL CAPS" fonts, because they have all nearly the same height and width, and are sometimes difficult to read. Emotionally, such fonts convey the impression of exclamation marks – which is one of the reasons why I use "SMALL CAPS" for some of the most relevant citations in Papers I–III, a choice that permits omission of leading and trailing quotation marks. Care should be taken that the legend really stands out on the background of the graph. Especially in the case of dark backgrounds (photographic negatives), one must make sure that all annotations stand out in proper contrast. Such textual elements (including arrows) should never interfere with the data elements (such as stars), and should never change the data content of the graphics. For an illustration, see Figure 38, where the labels for the 10–50% densities are dark on a relatively bright background, and the reverse for labels covering 50–100%. A similar change in label color is displayed in Figure 34, and in its color-transformed version in Figure 35.

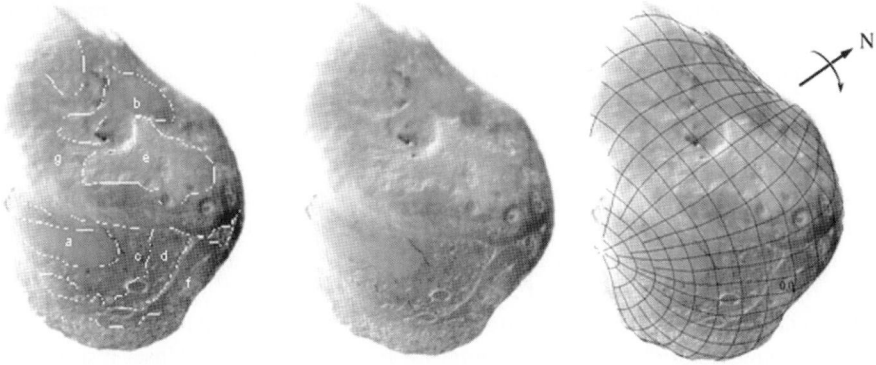

Fig. 22. Example of optimal graphical elements: Comet 9P/Tempel 1, as seen by the Deep Impact cameras. *Left:* with boundaries of layers marked, and identified by small letters. *Center*: without markings, for clear viewing of the surface features. *Right*: with the equatorial coordinate system superposed for easy reference in the text (after Belton *et al.* 2007, with permission).

Labels inside graphs are often added automatically, for example near contour lines. The software gives the user the possibility to define label placement, typically by specifying amounts above, below, left or right from the plotted point. But in some cases such automatic labeling leads to overlapping of labels and data information, and such graphs need more attention, even manual adjustment. Take for example Figure 33, where the wavelength labels along the horseshoe curve have been placed with three different settings – even including reversing font color – to avoid overlap of curves and labels. Labeling of all points in Figure 46 would lead to overlap of labels in the upper panel, and to overcrowding in the lower panel; therefore the two planets between Mercury and Mars were left unlabeled.

7.5.2 Scale factors and orientation

In astronomy (and in all geosciences) graphics frequently need two additional elements: the scale factor, and the orientation. As such, Figure 22 is a nice example of nearly perfect graphical communication: the image is of high resolution, and it provides an orientation arrow as well as a spin-direction arrow. The left image highlights the cometary surface features selected by the authors, the right image explicates the coordinate system, and the middle panel shows the natural and original view, so that the reader cannot be misled by the author's interpretation. For optimizing print contrast, I inverted the grayscale and, therefore, the caption mentions "after Belton *et al.* (2007)".

I obtained permission to reproduce from first author Belton, but not without first solving a most confusing issue: downloading Belton *et al.* (2007) through ADS delivers a PDF file carrying the bibliographic information "*Icarus* 191 (2007), 573–585" with the small print in the footnote mentioning *"This article originally appeared in Icarus Vol. 187/1, March 2007, pp. 332–344. Those citing this article should use the original publication details."* I first had overlooked this warning, fortunately the author informed me that my reference was not the correct one. We thus have here a very bad example of bibliographic confusion – with neither the author nor the reader being aware of it, causing needless email exchanges between the author and the user.

7.5.3 Insets

An inset graph is a graph within a graph, a smaller-scale picture inserted within the bounds of a larger one. Figure 3 illustrates the application of an inset: by using space without information content inside the larger graph, a vintage diagram is included at a smaller scale – though with preservation of resolution.

If an inset refers to a result or an argument from within the same paper, all its elements (labels, legends, ...) should be of appropriate font and size. If the inset is a reproduction of a graph published elsewhere, the size can be downscaled regardless of font size, provided the full reference to the original source is given (as is done in Fig. 3).

Figure 23 is a good example of proper combination of data and inset graphics: the dynamical orbit of the young massive binary system θ^1 Ori C is illustrated (Fig. 8 of Kraus *et al.* 2009). The available astrometric data and their error bars are shown, as well as $O - C$ vectors[88] (arrows). The original Figure shows multiple orbit solutions. The nine insets (interferometric images) illustrate the data, and contribute to a most decorative composition.

[88] $O - C$ stands for Observed *minus* Calculated.

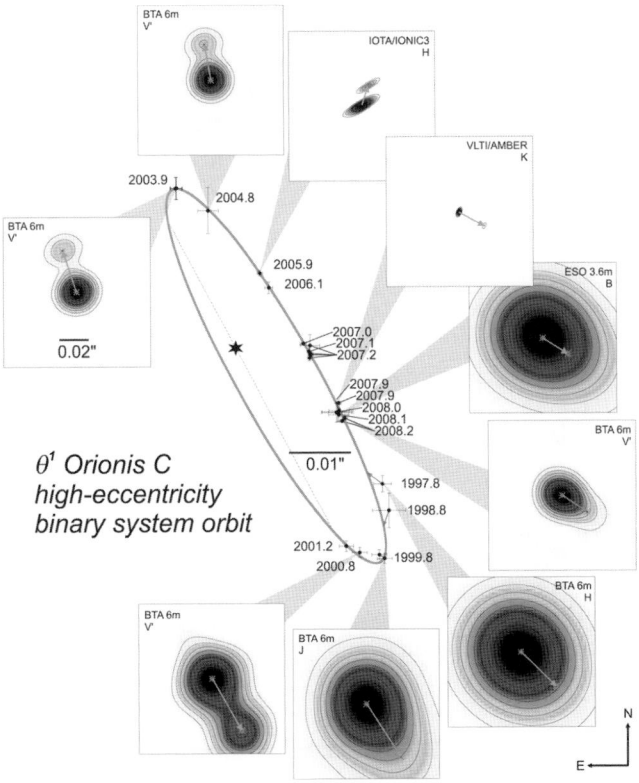

Fig. 23. Example of proper combination of data and decorative graphics: the dynamical orbit of a young massive binary system. Based on Kraus *et al.* (2009).

7.5.4 Grid lines

Grid lines are remnants from the gridded graph paper used in pre-computer days (see Fig. 40), and these grids were quite often printed with a special green or light blue ink that would not reproduce on a XEROX copy or a print. Grid lines are now a standard and automatic aspect in business graphics, but in scientific graphs of high data density, grids often form visual ballast that serves no purpose, simply because the reader will not try to derive precise quantitative information from graphs. The more so because the younger generation has never been accustomed to use grid lines, so their visual-mental perception system is not well developed in suppressing the grid lines and seeing only the data. In short, grid lines – especially heavy grids – unnecessarily clutter the diagram and get in the way of seeing the data: they are to be considered as chartjunk, unless they are faint, really helpful for reading off numerical data values, or for enhancing visualization, see Figure 21 and also Section 4.4.

Figure 24 shows Figure 1 of Holweger & Oertel (1971) in its original form, displaying 15 redundant grid lines, together with the same graph with the grids removed. Non-cartesian plots (polar and ternary diagrams) may, nevertheless, profit from light-shaded gridding because their axes are less commonly used, and the reader needs more effort to

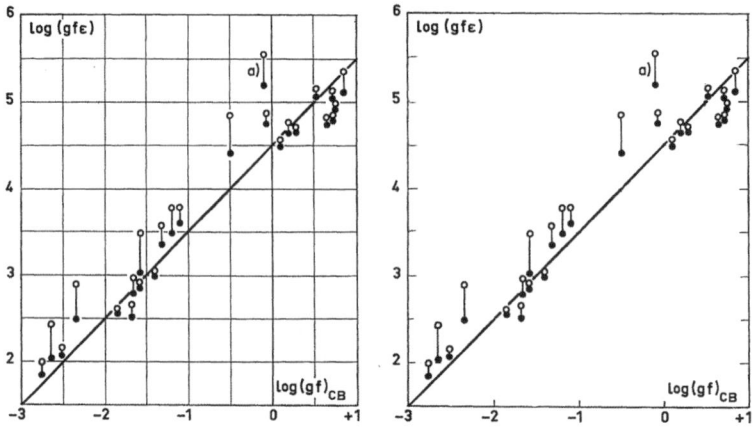

Fig. 24. Example of graphics with and without grids. Based on Figure 1 of Holweger & Oertel (1971).

assess the data. As mentioned already, 3-D surfaces are often overplotted with a mesh grid, though in Figure 9 the grid lines were omitted on purpose, with the result that the reader cannot tell whether a peak exceeds ordinate value 100 – the more so because the color code is not included.

These principles equally apply to images, graphs and tabular material. But functional grid lines can be useful. Figure 25 is taken from a paper by Usoskin *et al.* (2007), and illustrates another example of a very good diagram: sunspot activity through the Holocene is shown,[89] and the two horizontal lines represent crucial data levels, not just grid lines. The diagram cannot fail to direct the eye to a most important fact: the present-day grand maximum in the decadal mean sunspot index, the highest level in more than ten millennia. Beware, though, that this graph is based on sunspot data supplemented by proxy data records. This graph has a simple composition, nevertheless the scientific message stands out very clearly.

7.5.5 Decoration

In genuine data graphs there is not much reason to add decorations, but it is acceptable, as long as decorative elements do not overload the graph, and as long as they do not change any quantitative aspect of the data. An example of inclusion of a decorative banner is shown in Figure 5.

[89]Only the lower part of the original is reproduced. This diagram figured on the cover page of the first issue of A&A Vol. 471.

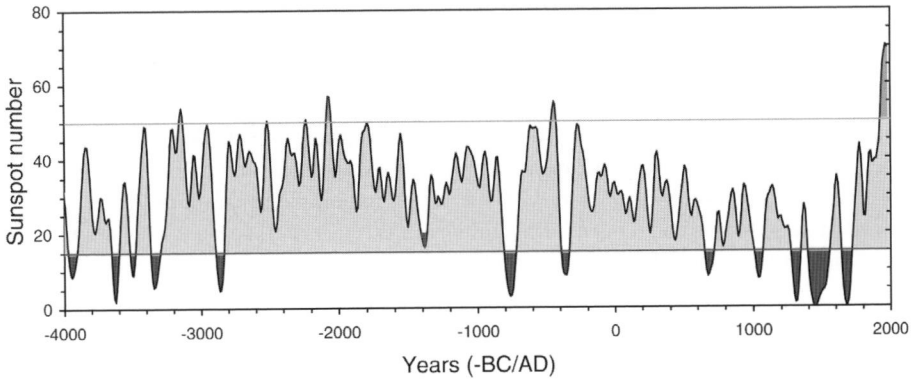

Fig. 25. Reconstruction of decadal average sunspot numbers through part of the Holocene. The x-axis label is explained on page 106. Source: Usoskin *et al.* (2007).

7.6 Aspects of composition

7.6.1 Galleries

Galleries are compositions – often ordered in an array – of several smaller images that deal with the same aspect of the data (for example, a set of finding charts or spectra). If possible, all panels in a gallery should have the same size, and the same amounts of white space between horizontal and vertical boxes, so that all elements of the gallery should render in a neatly tiled pattern. A simple case of gallery arrangement is Figure 22, which can be made as a single image file, or can be combined by compiling three separate image files side by side in the manuscript's LaTeX file. When doing so, one must make sure that excessive white spaces are trimmed off each image, otherwise these spaces will add up and create too large gaps between the image components. Such gaps are not only inaesthetic, they also render images too small in size. Images should thus be cropped, and preferably be oriented in the same way (portrait or landscape) as intended for print (for example, Figs. 27 and 43).

When building galleries of images (especially negative finding charts), the matrix of gray squares on a white background can produce the so-called *Hermann grid illusion*,[90] with dark areas appearing at the intersections of the white spacer lanes. The gallery in Figure 27 will – under specific ambient light – invoke such an illusion when viewed on a computer monitor.

[90]Observed by Hermann (1870), in a German translation of John Tyndall's lectures.

7.6.2 Size ratios

There are no rigid rules about the size ratio h/w (relative height and width) of graphics. Some aspect ratios[91] are "hardware-defined", for example visualizations of an imaging detector's field of view, or a map of an elongated celestial object. Scatter and line plots may also have aspect ratios that are conditioned by external factors: color–color diagrams, for example, are usually depicted with the same magnitude scale on both axes. LaTeX, unfortunately, forces (or seduces) the author to use the full page width for graphs. The following aspects should always be checked:

- tall and thin graphs are to be avoided, if possible;
- some designers swear by the *golden section* $\frac{h}{w} = \frac{1}{1.6}$, a so-called pleasing proportion,[92] though there is no good reason to apply this strong belief, except for the fact that such a ratio minimizes wasted space on both sides of illustrations that do not use the full page or column width (unless text can be set adjacent or wrapped around);
- changing the reproduction h/w ratio should be avoided, see the consequences in Figure 26;
- check the reduction ratio: for this book, the ratio is 1:1, hence Figure 16 should have correct metrics;
- take maximum benefit of the page-size ratio. For this book the ratio is $\frac{h}{w} = \frac{1}{1.6}$, for an A&A full page it is $\frac{1}{1.35}$, and for a single column it amounts to half as much.

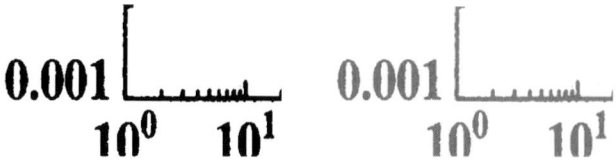

Fig. 26. Illustration of adverse effect of changing the h/w ratio: dimensions of axis labels and titles have changed along. Note the clipping of the lower part of the x-axis labels. *Left*: bitmap image, *Right*: grayscale rendering.

One element to check further is whether the aspect ratio unwittingly influences the reader's perception: high h/w ratios may leave the impression that a given slope or trend is much steeper than it would appear in a graph with a low h/w ratio. This facet is even more important when two adjacent graphs with different h/w ratios are compared.

Stretching and squashing graphs, when all objects are grouped, will distort all labels, and circular pie charts will become ovals. In single-object graphs, even modest compression will become visible because also the grid lines are affected. Undesired and sometimes undetected squinching is sometimes introduced by the scanning process, and therefore it is required to test your scanner at least once by scanning a page with grids only.

[91] The aspect ratio is the ratio of longest to the shortest image dimension.
[92] *i.e.*, movie-screen like.

7.6.3 Figure captions

The caption is a most important textual element – although captions are not composed within the graphical package, but in the manuscript itself. Captions should be brief, at the same time complete. The caption tells the reader how to read the graph, but should not list conclusions and interpretations, nor explain acronyms and abbreviations. A limited amount of data may be included, for example Figure 45 shows a Table embedded in the caption, a composition that allows to save space.

Captions are typeset in the manuscript, and, as such, correct fonts and font sizes are automatically applied. The house style of the journal should be consulted, especially concerning descriptions like *top, right, left,* ..., and whether these indicators should be followed by a colon. The caption is the place to list the permissions to reproduce, or the *"courtesy of ..."* or *"with permission from ..."* statements. If you reproduce existing artwork, the location (museum or archive) and the geometric dimensions of the original have to be given.

When a Figure contains color symbols, it makes no sense to refer in the Figure caption to the colored elements when the publication is in grayscale. As this book is printed in black-and-white, but appears in a full-color digital version, different graphic symbols were chosen to represent colors in such a way that the reference to a symbol allows unambiguous decoding, whatever the color mode of the consulted page.[93] This approach is crucial if you want your graphics to be understood by persons who suffer from color-vision deficiencies, see Section 9.4.5.

7.6.4 Figure numbering and placement

Graphics should be placed immediately after the first mentioning in the text, and may precede the citation only if it is placed on the same page or on the page opposite to the reference. The graphics placement in the present paper deviates somewhat from this rule, because several graphs contain design elements that refer to descriptions in different places, often located a dozen or more pages apart.

If a graph extends over more than one page, the Figure count should be interrupted so that all parts of the Figure refer to the same number, and the caption should clearly indicate this with a statement like "Figure 1, continued".

Avoid large empty spaces above and below the plotted information: such spaces are better used for scaling up the graph to show more detail. These spaces are often produced by an improper bounding box of a POSTSCRIPT Figure, which may even result in partly covering lines of text (or the running title or page number) above the Figure. LaTeX cannot handle Figures with text wrapped around, though unaesthetic empty spaces can be avoided by placing the caption adjacent to the Figure, see for example Figure 1 and, especially, Figure 36.

[93] The most significant graphs of this paper are also reproduced in quadri on 16 unnumbered pages after page 170.

8 Image quality: high resolution and low resolution

8.1 Image resolution

Every Editor of a proceedings book knows that deficient image resolution is a very common problem. Image resolution refers to the amount of detail, sharpness and clarity of an image, and is quantified in terms of pixels, dots, or lines per unit of length (usually the inch). In a way similar to the confusion between typeface and font, resolution is often referred to with incorrect terminology. Experience shows that the relation between the intrinsic resolution of an image and its print resolution, is often not well understood (*i.e.*, the difference between resolution of the capturing device and the display device, see the discussion of Fig. 27 below). The human eye, for example, has an optical resolution of 0.1 mm or ~ 250 dots inch^{-1}, see the small bars on the upper right of Figure 16 with tick marks at 1-pt interval,[94] illustrating that it is very hard to achieve such a resolution in desktop printing. The terminology on resolution varies according to output device,[95] see Table 6.

Table 6. Image resolution according to output device.

device	typical value	unit	abbreviation
human eye	250	dots inch^{-1}	dpi
screen (websites)	72–100	pixels inch^{-1}	ppi
laserprint	300–600	dots inch^{-1}	dpi
inkjet	up to 1200	dots inch^{-1}	dpi
imagesetter*	2500–5000	dots inch^{-1}	dpi
scanner	at least 150	samples inch^{-1}	spi
halftone	150 and higher	lines inch^{-1}	lpi

*Highest quality printers for professional publishing.

High-resolution graphics was a problem in the past, and computer-generated data have often been printed at a resolving power not better than the line and character spacing of the typewriter/printer (3–4 pt, see Fig. 52 that very much resembles line-printer graphical output resolution of the 1960–70s). This hardware problem is extinct now, nevertheless an amazing number of sloppy-resolution artwork still populates journals and books, mainly because properly-generated computer plots are simply saved to an output file at screen resolution, and printed as such.

Optical resolution is the resolution determined by the optical and mechanical hardware of the capturing device (scanner or camera), and is the true resolution that the device can handle. *Digital resolution*, on the other hand, is an artificial concept referring to resolution arrived at via pixel interpolation, or by means of another transformation. The frequency of lines in a halftone screen, lpi (introduced in Sect. 6.1) also refers to

[94] Thus, the average human eye can resolve graphics that are about three times better than this example.

[95] Digital-camera manufacturers sometimes call *capture resolution* the number of pixels of the image sensor.

resolution. That concept, however, is not relevant to most authors, since most of us do not even know the lpi specification of our desktop printers, simply because lpi is not a matter of hardware,[96] but a software issue that is part of the halftone creation process. The three concepts are summarized in the box below.[97]

ppi = image resolution: includes only image information
dpi = output resolution: includes output device information
lpi = frequency of lines in a halftone screen or scan

The pixel is a unit of information, and **has no meaning unless its size is given**, *i.e.*, by specifying the number of pixels per unit of length. Monochrome images require much higher resolution than halftones in order to prevent aliasing (staircase appearance) of diagonal lines. Combinations of shades of gray and text or lines (for example, finder charts with annotations), must reconcile aliasing with file size, and are typically set to 500–900 dpi. Images should also be scaled to the size planned for print, but care should be exercised when resampling: adding pixels by interpolation (resampling up) leads to image degradation, while resampling down leads to deleting information. Resampling also plays a role in image rotation: where rotation over 90° does not change image content (because pixels need not be repositioned over the border between two pixels – that is, when pixels are square), every rotation over other angles will degrade image quality.

Remember that sending 72 dpi to a 600 dpi printer produces output that is no better than that of a 72 dpi printer. Figure 27 reproduces a detail from the book cover illustration: the panels have image resolutions of, respectively, 600, 300, 100 and 72 ppi, and are printed on an approximate size of 51 × 28 mm (2.0 × 1.1 inch). So the print resolutions are, respectively, 300, 150, 50 and 36 dpi. The example shows that the 300 ppi image can be printed at an acceptable resolution if the physical dimension of the image is scaled down by a factor of two (thus a 4× smaller print).

The consistency in resolution inside artwork images must always be considered: if you mount images in one single composition or gallery, you must make sure that the resolutions of all sub-images match – at least when the original images have comparable resolutions. But not only the coherence in resolution in a gallery is important: the viewing resolution of each sub-image must be acceptable. For example, if an online publication offers a 5 × 5-cm 600-dpi graph (mostly JPEG or BMP) with 5 × 10 sub-images, then printing the complete graph at 2× magnification will yield a comfortable 300-dpi print, but a same-size print of one gallery element will result in 60×30 dpi, totally inadequate for evaluating or digitizing the data.

[96]Except for the tapestry artwork discussed in Section 3.2, with 0.46 lines mm^{-1} in the vertical direction (see page 74), and thus 11.7 lpi.

[97]Dpi is not the size of the dot that the output device produces, but describes the spacing of the grid of dots, also called the *addressability*. Resolution is the smallest feature that can be printed, and is a combination of dot size and addressability. Since dot size is often much smaller than pixel size, dpi must include device information, whereas ppi only has image information.

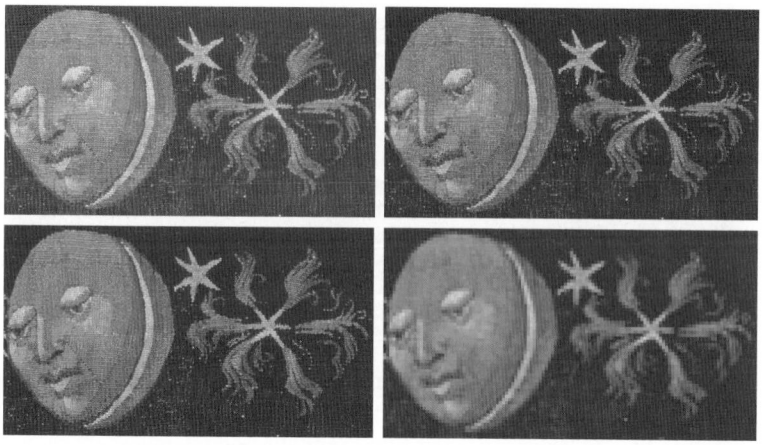

Fig. 27. *Top left:* detail of cover illustration at 600 dpi. *Top right:* same at 100 dpi. *Bottom left:* detail of cover illustration at 300 dpi. *Bottom right:* same at 72 dpi. The lowest-resolution images clearly show ragged outlines and also *pixelation*, *i.e.*, individual image pixels becoming visible.

8.2 Sampling frequency and scanning

Sampling is the technique of selecting an appropriate sample of a population that is suitable to represent the complete population. In the context of this paper, sampling refers to collecting (image) data at equidistant points. The sampling theorem[98] states that

> IF A FUNCTION $f(t)$ CONTAINS NO FREQUENCIES HIGHER THAN W COUNTS PER SECOND, IT IS COMPLETELY DETERMINED BY GIVING ITS ORDINATES AT A SERIES OF POINTS SPACED $\frac{1}{2}W$ SECONDS APART

(Shannon 1949). In other words, **the sampling frequency must be more than twice the frequency of the original signal**, and consequently one cannot extract from a continuously sampled dataset any frequencies greater than $\frac{1}{2}\,\Delta$ (where Δ is the sampling interval), because aliases arise due to beating between the sampling frequency and any real frequency present in the data (see also Kurtz 1983 for a discussion related to Fourier analysis in astronomy). Thus, when scanning a photograph intended for 1:1 reproduction, you must sample with twice the frequency of the output linescreen:

$$\mathbf{ppi_{scanning} = 2 \times lpi_{output\ printing}}$$

When reproduction is on a different scale, the above-mentioned value must be multiplied with the reproduction scale factor (*reduction scale factor* $<1<$ *enlargement scale factor*).

[98] Also called the Nyquist–Shannon sampling theorem.

The scanner's resolution qualification is determined by the x-direction sampling rate (the number of sensors in the detector array) and by the y-direction sampling rate (the interval of the stepper motor). Most desktop flatbed scanners have a true hardware (optical) resolution[99] of at least 300×300 ppi. Below are a number of rules-of-thumb for good scanning practice:

- scanning line art (drawings, etc.) and texts must always be done at the output resolution of the final output device, in 2-bit mode, because 8-bit halftones or any other scanner color option will make the image appear smoother – just compare the 8-bit and the 2-bit scan of lineart in Figure 26: the 2-bit result appears crispier;
- consequently, scan at grayscale resolution only when the original is a true grayscale (or color) graphic;
- scan at the target output resolution ($2 \times$ lpi) of the printer, or about half the printer dpi: you can always decrease resolution after scanning, but not the other way around;
- at the same time, make sure that the settings of the video board match the capturing image resolution, *i.e.*, the video settings must be 24-bit and not 16-bit mode;
- images for screen used to be 72 dpi,[100] but nowadays screen resolutions exceed 100 dpi, and thus scanned graphics for the web are increasingly produced at 100 dpi;
- start with high-quality originals and avoid images from books or magazines;
- take care with historical drawings: they may be so faint and specked that no amount of effort can improve them (see Fig. 30, which also shows lines that overlap text);
- scan at the target output size;
- save/compress with minimum loss, better even compress lossless, and
- verify the relation between the scanner output and the image brightness input, using test strips or grayscale wedges.

Figure 28 illustrates a quite instructive example of a scanning experiment. The first version of this book's cover image scan was delivered at 300 dpi resolution, but because the grayscale patch was not very well resolved, I requested a new scan at 600 dpi. Figure 28 shows the normalized histograms for the B channel:[101] whereas doubling the resolution should only change the number of pixels (and the file size) by a factor of four (and hence make the distribution appear smoother), the histograms are markedly different – a fact already obvious from visual inspection since the second image was much darker than the first, see Section 11.4 for more details. Figure 28 is also an example of how straightforwardly decodable color graphics (through its legend), may become less obvious when printed in black-and-white.

[99] That is, not the software-enhanced interpolated resolution through the creation of extra pixels in between the scanned ones.

[100] Because the original Macintosh 14″ monitors had a fixed 72 dpi resolution.

[101] See Section 9.4.2 for the meaning of B.

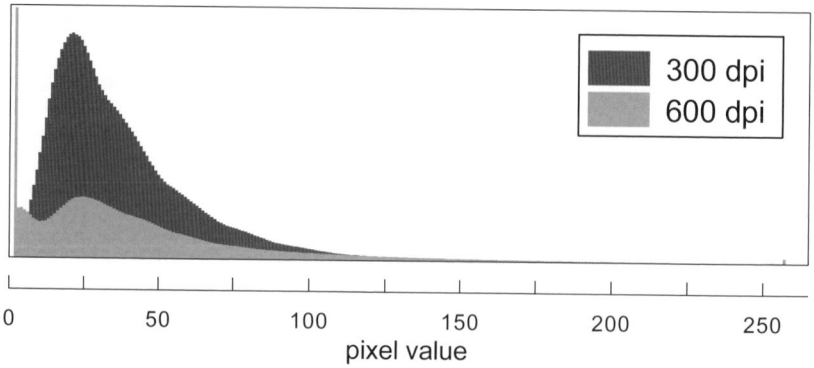

Fig. 28. Normalized histograms of the cover image, based on scans at 300 and 600 dpi resolutions.

9 The complexities of handling color

Most people can see 20 000 to 100 000 colors (in the human visual range, 380–800 nm), and today's computer artists have a palette of over 16 000 000 (*i.e.*, 255^3) colors at their disposal. All these colors come to us as emission spectra of light sources, or reflection[102] (but also transmission and absorption) spectra of materials (paper, glass, plastics, ...). Yet, using more than 2 dozen hues may already lead to bad graphics. Let us first see why we want to use color in our designs, and if all of us are really much helped by doing so.

9.1 Why use color in graphics communication?

Technically spoken, white and gray differ only in level of brightness: gray is not a true color[103] because different grays only differ in luminance, and not in spectral signature. Black, on the other hand, can be described as "zero color" – *i.e.*, no signal at all. Figure 4, for example, uses one color in various shades, and this color view can easily be replaced by a grayscale shading without any loss of information, though with perhaps a less pleasing visual outcome. Some authors are convinced that black-and-white or grayscale graphics can convey any message as well as color graphics can – that is, if the design is well done.

Color simply is an additional – though very powerful – element of information, and if well conceived, color graphs may help your message to better stand out. However, color is expensive, both in printing cost and in memory storage, and haphazard addition of some colors just cannot improve your graphics: **sloppily designed color work is always worse than any black-and-white graphics**. Bad color work can lead to wrong emphasis and misinterpretation, especially for color-blind viewers.

[102]See Figure 48 for an illustration of the impact of ambient light on the perceived color of the substrate.
[103]For example, in the four-color-map problem in cartography, gray is not one of the operational colors.

The choice of the color coding is entirely free, although there exist some traditions, for example, using blue for hot and red for cool in astronomical spectra, or the typical color coding of bathymetric and topological maps that use light to dark blue for ocean depth, and green to brown hues for height of landmasses. There are several reasons for using color in graphs:

1. to label items in a graph (color as symbol);
2. to separate distinct data (color as identifier);
3. to emphasize specific parts of a graph (color as organizer);
4. to measure (color as quantitative information);
5. to facilitate teaching (color as didactic tool);
6. to represent the real world (color photography), and
7. to decorate graphics (color as decoration).

There is a basic principle[104] of graphic design:

DESIGN IN BLACK AND WHITE
ADD COLOR WHEN THE DESIGN IS COMPLETE

In other words: use color primarily for emphasis and for conveying additional information, but **do not let color be the only visual differentiation in your graphs**.

9.2 Different colors for everyone

We actually do not "see" colors: a specific visual stimulus activates our color detectors (the photoreceptors called cones) in the retina of the eye, and color is nothing else than a subjective biological sensation provoked by the reaction of the brain to that amount of stimulus. Most human eyes have cone receptors with different sensitivities for short-, medium-, and long-wavelength light, and it is exactly this wonderful physiological property of the eye that is used in all principles of color reproduction in painting, dyeing and printing – even in weaving, see the footnote on James Le Blon on page 75. The color of an object or a light depends on:

1. the spectrum of the light emitted by or reflected from it (see Fig. 48);
2. the ambient light;
3. the observer, and
4. what the observer was previously looking at.

[104]This principle is also applied in photography courses and schools: students first follow courses in black-and-white photography before entering the color photography curriculum.

A complete physical description of a color would involve full knowledge of the spectrum of all involved light sources, in addition to the spectral response of the observer's own eyes.

I recently gave a talk at a conference, and that same day, during the closing dinner, a gentle young man came to me and said *"That was a nice talk, with good graphics, but never again use red characters on a green background: I am color blind."* He explained that he could, however, read some of the red characters on green background because I had used large-size shadowed or embossed fonts. Very few people are completely color blind, and most color-vision deficient persons can see colors, though they have difficulty differentiating between particular shades, see Section 9.4.5 for a more profound discussion.

9.3 An early generic mathematical color model

The foundation of the theory of color composition was formulated by Isaac Newton in 1704 in his *"Opticks"*: every color in nature is produced by mixing different colors of the spectrum (of sunlight). Newton divided the circle in seven parts[105] *"proportional to the seven musical Tones or Intervals of the eight Sounds"*,[106] and wrote that the spectrum (from blue to red) appears

> ...TINGED WITH THIS SERIES OF COLOURS, VIOLET, INDICO, BLUE, GREEN, YELLOW, ORANGE, RED, TOGETHER WITH ALL THEIR IMMEDIATE DEGREES IN A CONTINUAL SUCCESSION PERPETUALLY VARYING: SO THAT THERE APPEARED AS MANY DEGREES OF COLOURS, AS THERE WERE SORTS OF RAYS DIFFERING IN REFRANGIBILITY.[107]

Nowhere does Newton put forward that the spectrum contains *only* seven colors (see Fig. 29, which actually is a polar diagram, and also a pie chart) – to the contrary, he explicitly stated that there exists a wide sequence of spectrally pure (*homogeneal*, as he called them) hues that cannot be further split by refraction: monochromatic light as we call it today. This leads to the two basic questions in colorimetry, *viz.*,

1. how many different colors are needed to produce *any* color by mixing, and
2. in what amounts must these spectral hues be mixed?

Of course, artists have long known: mixing blue and yellow yields green, and by mingling red, blue and yellow dyes in different proportions, painters could make almost any color they wanted.

---

[105] Scientific records before 1611 trace four aspects of color diagrams: circles based on seven colors, circles based on four colors, circles organized for a red-yellow-blue mixing system, and historical relationships between pigment primaries and spectral primaries. Parkhurst & Feller (2007) conclude that color organization diagrams are at least as old as the 13th century, with the seven-color circle possibly the oldest found, the four-color next (14th–15th centuries), and the red-yellow-blue circle the most recent (18th century). No graphic records of circular systems earlier than 1611 have yet been identified.

[106] Newton associates color with *intervals of sounds*, but nowadays many sources mention frequencies of *tones*, whereas it is the ratio of *sensations* to which Newton referred.

[107] *Opticks* (1704), Prop. II Theor. II, page 87. Most people can see six colors, and the splitting of purple in violet and indigo is not very widely perceived, as remarked by John Dalton, see page 135. Indigo comes from Latin *indicum*, and refers to India as a supplier of indigo dye.

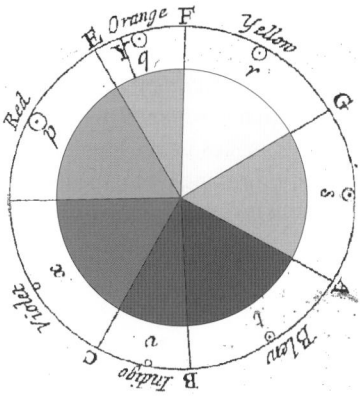

Fig. 29. Newton's color circle (Fig. 11 from *Opticks* 1704, Book I, Part II, Plate III), displaying seven colors selected by Isaac Newton. Probably the first graphical record of the concept of circular color systems. Newton's basic drawing is overplotted here with the 7 *roygbiv* colors: red, orange, yellow, green, blue, indigo and violet.

We owe a lot about color theory to the experiments of James Clerk Maxwell, and especially to his 1855 and 1860 papers. Maxwell's (1855) *Experiments on colour, as perceived by the eye, with remarks on colour-blindness* is even very remarkable as a scientific paper from the viewpoint of the principles explained in Paper I. The title of his paper is succinct, though very complete. The paper has no IMRaD structure whatsoever,[108] but its opening paragraph

> THE OBJECT OF THE FOLLOWING COMMUNICATION IS TO DESCRIBE A METHOD BY WHICH EVERY VARIETY OF VISIBLE COLOUR MAY BE EXHIBITED TO THE EYE IN SUCH A FORM AS TO ADMIT OF ACCURATE COMPARISON; TO SHOW HOW EXPERIMENTS SO MADE MAY BE REGISTERED NUMERICALLY; AND TO DEDUCE FROM THESE NUMERICAL RESULTS CERTAIN LAWS OF VISION.

explains what the paper is about, including all elements needed in a structured abstract – objective, methods, results – in fact, a structured abstract *avant la lettre*.[109] Maxwell presents error discussions, describes tossing out outlying data,[110] because two observers *"...namely, Mr R. and Mr S., were not sufficiently critical in their observations to afford any results consistent within 10 divisions..."*, explains boundary applications of the theory (related to the color-defective eye), and the corrections needed for such applications. In this paper he describes his color-mixing experiments that are based on rapidly-rotating color-paper sections,[111] from which he concludes that color perception is not due to the real identity of the colors, but on account of the constitution of the eye of the observer. And that the law of color vision is identical for all "ordinary eyes".

Young (1802) had put forward[112] that *"each sensitive point of the retina contains 3 particles matching the three principal colors, red, yellow, and blue..."* enabling the eye

[108] See Paper I, Section 9.

[109] Idiom meaning *before the concept existed* – not to be translated when used in another language.

[110] So-called outliers, see Section 11.4.

[111] *Section* refers to a geometric section, not to a part of a paper.

[112] The Bakerian Lecture, *Scholium*, page 20.

to react to "every possible undulation".[113] Such principal colors are now called *primary colors* or *primaries*,[114] the corresponding receptor types are generically named short, medium and long-wavelength cones, and form what is called a *trichromatic color system*. Maxwell, however, struggled with the fundamental difference between mixing light rays and mixing lights reflected by pigments: mixing dyes[115] makes us see any light that the colorants do not absorb, or do not extract – therefore, color mixing by pigments or by dyes is called a *subtractive process*. In combining light sources, there is no extraction, and all hues combine in what is called an *additive process*.

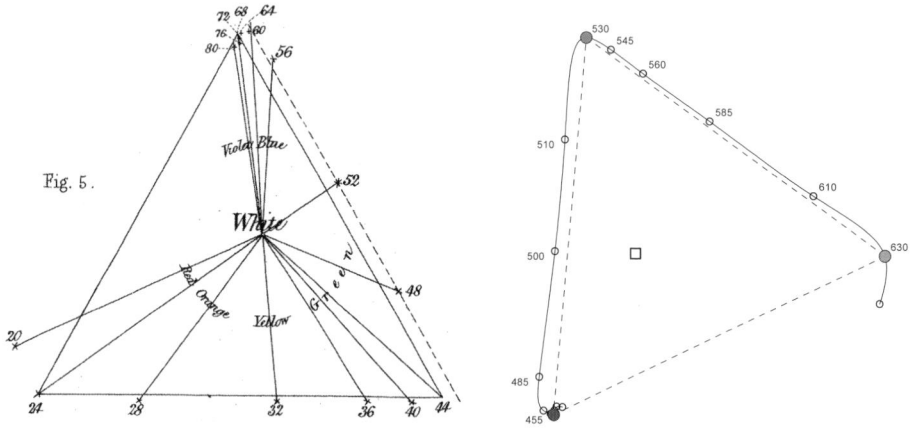

Fig. 30. *Left:* Figure 5 from Maxwell (1860). Channels 24, 44 and 68 correspond, respectively, to **his** selection of red, green and blue primaries. *Right:* rendering of Maxwell's triangular graph, rotated by 145° for comparison with the chromaticity diagram in Figure 33. The central point (□) is the position of "white" as derived from twenty observations. The numbers along the curve are wavelengths of pure spectral hues (in nm), and correspond to Maxwell's channels numbered 20–80.

Maxwell's great discovery was to describe the system that relates any color to the set of primaries, in the so-called *Maxwell triangle*, a ternary diagram where each primary color (red, green and blue) is placed at the vertex of an equilateral triangle, as shown in the left diagram of Figure 30. Inside such a triangle, every color can be made by mixing different amounts of each primary,[116] and white (also called the white point) is obtained by adding maximal amounts of each primary (for an example, see Fig. 48).

Mathematically that was clear, but psychophysically there are two complications that make it impossible to construct *any* color from the three primaries:

[113] *Undulation*: hue or wavelength – note that this statement is not strictly correct: not every receptive cell of the retina can distinguish color; some cells (rods) do not have any color sensitivity at all.

[114] Primary colors are colors from which all other colors can be derived by mixing in different amounts.

[115] Dyes are chemicals dissolved in a fluid, pigments are small particles that are not dissolved in liquid.

[116] That is, primaries should be independent: no additive mixture of two primaries should yield the third.

1. the primaries are not (in fact, need not be) exactly known, and
2. primaries are not monochromatic, but have a substantial *bandwidth* that includes neighboring hues (see Fig. 31).

Maxwell's concept is called a *trichromatic color space*. Let me just make a sidestep, and remark that the definition of such a trichromatic color space is very similar to the definition of a photometric system in astronomy, such as the Johnson *UBV* system (that also has a set of three overlapping wide passbands).

Maxwell (1860), on the other hand, is a fully-structured paper with introduction, history, method, description of instrument,[117] method of observation, data reduction, error discussion, results, comparisons, conclusions, and even a postcript note on color-deficient observers (but no Abstract, although the last Section of the Introduction lists the structure of the paper). His paper refers in detail to previous work by Newton, Young, Helmholtz, Fraunhofer and Poggendorff, and sets the stage for his own research: his experiments demonstrate that the chromatic observations leading to color equations can be made with great accuracy, both by normal and by color-deficient eyes. In a closing note, he explains the difference between Newton's theorem on the mixture of colors[118] (Newton's color circle divided into seven parts, Fig. 29), and Maxwell's own model.

For this investigation he does not work with rays reflected from a spinning color wheel, but with a purpose-designed optical device that allows him to combine red,[119] green and blue rays to render shades of white. From numerous measurements of the whites at 16 spectral locations, he constructs color equations that yield white light, and obtains intensities of the three standard colors of *"observer J"* (James himself), which he subsequently marks as R, G and B. Amazingly, he arrives at describing all experiments and their results in words, and in pure black-and-white graphics.

Maxwell's experiments led to the first mathematical (trichromatic) color model. Figure 30 reproduces the results for his own eyes in a ternary plot, where each of his selected primary colors (R for scarlet, G for green and B for blue) is a corner of the triangle.[120] Points outside the triangle result from negative components of one primary; measurements in the blue-violet corner of the triangle are quite difficult, and led to more erratic results. The rightmost panel is exactly the same diagram, but with the location of Maxwell's pure spectral hues connected by a spline curve (see Sect. 11.7). Today we call such a diagram a *chromaticity diagram*, and the 3-dimensional space it describes is called a *color space*. A chromaticity diagram tells us only which mixtures of light will produce a specific color. Figure 30 thus shows James Clerk Maxwell's personal color space, or the area delineated by the triangle of the primaries that he selected. The range of colors that such a model can represent is called the *color gamut*.[121] His 1860 paper also contains the numerical gamut for another person (*"observer K"*), and Maxwell comments on the

[117] Maxwell expertly used standard equipment, but also developed his own apparatus.

[118] *In a mixture of primary colours, the quantity and quality of each being given to know the colour of the compound.* Newton, *Opticks*, Book I, Pt. II, Prop. VI.

[119] Maxwell notes *bright scarlet* (vermillion).

[120] Though not immediately obvious, this drawing is also sort of a pie chart.

[121] The word *gamut* originated in music, and nowadays refers to the whole range of pitches an instrument or a voice can reach.

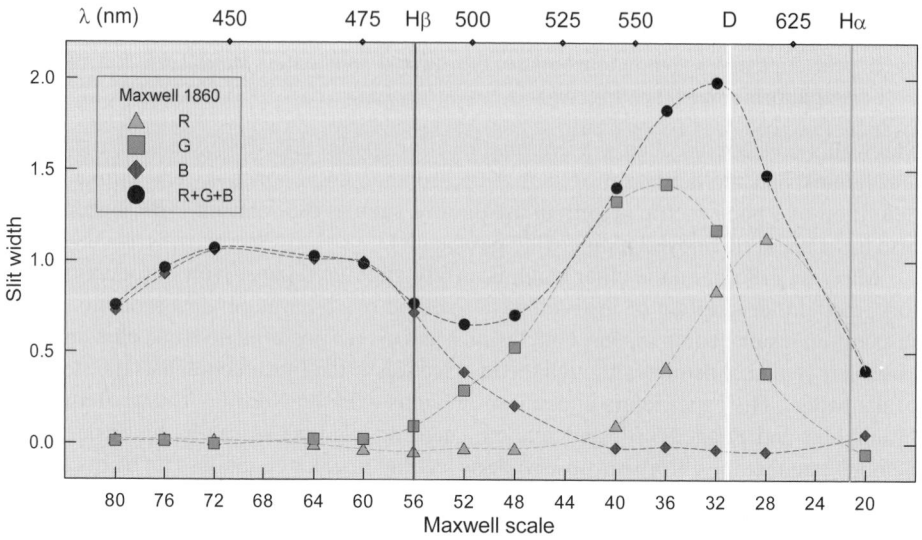

Fig. 31. *RGB* spectral response curves for James Maxwell's own eyes. Dashed lines are spline curve fits. The vertical line near scale value 31 is the doublet of the sodium spectrum near 589 nm (rendered in yellow and thus almost invisible on a grayscale print). Hα and Hβ (Fraunhofer C and F, respectively), are also indicated. The linear scales of the x and y axes are in laboratory units. The duplicate x-axis is expressed in wavelength on an approximate (nonlinear) scale, with small ♦ symbols instead of tick marks.

"blindness" of his very own eyes to parts of the spectrum: *"the feeble intensity of certain kinds of light as seen by the eyes of J . . . "*.

Thus, *trichromacy* is a property of ourselves, and the color space as shown in Figure 30 is unique, and James Maxwell's only. Or in modern terms, *device-dependent*: every person has her/his own color gamut, just like every input or output device also has its own gamut,[122] in fact one could say that each device has *its own language for colors*. Color – like beauty – is in the eye of the beholder,[123] and is not a synonym for wavelength.

Figure 31 shows Maxwell's empirical *color-matching functions*, as well as the sum for each retinal receptor type. His general conclusions list that his experiments show that the number of primary elements of color (for the normal eye) is **only** three, and that the *exact* wavelength-positions of the primaries are not ascertained.

9.4 Color models, color spaces and color palettes

A *color model* is the way color is mathematically described with 1, 2 or 3 primaries. A color space is the set of all the colors that can be represented using a specific color model.

[122] More precisely: a subset of visible colors that the device can display, or the area enclosed by a color space.
[123] *Beauty is in the eye of the beholder*, quoted by Marcia in *Molly Bawn*, Margaret W. Hungerford (1878).

Once a color system is specified, the amount of each component needed to create a color is given as a number from 0 to 255, and colors become coordinates in that specific color system. Dark corresponds to (0,0,0) and white has three components with value 255. Any other color with three equal coordinates corresponds to a shade of gray (see also Fig. 48).

From the previous Section we know that:

1. a human color space is described in three dimensions;
2. the exact function of the primaries is not critical, as long as they are different and not co-linear, and
3. the color-mixing process can be additive (red, green and blue for video and computer monitors) or subtractive (cyan, magenta and yellow for printing).

The remainder of this Section briefly describes the color spaces that are most important for scientific writing, *viz.*, the generic human-observer color space (XYZ), the color space of computer and video screens (RGB and sRGB) and the color space of our printed scientific writings ($CMYK$). At the same time, the concept of *color palette* is introduced: a color palette is a subspace of the color space, and is defined by the graphing software's options.

9.4.1 The CIE XYZ color space

The CIE[124] XYZ color model was developed in the 1930s as a perceptual color space that approximates the human visual response to different colors – that is, two colors with same XYZ will look the same to a CIE standard human observer.[125] Figure 32 shows the CIE xyz standard observer functions. The CIE XYZ color space contains the human gamut of all perceivable colors, and is mostly visualized as a projection x, y, Y. The Y component (called the luminance of the color) characterizes the intensity of light, and the chromaticity coordinates x, y (defined by the dimensionless ratios $x = \frac{X}{X+Y+Z}$ and $y = \frac{Y}{X+Y+Z}$) stand for an objective measurement of the color of an object, leaving aside all brightness information.

XYZ colors with the same chromaticity values have the same hue, but can differ in absolute brightness Y. Hence, colors can be described either by XYZ coordinates or by xyY (chromaticity plus luminance Y). Figure 33 shows the x, y *chromaticity diagram* (for a given Y). The straight line connecting the two end points of the horseshoe curve is referred to as the *purple line*, which corresponds to the singularity in the Newton wheel when leaping counterclockwise from red to violet. The point of equal energy is the CIE standard for white light. The pure (*i.e.*, fully saturated) colors lie on the ridge of the curve, and any line joining any two points in the diagram defines all colors that can be obtained by combining these two colors.

In daily life there is not much reason to worry about the XYZ system: its role is, just like halftoning, embedded in the graphical rendering software.

[124] *Commission Internationale de l'Éclairage*, organization that develops color standards: `www.cie.co.at/`

[125] The XYZ primaries are not material primaries, but invisible mathematical concepts, because they lie outside the gamut of the real colors.

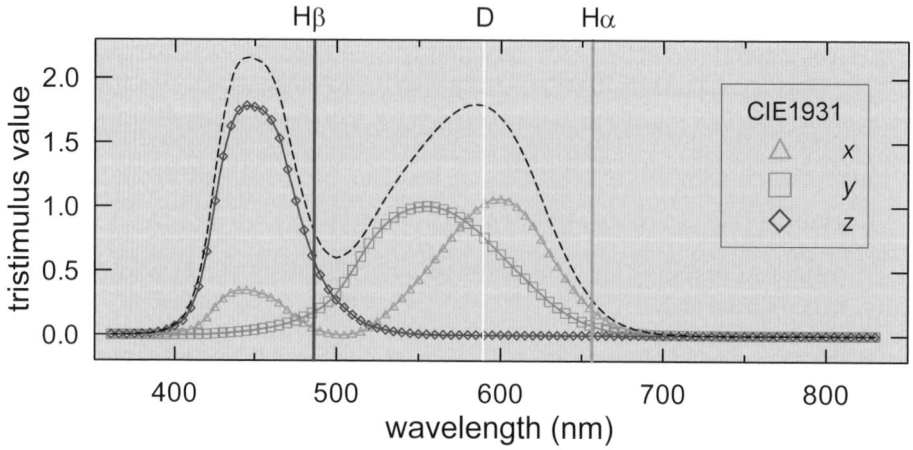

Fig. 32. CIE *xyz* standard observer functions, and their sum (dashed line). The top *x*-axis gives the approximate positions of the spectral lines shown in Figure 31. Both wavelength scales are linear.

9.4.2 The *RGB* and *sRGB* color spaces

The *RGB* color space is a color model for screen colors, *i.e.*, colors related to the emission of light; each number in the range 0–255 represents the intensity in each channel of the additive primaries. The *RGB* gamut is a triangle (Fig. 33) in the x, y plane.[126] The *sRGB* variant is a standard color space (designed for a typical CRT display) that has become the default color space for WINDOWS. The *RGB* triangle envelopes the gamut of colors that can be obtained by combining the three primaries,[127] but its gamut does not enclose all visible colors. Note the differences with Maxwell's triangle and gamut shown in Figure 30.

9.4.3 The *CMYK* color space

The cyan–magenta–yellow–black *CMYK* color space is not used for emission-light colors, but for reflected-light colors when printing on paper, cloth or other stuff. Cyan, in the subtractive color system, is one of the three primary colors, in the additive system it is a secondary color. Cyan absorbs red, Magenta absorbs green, and Yellow absorbs blue. The reason for adding the black *K* ink is that the subtractive primaries *C*, *M* and *Y* do not entirely remove their complementary *R*, *G* and *B* colors, hence their mixture produces a brownish tint.

The five-sided polygon in Figure 33 shows the gamut of the *US Web-coated SWOP*[128], a generic *CMYK* color profile. There are, however, additional factors to be taken into account, *viz.*:

[126] *i.e.*, for a given Y

[127] Only the relative amount of each primary is needed to quantify the color.

[128] SWOP: Specifications for Web Offset Publications.

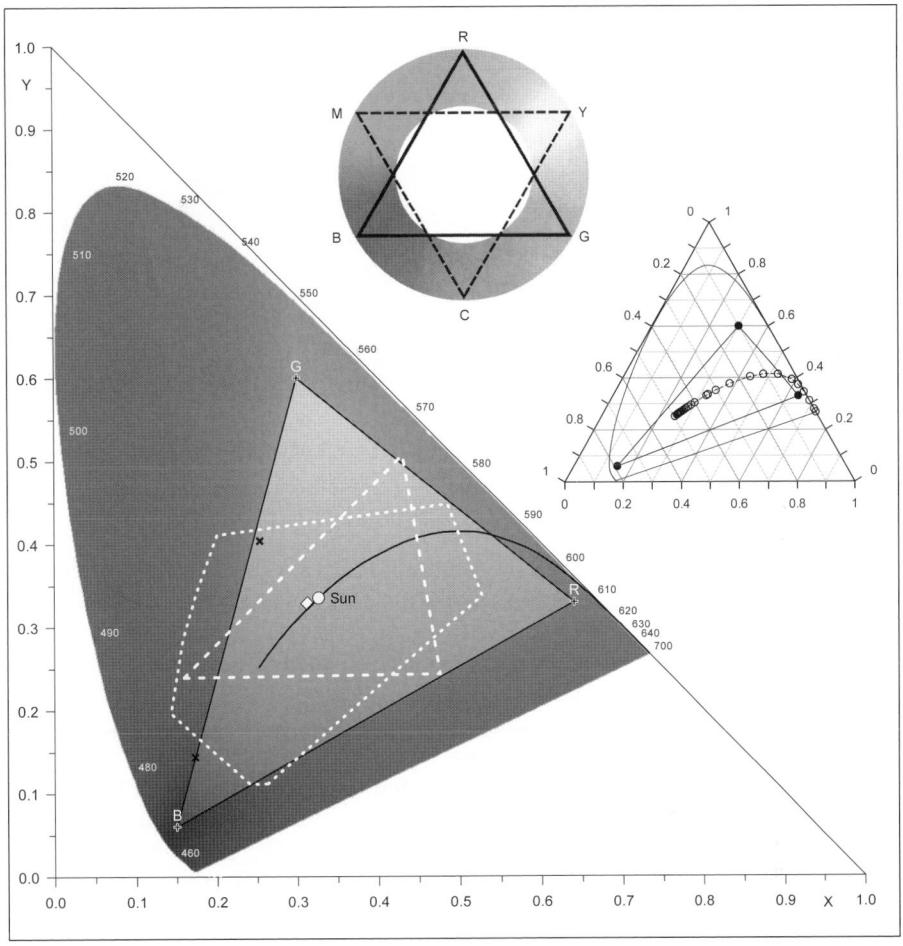

Fig. 33. CIE *xy* chromaticity diagram for the average person, with the *RGB* gamut triangle (typical color CRT monitor) and approximate *CMY* gamut (for color printing). The five-sided polygon is the gamut of a generic *CMYK* color space. The curved black line is the locus of black bodies (500–25 000 K). The pure (fully saturated) colors lie on the ridge of the horse-shoe curve. The position of the Sun is indicated, as well as the white point for a CRT monitor (◇: point of equal energy corresponding to the CIE standard for white light). × represent both A&A blue colors in *RGB* (coordinates from Table 7). The triangular inset shows some of the same data elements in a classic ternary diagram triangle, with some positions of black-body radiators (○) in the stellar-temperature range up to 25 000 K. The colored circle is a rendering of the color wheel representing the visible light spectrum with the positions of the *RGB* and *CMY* colors (Source: Kodak Handbook 1979). Primary colors are separated by 120°, and complementary colors are located directly opposite each other. The colors inside the horseshoe area were adapted from Steer (2004), with permission.

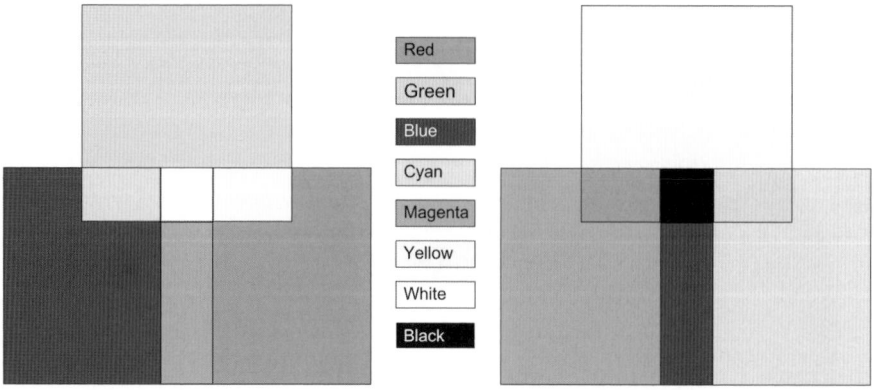

Fig. 34. Primary additive *RGB* (*left*) and subtractive *CMY* (*right*) colors. One graph follows from the other by permutation of colors and swapping white to black.

1. the reflective properties of the print substrate play a significant role: printing on yellowish paper, for example, will yellow the whites, and modify all colors;

2. the environment lighting plays an important role: ambient light of a different color will change the perceived hues – see the horizontal band in Figure 48.

CMYK is very device dependent, and covers fewer colors (a smaller gamut) than *RGB*, with the consequence that **some *RGB* colors cannot be represented in *CMYK***. Though *RGB* files contain more color data than *CMYK*, they nevertheless measure less in disk space.

The A&A cover and logo uses two hues of blue – these are also used in the timeline shown in Figure 5, and in the central square in Figure 2. Table 7 lists their *CMYK* values and corresponding *RGB* coordinates (with their positions marked in the chromaticity diagram of Fig. 33).

Table 7. The A&A cover and logo uses two hues of blue (no *R* involved).

Color	*C*	*M*	*Y*	*K*	*R*	*G*	*B*	Examples
light blue	72	0	23	0	0	188	201	Figure 5
dark blue	100	65	23	15	0	83	130	Figures 5, 17 and 33

There are basically two types of inks for printing: dye-based inks (*i.e.*, the more common types), and pigment-based inks (which are chemically stable over many decades, are much more *lightfast*[129] than dye inks, and are also much more expensive).

[129]Lightfastness is the degree to which a print resists fading when exposed to light.

9.4.4 Device-independent colors: spot colors

When colors cannot be made by mixing primaries, the graphic arts industry uses *spot color*[130], *i.e.*, industrially prepared special inks that are applied one by one. The PANTONE[131] Matching System, for example, uses 14 base pigments to produce each of the more than 1000 spot colors in their list, using a proprietary and unique ink mixing formula.[132] This proprietary color space assures a perfect match of the colors on the customer's monitor with those on the vendor's screen. The lists of color numbers and pigment values are the intellectual property of PANTONE and are thus copyright-protected, therefore they are only provided within high-cost software packages.

9.4.5 Degenerate color spaces

Section 9.3 referred to Maxwell's papers that describe in detail all the intricacies of color vision using black-and-white graphics: he just had no choice, because color printing was simply not available. Society, in those days, was not so colorful as it is now: except for the colors in nature, and the colors in paintings, tapestries, buildings and clothes, people were not much exposed to color. Nowadays, color is ubiquitously used in projection, in monitor viewing and in print, and quite often more for purposes of coding than for aesthetics.

All previous Sections dealt with trichromaticity, the three-dimensional color model in which every color stimulus can be described with three numbers referring to three distinct primary colors. The extreme apex is the *black-and-white color space*, with only two levels of *shade*. Then comes grayscale, a monotone space with 255 gray levels, followed by duotone, a mode with just two colors where the color triangle becomes singular.

On page 126, an allusion was made to persons with red-green color blindness. Such dichromatic[133] form of color blindness is called *deuteranopia*, from the Greek expression for "second does not see", referring to the second cones in the retina (the G channel) that are defective. This does not mean that green colors are invisible to people suffering from deuteranopia: most people with this condition can see quite well in the middle region of the spectrum, though they do not discriminate color differences. In about 5–6% of the (mainly male) population, at least one of the receptor types is affected with this anomaly, which means that about 7 to 8 students that attended the first and second SWYA schools are expected to be affected in one way or another. Such reduced color vision is described by a two-dimensional color model.

Maxwell was not the first to comment on different observers' abilities to discriminate colors. John Dalton (1798), already in 1794 remarked[134] that, unlike persons in general who can distinguish six kinds of color, he can see only two or at most three colors: yellow,

---

[130] An ink that is printed in a single run, like the individual C, M, Y as opposed to *process color*, which combines primary inks.

[131] PANTONE – founded in 1963 – is the developer of the most widespread color communication standard in the printing industry, and is a leading source in the technique of accurate communication of color in printing, publishing and textile industries.

[132] Instead of words or coordinates for describing colors, standardized labels and numbers are used (for example PMS 288, a bluish tint).

[133] A dichromat is a viewer with one of the RGB primaries missing.

[134] His lecture was read on October 3^{1ST}, 1794 (note the confusing format of this date).

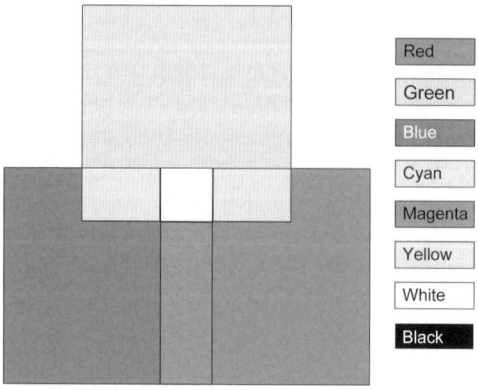

Fig. 35. Simulated *RGB* colors as perceived by a viewer suffering from protanopia. Black and white were untouched, as well as the labels in the legend.

blue and purple, and that yellow and green seem to him one color (for a full citation, see the caption of Fig. 36). Therefore, *Daltonism* is the term referring to dichromacy characterized by a lowered sensitivity to green light, *i.e.*, by deuteranopia. When the red cones are defective, then reds, yellows and greens are indistinguishable, and this deficiency is called *protanopia*. *Tritanopia* – the inability to see the third color (*B*) is quite rare.

Figure 35 reproduces the left panel of Figure 34 with *RGB* colors transformed[135] as if perceived by a protanope. The large majority of readers (about 94% of the population) will thus recognize the importance of selecting appropriate colors when designing graphics. This consideration is not a matter of scientific writing alone, but also a major preoccupation in business graphics: Cole (2004), for example, mentions having received an enquiry from a major bank that sought advice on the use of color in their graphs, because a customer with defective color vision had complained of his inability to distinguish color-coded data in their reports.

Web designers have defined a color palette of 216 colors for use in 8-bit (256-color) graphics mode – called the *Websafe palette* – and other colors are then dithered[136] to that color palette. The scheme is based on six steps of each of the *RGB* display ranges, hence the *RGB* coordinates of each color in this palette differ from the coordinates of each other color by 51 units, or by a multiple of this value. Figure 36 shows the positions of these colors in an *RGB* chromaticity diagram. The first aspect that draws immediate attention is that, though the color coordinates are regularly spaced (*i.e.*, relatively equidistant) in the *RGB* lattice, the spatial equidistance is not conserved in the ternary diagram.

Figure 36 shows that the color perception of people suffering from deuteranopia is very close to a duotone color space. Hence, **your graphical message will stand out when you avoid the confusion locus – common to deuteranopia and protanopia – along the *RYG* line of the spectrum, and when you select blue or white hues** (*e.g.*, for the backgrounds of powerpoint presentations).

Dalton's citation, partly mentioned in the caption of Figure 36, though, comprises a most confusing error in the subsequent sentence: *"The difference between the green part*

[135]Transformations based on information provided by the *British Telecom* website, `http://www.btplc.com/`

[136]Dithering: randomly juxtaposing pixels of two colors to create the illusion of a third color.

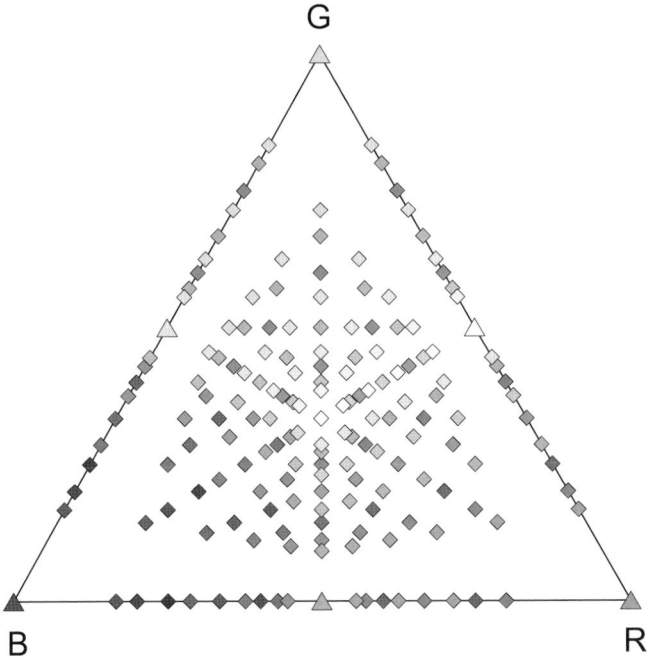

Fig. 36. *Top*: WEBSAFE palette of 256 regularly-spaced colors. The △ symbols for the *RGB* and *CMY* primaries are larger than the ◇ for the composite colors. Black, and the shades of gray (20%, 40%, 60%, 80%), are not shown, since they coincide with the white point with coordinates (1, 1, 1). *Bottom*: simulation of WEBSAFE colors as perceived by a deuteranope, accompanied by Dalton's *verbatim* description of his own color gamut.

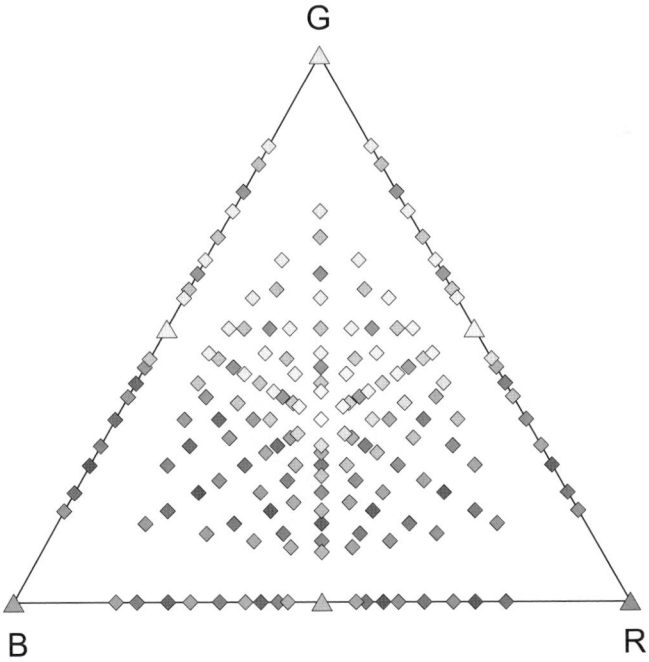

THAT PART OF THE IMAGE WHICH OTHERS CALL RED, APPEARS TO ME LITTLE MORE THAN A SHADE OR DEFECT OF LIGHT; AFTER THAT THE ORANGE, YELLOW AND GREEN SEEM *one* COLOUR, WHICH DESCENDS PRETTY UNIFORMLY FROM AN INTENSE TO A RARE YELLOW, MAKING WHAT I SHOULD CALL DIFFERENT SHADES OF YELLOW (DALTON 1798).

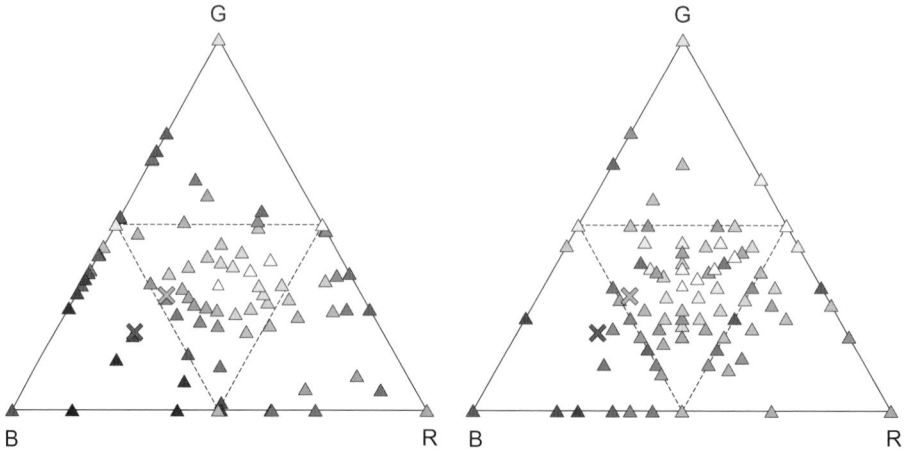

Fig. 37. Two standard color palettes. *Left*: PHOTOSHOP, *right*: GRAPHER. The × signs indicate the positions of the "A&A blues". Axis tick marks and labels have been omitted. Each apex of the triangle has the corresponding tristimulus value 1 with the two other values 0.

and the yellow part is very striking to my eye: they seem to be strongly contrasted." – he of course meant to say "between the green part and the blue part", a regrettable *lapsus*[137] most certainly overlooked during shallow proofreading.

9.5 Color workflow

9.5.1 Color management

Most graphical design software will open in the s*RGB* default color space, which is quite poor for evaluating the printed outcome. As an alternative, an *RGB* working space with a maximum overlap with *CMYK* can be selected. The monitor must be tuned to a contrast setting that is as close as possible to print, by applying an appropriate *gamma correction* that characterizes the nonlinear relationship between image pixels and monitor brightness. Gamma (Γ) is a parameter that assigns the relative brightness (input to output) of midtones: when $\Gamma = 1$, there is no adjustment (*i.e.*, the default), with $\Gamma > 1$ the dark areas receive more detail, with $\Gamma < 1$ the light areas get more tone resolution. Γ variation is nonlinear, as it conserves the responses at the white and black ends. For optimal prepress work, a gamma of 1.8 should be selected. In other words: **a color space always includes a gamma specification**.

Color management is the control of all aspects of color from the color space of the conception and design stage, up to the color space of the final printing – again, I must refer

[137] A slip or lapse.

here to Shannon's model in Figure 2. Color management systems try to take care of the different color reproduction characteristics of monitors, cameras, printers and scanners, so that the colors look the same, regardless of the hardware or the operating system used to reproduce them. Without color management one cannot hope to accurately predict output color on the basis of input color. Color management relies on ICC[138] profiles (file extensions `.icm` or `.icc`), which are files that contain the data of the color characteristics of the output device. This information describes how the scanner, monitor or printer reproduces input colors.

An element of the color management is the choice of the *color palette*, a subset of preselected colors provided by your graphical software: finding the right color palette is almost as important as drawing the graph. Figure 37 illustrates two such standard palettes in a ternary *RGB* diagram. As the graph illustrates, both examples cover different areas and color ranges, and even reveal color anomalies. One can usually extend the palette through the use of a *color picker* that allows you to select any *RGB*-additive mix.

9.5.2 Proper color balance and gray tonality

The Publisher/Printer can adjust color balance and exposure only if you provide calibration exposures. The front cover illustration, for example, features a *Kodak Color Separation Guide and Gray Scale*, a reflection step chart of 20 stepped, gray values[139] for comparing tone values of the original with the reproduced image (see Fig. 38, and also the back cover). The color-control patches strip contains standard colors used in *RGB* and *CMYK* to help match the colors in the printout with those of the original. When such an exposure is included in a photographic reproduction, it is possible to faithfully render the original colors in the printed image.

Let us now see how the technical principles apply to real life. Figure 38 shows the calibration strips for the cover illustration in *RGB*. The Publisher's prepress adjusts the grayscales and transforms the *RGB* to *CMYK*, applying in-house color and grayscale profiles that support their calibrated printers.[140] Figure 39 shows measured brightness level (on the left y axis) and normalized pixel value (on the duplicate y axis) as a function of reflectance or density[141] for three different exposure types: the middle grayscale shown in Figure 38, a sample scanned Kodak grayscale, and the Kodak grayscale photographed together with the book cover illustration. The two lines have different slopes, which means that applying the upper curve will result in an image with less contrast and lower dynamic range,[142] than when applying the lower line. The shape of the Kodak grayscale curve is nonlinear as a result of using a nonlinear detector, the photographic film.[143]

---

[138] *International Color Consortium*, `http://www.color.org`

[139] 20 steps in 0.10 density increments ($\frac{1}{3}f$-stop) between 0.0 [white] and printing black. Label 0 (A) corresponds to density 0.05, label 19 corresponds to 1.95.

[140] Note that the cover and the pages of the book are not printed on the same printers.

[141] Density equals $-\log_{10}\frac{\text{reflected light}}{\text{incident light}}$, and density 0.05 is the reflectivity of white paper ($\sim 90\%$) because typical paper increases density by about 0.05.

[142] The dynamic range of an image is the range between the lightest and darkest pixels in the image. Beware that recordable dynamic range and displayable dynamic range can differ.

[143] This curve also embeds a basically linear component due to the scan of the photographic positive print.

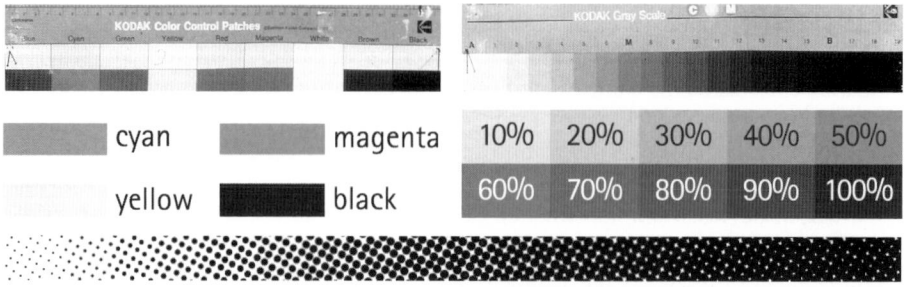

Fig. 38. *Top left:* Kodak control color patches for the cover illustration. *Top right:* Kodak grayscale strip. *Middle*: *CMYK* patches and gray scale used for printing the cover (both middle images are also reproduced in full color on the back cover of this book). Note that the labels for densities 10–50% are dark on a relatively bright background, and *vice versa* for the labels 50–100%. *Bottom*: halftone image of the Kodak grayscale strip, maximum dot size 7 pixels, screen angle 120°.

The point I make here is not so much the technical aspect by itself, but the fact that the author or designer (leftmost box in Fig. 2) has not the slightest control of what the reader (rightmost box in Fig. 2) will see if there is no **accurate communication with the Publisher's prepress staff**: this connection thus becomes one of the greater sources of *communication noise* alluded to on page 69. Just check for yourself: compare Figure 27 (in the color pages) with the reproduction on the back cover of the book – at the moment of writing, I had no control whatsoever on the degree of color matching of both prints.

9.5.3 Color transformations

But if every person and every output device has its own color space, how can we ever hope to communicate, especially since all these spaces evolve with time?[144] Just like language translation, color conversion is not always a straightforward one-to-one mapping.

The first type of transformation is related to the person who views the screen. Such transformations consist of matrix multiplication plus vector translation of the form[145]

$$\begin{pmatrix} R \\ G \\ B \end{pmatrix} = \begin{pmatrix} 3.2209 & -1.5392 & -0.4979 \\ -0.9696 & 1.8762 & 0.0408 \\ 0.0554 & -0.2047 & 1.0573 \end{pmatrix} \begin{pmatrix} X \\ Y \\ Z \end{pmatrix}$$

Each monitor has its own linear *XYZ*-to-*RGB* transformation matrix, and bringing two adjacent *RGB* monitors to the same standard involves multiplying the color matrix of one monitor with the inverse color matrix of the other. When color-calibrating monitors and printers, you should keep in mind that

[144]The eye ages with time, computer screens age with time, even ambient temperature plays a role.
[145]The coefficients of this matrix are just one possible set of coefficients.

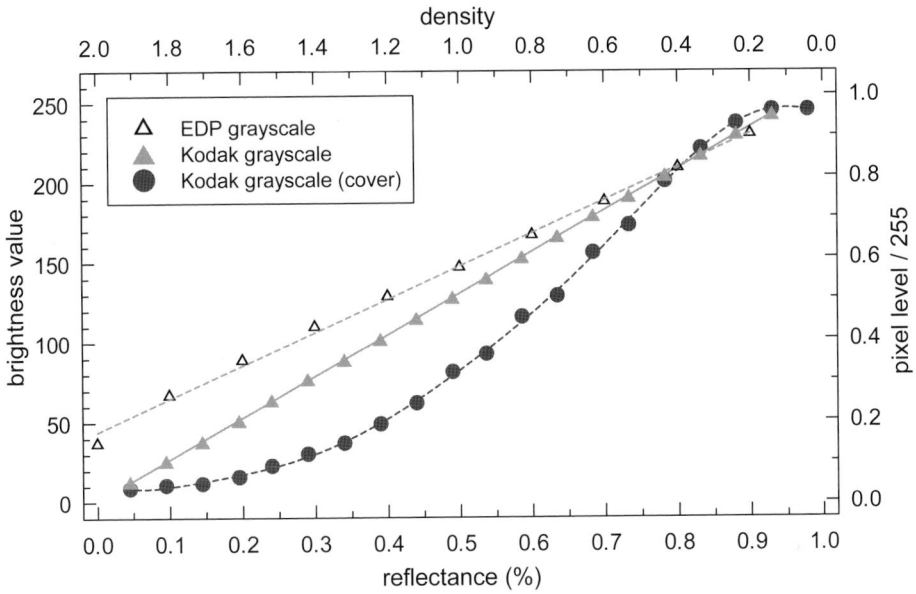

Fig. 39. Brightness as a function of reflectance or density for three different exposure types. EDP grayscale △: measured pixel level as a function of reflectance for the middle grayscale in Figure 38 (reflectance data provided by the Publisher of A&A). Kodak grayscale ▲: pixel measurements from a scanned Kodak grayscale as a function of density. •: pixel values from the Kodak grayscale photographed together with the book cover illustration as a function of density (see text). Curves are linear and polynomial fits. Note the symbol sizes in this graph: triangles and circles all have the same nominal 8-pt size, yet they are not perceived as of equal size.

1. the above relations lead, mathematically speaking, to a perfect fit as soon as three independent tristimulus values are provided. To be robust, however, many more empirical data must be acquired in order to secure accurate color coefficients from a least-squares solution;

2. the entries should have a reasonable extent of color diversity: calibrating with a handful of points that are close in the *RGB* space will always yield a poor calibration with a very limited range of validity (just like in astronomical photometry as referred to on page 104);

3. monitors age with time, and their gamuts change along, hence a good calibration setup of today will not necessarily work out well one month later, and

4. when adjusting several monitors to the same scale, each monitor must be separately calibrated. A matrix with coefficients that are averages from several other matrices, will never be a proper calibration.

As mentioned already, *RGB* has a larger gamut than *CMYK*. Therefore, although one can convert any *RGB* to *CMYK* without losing information, the inverse is not the case:

when a *CMYK* color is converted back to *RGB*, information is lost. For the conversion to *CMYK*, the A&A Publisher uses the *ISO coated V2* profile[146] to generate the PDF file of the illustration that is printed on the cover of each Volume of A&A (Hosotte 2009).

It is very important to understand what happens to your graphics during the journey from your desk to a digital or printed publication. The process is not very widely understood, even by people who work in the printing and in the graphics industry. For Sterken (2009), I asked to print the same 12th-century world map on the cover, and also inside the same book. At proof stage, I expressed my unhappiness about substantial color differences between the prints of the same map. The technician told me that it was *"impossible to equalize the colors since the production involves two different printers"*. The problem really was that one person had transformed my manuscript to a *CMYK*-based PDF file, whereas I had delivered the cover as an *RGB*-based PDF. A publishing or printing company with a good prepress service should anticipate such situations, moreover the color profiles of their respective printers should be so well maintained that all their equipment operates in concert, without the client even having to think about the issue. The really good companies calibrate their monitors and printers on a weekly basis. Unfortunately, *canned* profiles – generic profiles for a specific brand and model – are too often used instead of custom profiles.[147] This example illustrates, again, the role of the Publisher's prepress services, as described in Section 3.2 in the paper by Agnès Henri in Part 1 of these Proceedings.

9.5.4 When to change color space?

The basic workflow, in principle, is to design all artwork **in the color space with the largest possible gamut**, and to convert to a more restricted gamut at the print-ready moment only, *i.e.*, **make all retouches and changes always in the same wide-gamut system, and transform only when the job is ready to be sent to the printer**. Moreover – since *RGB* files cover a larger gamut than *CMYK* files, at the same time they weigh less in megabytes – keeping an *RGB*-based workflow is the most versatile and the most economic approach. Using the language metaphor on gamuts (page 130), it is obvious that making repeated changes by **transforming back and forth between color systems is like translating the same sentence back and forth between languages**: every translation destroys information, and increases communication noise.

The commercial printer at the end stage uses profiles to convert your *RGB* space to a very specific *CMYK* color space that takes into account the press, the inks and the paper stock. This transformation should happen only once: **at the very final stage, where most of the money is spent**, and this transformation is done by the company that prints your work, in the same automatic and invisible way as the halftoning is performed. The only way to evaluate the color print, is by proofprinting on the final device, which is a costly affair. Alternatively, a well-calibrated printer will do, and a lot of trouble can be avoided by "proofing" via previewing on screen in *CMYK* mode.

---

[146] ICC standard color profile created by the *European Colour Initiative* (ECI) `www.eci.org`

[147] Not to speak of variations in color that are caused by variations in ink supply, room temperature and humidity, the batch of paper, and whether the print run has just commenced, or is at its end.

10 Principles of graphical excellence

Edward Tufte published a series of exquisite books (Tufte 1983, 1990, 1997) on graphical design, with some of his material presented already for the first time by Funkhouser (1937). It is most interesting to compare the print quality of Tufte's and Funkhouser's publications, and see the enormous advances in print technology over half a century, in all aspects of resolution, font and (lack of) color. Though the content of these books and papers is only partially useful for astronomy communication, the general principles are valid for about all aspects of our graphical work. Tufte (1983) phrases the principles of graphical excellence in the following terms:

> **GIVE TO THE VIEWER:**
> **THE *greatest* NUMBER OF IDEAS**
> **IN THE *shortest* TIME**
> **WITH THE *least* INK**
> **IN THE *smallest* SPACE**

In other words: optimize in all possible directions in order to obtain the highest possible information content characterized by clarity, precision and efficiency. He also formulates five rules:

1. above all else show the data;
2. maximize the data-ink ratio;
3. erase non-data ink;
4. erase redundant data-ink, and
5. revise and edit.

By way of experiment, I checked all graphics of the first three years of A&A (1969–1971) and compared them with graphics appearing during the period 2007–2009. Although the recent graphs look a lot more professional than those of the early days, some classes of errors commonly made in the 1970s are still committed today! In the subsequent Sections I concentrate on some of these errors.

10.1 Managing ink

Tufte defines the *data-ink ratio* as the ratio of data-ink to the total ink present in the graph, *i.e., the percentage of ink used for displaying non-redundant information*. In particular, thick grids, crosshatching, dark shading, labels and ticks, oversize legends and rubble of computer plotting (non-specified characters or identifications[148] at the top or bottom of a graph) decrease this ratio, and render graphs unaesthetically overladen. In other words: **whatever ink you add to a graph should have a reason**. This also refers to his concept of *data density*.

[148]The kind of identifier, also known as chart junk.

$$\text{DATA-INK RATIO} = \frac{\text{DATA-INK}}{\text{TOTAL INK PRESENT IN THE GRAPH}}$$

$$\text{DATA DENSITY} = \frac{\text{NUMBER OF ENTRIES IN THE DATA BASE}}{\text{AREA OF DATA GRAPHIC}}$$

Figure 40 is a typical example of a vintage graph hand-plotted on a printed polar-coordinate grid. Here you see that the heavy grid yields a very low data-ink ratio because of the closely-spaced grid cells, and because of the substantial weight of the grid lines. Moreover, the data area encompassed by the three line graphs identified in the figure legend, is less than 15% of the complete graph area. The "above all else show the data" principle could be implemented by restricting the graphics to a rectangular part delineated by the superimposed central square that I added for the sake of argument. The purpose of the square is also to illustrate an optical illusion that comes along when applying heavy raster lines: the sides of the square – although strictly rectilineal – appear slightly curved when viewed on the background of the circular grid.

Another example is shown in Figure 41, based on data kindly provided by Chanut *et al.* (2008), illustrating the temporal evolution of the longitude of a planetesimal. The left diagram shows the complete dataset. The huge block of ink covers more than 100 000 datapoints, yet to the reader's eye it is only one single pixel of information that covers 65% of the graph. Since the dependent variable is an angle, a non-data line is automatically added each time the argument passes from 360° to 0°, and a substantial amount of non-data ink is thus added. Nevertheless, it is not easy to find a good solution to present such a cyclic dataset: a polar graph offers some relief, but is not a really good method to help the viewer understand the process. One possible remedy is depicted in the right panel of Figure 41: applying an axis break removes the bulk of the disturbing ink, at the same time this gives more focus to the real behavior of the independent variable "trapped" in a confined interval after $t = 65\,000$. Remember, though, that a poorly-documented axis break may lead to confusion: in this case the eye may not readily recognize the removal of an interval of more than 60 000 years, although in Figure 41 the scale problem is documented by including proper tick marks on both sides of the interrupted axis. Graphs with similarly high ink density are very common when displaying frequency spectra of cyclic or periodic data.

The effect of non-data ink depends on the situation: Figure 18, for example, is quite clear when viewed in color mode, but its grayscale rendering of the non-data ink is somewhat confusing.

10.2 Revise, edit, revise, edit, and revise, edit

Revision of graphics is a most important act, that very well matches the phase of revising and proofreading texts as discussed in Section 10.1 in Paper I. Let it be clear: the first draft is only a working model – just like in the writing process – and the subsequent

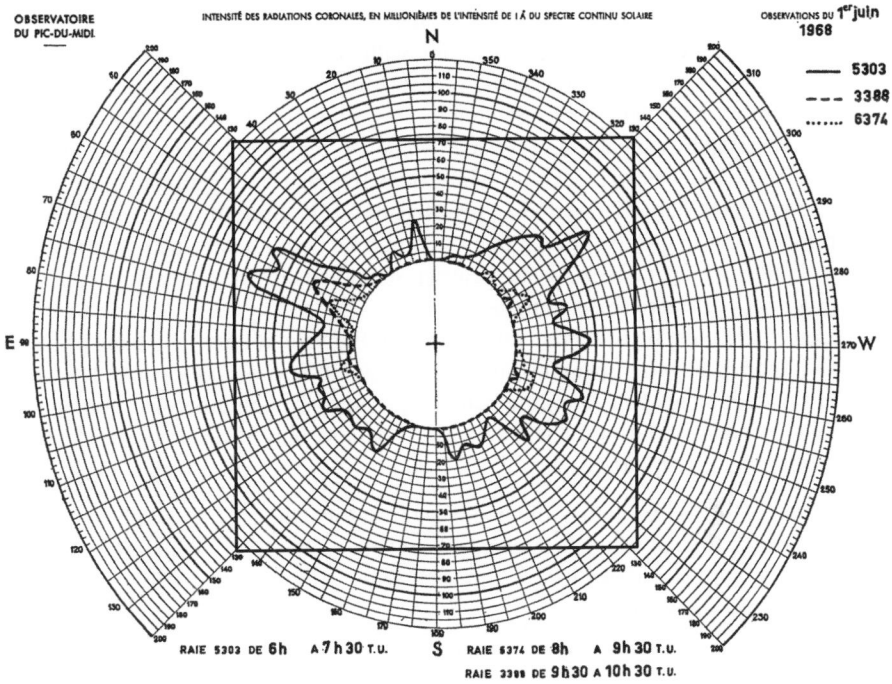

Fig. 40. Vintage graph hand-plotted on a printed polar-coordinate grid. The central square has been added for emphasis (see text). The title and legends provide, besides identification of the data, also information on the time and location where the data were obtained. Based on Rozelot (1969).

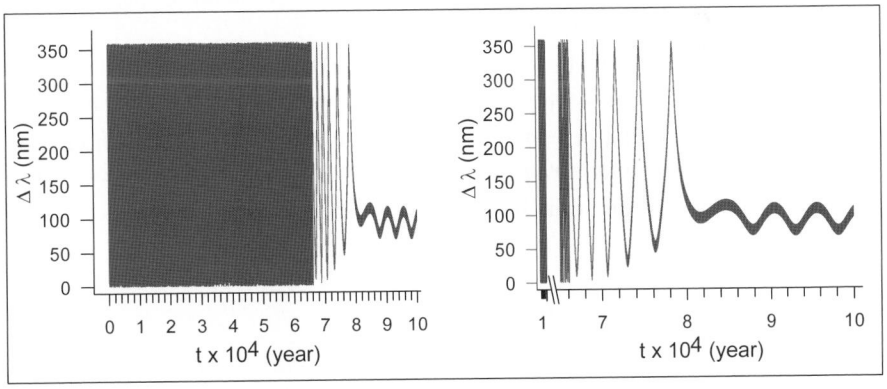

Fig. 41. Examples of graphics that is difficult to read. *Left*: the complete dataset. *Right*: same, with axis break. Duplicate axes were omitted for clarity. Line color for this graph was set to blue, in order to obtain a gray rendering in black-and-white print. Based on data from Chanut *et al.* (2008).

drafts are for optimizing and for colorizing. There is an absolute need for multiple drafts: experiment with several types of graphs, play with colors, change aspect ratios, and so on. After the last revision, a thorough "proofreading" of all graphs is necessary, with special attention to all elements of Section 7. A special act of proofreading is to verify how the color graphics appear in grayscale by converting the graphs to monochrome (or by setting the `dvi` viewer to black-and-white, or simply by printing all graphs on a black-and-white printer). Paper I referred to **proofreading for content** and to **proofreading for errors** – likewise proofreading graphics comprises both aspects, and your last check before submitting a manuscript for publication, should equally be a deep proofreading of all graphics for errors and for inconsistencies.

Extreme care must be observed with respect to the validation of any tabular material that was imported through scanning and subsequent OCR, especially when data are extracted from documents that date from before typesetting by computer was implemented. Such OCR is not without obstacles, as various defects may deteriorate the data. The following problems regularly occur:

1. when dealing with century-old books and catalogs, the user is not always allowed to xerox or to scan valuable documents, and hence must work with image files from online libraries or repositories, and such scans are often of too low resolution to adequately apply OCR;
2. older manuscripts (for example, Maunder 1904) had to save on typesetting labor, hence leading decimals and + or – signs were given at the beginning of a column only, unless they change midst a column, and only then the change was indicated;
3. confusion between o, 0 and °, between l, [, ! and 1, between 5 and 6, etc.;
4. obsolete characters – such as the ø for 0 – leading to grave errors;
5. extraneous spurious characters resulting from blobs of ink on the paper, to specks of dirt on the scanner's glass plate;
6. broken letters and symbols, for example part of a 0 may be missing and the character is then read as (, or a missing segment of 6 may lead to recognition as 0;
7. columns with missing or non-available data are sometimes filled with blanks, or with a sequence of dots, rendering such OCR products not strictly machine-readable;
8. genuine typographical errors by the typesetter, and missing characters (especially the decimal dot).

10.3 Layout: the typeset page as an image

Tufte (1983) says *"data graphics are paragraphs about data"*, hence the placement of these paragraphs is of prime importance: a typeset page, when considered in its entirety, simply is a composite image compiled in a semi-automatic way. Because we look at a printed page in one glimpse, we also see the aesthetics of the graphics paragraphs and the text paragraphs at the same glance. But each page can display several unaesthetic and even disturbing artifacts, as described in the list below. In addition, Section 13 in Paper III also discusses the possible effect of page layout on the *interpretation* of the text (*i.e.*, the case of Robert Millikan's alleged data cooking).

1. **Widows and orphans** are dangling words at the beginning or end of a paragraph. A widow is *a short line appearing at the top of a page, and is at the end of a paragraph*; an orphan is *a short line at the bottom of a page, or a word or part of a word on a line at the end of a paragraph.* Figure and Table captions can also produce such unaesthetic dangling words. Such artifacts result in excessive white space between paragraphs or at the top or bottom of a page, and can be avoided by changes in wording. I am often astonished to see widow-type appearance of one or two references appearing on a blank page at the end of a paper. Such occurrences involuntarily create the impression of an extra page (especially in proceedings books where space is limited), though they can so easily be avoided by technical tricks (slight resizing of one Figure, using abbreviations for journals, etc.).

2. **Alignment and right-margin justification**: *right justified* is often used instead of the term *right aligned.* Right justified is type aligned both right and left, right aligned is flush right and ragged on the left. Two-column page layouts often lead to unaesthetic alignments with lots of white space; an example is shown in the caption of Figure 13.

3. **The aspect ratio of graphics** (as already discussed on page 118) has two facets: the internal aspect ratio, *i.e.*, the optimal ratio that shows the data in the best way, and the page aspect ratio, as defined by the Publisher's rules and guidelines. It is the latter that causes most of the unaesthetic page layouts in journals. Since Publisher's rules cannot be changed, the internal aspect ratio should be optimized.

4. **The composition of graphical elements**, as illustrated in Figure 36. The most straightforward approach would be to place the triangles next to each other, but that requires downsizing, and results in loss of color perception. The more aesthetic solution is vertical arrangement above a caption over the full width of the page. The actual presentation is perhaps less pleasing at first sight, though it permits a direct explanation, in John Dalton's own words as published more than two centuries ago.

10.4 Style

Paper I stated that, in contrast to prosaic style, *"scientific style is direct, plain, assertive, context-independent, and brief ... "*. The same applies to graphics: your graphs will develop in a certain style direction, just like architects' houses bear the signature of their designers. That is why it is important to develop a good style as soon as possible, and I refer to Section 13 for a list of errors to avoid. Your graphical design skills will equally benefit from looking at all aspects of graphics that you encounter in books, journals – even in the daily environment around you. The most challenging task is to **remain consistent inside the same paper**, a goal that is quite difficult to achieve in a multi-authored work.

11 Lying with graphs

Some images, by definition, always "lie": *false-color*[149] representations of celestial photographs do mislead, especially images taken in non-optical windows of the electromagnetic spectrum. For example, only part of the Hubble Space Telescope wavelength window overlaps with the gamut of human vision, and other space telescopes even observe only in the ultraviolet or in the infrared spectral region. All the same, infrared space images acquired in a 75-μm band are combined with 150-μm and 300-μm band data and rendered with selected *RGB* primaries to yield pleasing and informative pictures.

The last paragraph of the Preamble in Paper I stated that we proceed *"on the premiss that the reader is strongly committed to truth in science, and even more so to the truthful communication of scientific findings"*. Still, good intentions do not always suffice: graphics may – inadvertently or on purpose – mislead the reader.[150] Images are just ideal for seducing or deceiving, and that particular power is the core force of all advertisement and publicity messages. However, an awesome drawing or a fantastic picture can never be the substitute for a proof or for cogent evidence, as Ronald A. Fisher (1925) so clearly stated:

> DIAGRAMS PROVE NOTHING, BUT BRING OUTSTANDING FEATURES READILY TO THE EYE; THEY ARE THEREFORE NO SUBSTITUTE FOR SUCH CRITICAL TESTS AS MAY BE APPLIED TO THE DATA, BUT ARE VALUABLE IN SUGGESTING SUCH TESTS, AND IN EXPLAINING THE CONCLUSIONS FOUNDED UPON THEM.

11.1 The copyright

The same rules on copyright as explained in Papers I and III apply to the use of graphics: you must always mention the source, and whenever it is obligatory, ask for permission to reproduce. This chapter contains several reproductions of copyrighted material with the copyholder's permission (granted for free, or for a fee). Figures 29 and 30 are copyright-expired images, since their authors have passed away more than 50 years ago, hence these images are in the public domain. But remember that, though the tapestry reproduced on the front cover of this book is more than 500 years old, it is the artwork's present owner (*Röhsska Museet of Fashion, Design and Decorative Arts*) that must give or sell the right to copy. Even if this artwork had been published in an artbook, it would still be protected through the copyright clauses in that book.

A special warning is in place about copyrights of internet images. Many images on the web have been uploaded without the slightest worry about copyright violations. Nevertheless, you do not have the right to copy and publish any web images without seeking permission from the copyright owner. Beware that some copyrighted material embeds well-hidden digital marks that allow tracing back the origins. Some publicly posted files even have digitally changed content (an intentional graphical bug) that will reveal the source of the image. Some graphics presented by Wikipedia are licensed under the *GNU*

[149]False-color differs from pseudo-color: the former refers to an *RGB* rendering of images acquired with non-*RGB* filters, whereas the latter refers to a monochromatic image represented in color, see Figure 4.

[150]Wilful deceit with graphs is discussed in Section 6.7 in Paper III.

Free Documentation License,[151] and may thus be reproduced without explicitly asking for permission. Again, **permission to reproduce is not the same as transfer of copyright**.

11.2 Retracing graphs

But what about so-called *derivatives*: are you, for example, allowed to scan Figure 13, and retrace[152] it? And if the retraced version is modified (for example by adding a branch to the tree in Fig. 13), is that enough to claim new copyrights? If you reproduce either a scanned drawing or a traced version, you cannot claim copyright on the image since it is a slavish reproduction. Stronger even, nobody else may reproduce your version, whether or not you forgot to reveal its source. However, if you introduce original variations in content, the situation becomes more complex, take for example Figures 1 and 2, where the former is a slavish copy, whereas the latter is an expanded version with added interpretation. Note that adding only color is never enough for claiming copyright, so Figure 30 in Paper III – though not a slavish copy – is a derivative and thus, not new.

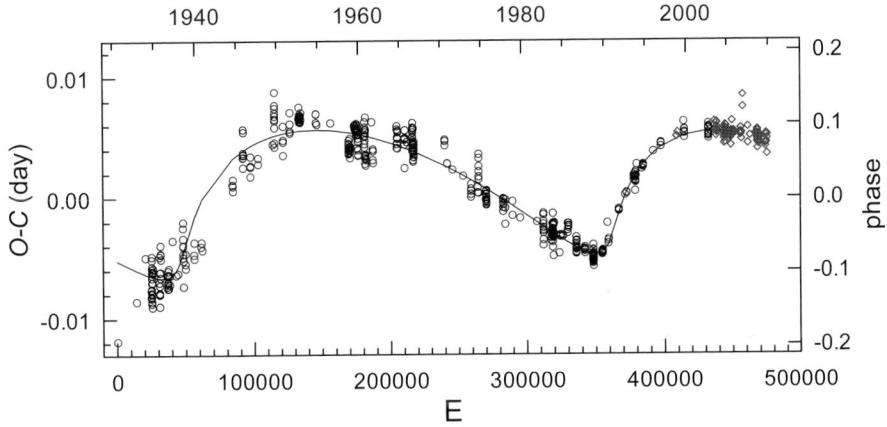

Fig. 42. $O - C$ curve of CY Aquarii with respect to the ephemeris of Fu & Sterken (2003). The continuous curve is the light-time orbit solution; ○: data from Fu & Sterken (2003); ◇: new data. Based on Figure 4 of Tuvikene *et al.* (2010).

One problem with retracing and reproducing is that you should avoid to change the content of the original in such a way that the original message gets distorted. Figure 42 is an example of a reproduction of an $O - C$ diagram of the pulsating variable CY Aquarii adapted from a graph from Tuvikene *et al.* (2010), which, in turn, is a Figure from Fu & Sterken (2003), overplotted with new data (the ◇ symbols) and an axis extended to fit the original. Note that the four axes of this graph have different units, and that the opposite

[151]The *GNU General Public License* is a free, copyleft license for software and other kinds of works: `www.gnu.org/licenses/gpl.html`

[152]Retracing is a kind of reverse-engineering, in fact the twin of text OCR. In the past, tracing was done on transparent (and often gridded) *tracing paper*.

axes are redundant, but help the reader to interpret the data. The lower x axis is the cycle number (commonly indicated by E), the top axis shows the equivalent time baseline, and reveals that more than 80 years of data are involved (the star's pulsation period is $0\overset{\mathrm{d}}{.}0610$, hence the large E numbers). The left ordinate axis gives the residuals in days, whereas the opposite ordinate axis uses dimensionless phase (residuals divided by the period). This axis also gives additional information: it tells the reader that the meandering of the period covers almost 40% of the cycle length. So, Figure 42 barely differs from its source in Tuvikene *et al.* (2010), and cannot possibly be called a new graph, simply because the *added* information content is *nihil*[153]. The Tuvikene *et al.* (2010) Figure, however, significantly differs from the source in Fu & Sterken (2003), and can be considered new – although more than 75% of its content was retraced – because the *added* information content is significant.

11.3 The lie factor

When visual representation is inconsistent with the numerical content, there is distortion. Tufte defines the concept *lie factor* as the ratio of the size of the effect shown in the graph to the size of the effect in the data. Lie factors of 2–5, unfortunately, seem to be quite common.

$$\text{LIE FACTOR} = \frac{\text{SIZE OF EFFECT SHOWN IN THE GRAPHIC}}{\text{SIZE OF EFFECT IN THE DATA}}$$

In particular, one must **make sure that interpretation of graphics does not depend on any variation in graphic design**: only those graphical methods should be applied that do not modify data content, as is stated by Butler (1979):

> ...IT IS INCUMBENT UPON THE USER TO ASCERTAIN WHETHER THE MANIPULATIONS REQUIRED TO PREPARE THE GRAPHICAL DEVICE HAVE IN ANY WAY MODIFIED THE RELATIONS AMONG THE SELECTED VARIABLES.

Auto clipping, auto scaling, auto binning and auto axis-breaking are some of the automatic features that can lead from misrepresentation to outright lies. Remember, though, that some of these variations are often enforced in an automatic way by the addition of a dimension, see Section 4.3.4 on representing 1-dimensional data by 2 or 3-dimensional quantities. Especially fitting a third dimension to histograms and bar charts is like outfitting a 2-D movie with an extra 3-D dimension: it adds nothing to the information content, although it can distract.

[153]From Latin, used for "nothing" or "null".

11.4 Outliers

An outlier in a set of data is defined by Barnett & Lewis (1984) as *"an observation which **appears to** be inconsistent with the remainder of that set of data"*, or *"one that **appears to** deviate markedly from other members of the sample in which it occurs"*.[154] These authors underline that the phrasings **"appears to ..."** are crucial: they emphasize that it is a matter of subjective judgment on the part of the observer whether or not to pick out some discordant data for scrutiny. Daniel Bernoulli, in 1777 already, dealt with the problem of outliers in astronomy, and wrote[155]

> IF THERE IS NO SUCH REASON FOR DISSATISFACTION I THINK EACH AND EVERY OBSERVATION SHOULD BE ADMITTED WHATEVER ITS QUALITY, AS LONG AS THE OBSERVER IS CONSCIOUS THAT HE HAS TAKEN EVERY CARE.

A similar statement was made by Delambre & Méchain (1806), see Section 6.4 in Paper III.

Some experimental results can just be plainly impossible, but others can appear quite unlikely, though still possible – remember that in a normal frequency distribution 0.2% of the scores may deviate from the mean by more than 3 standard deviations – which makes rejection of outlying values difficult to justify.

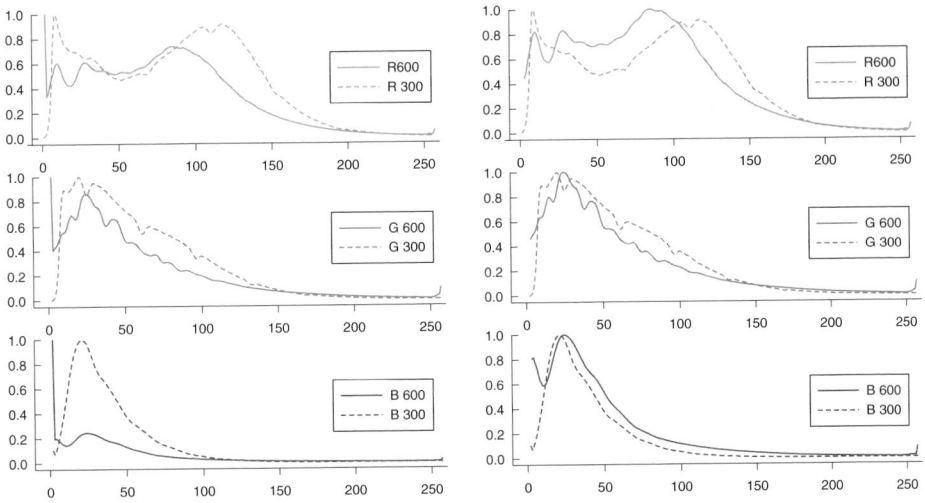

Fig. 43. *RGB* histograms of the cover image based on 300 and 600 dpi scans. *Left*: unedited histogram data. *Right*: after removal of an outlier at pixel value 0 in all channels.

Outliers can easily fool you during the data analysis. Take the odd histogram of Figure 28, and compare it to the same information presented in Figure 43. The left stack

[154]The `boldface` was not part of the original statement, but the emphasis was expressed in the text.
[155]Translation from Allen (1961).

of curves in Figure 43 – actually histograms drawn as curves – are from the same source as the data shown in Figure 28 (the *B* panel actually is identical to the one in Fig. 28). The right stack is derived from exactly the same data, however one outlier – a huge spike at pixel 0 in each band – was removed. The outcome is astounding: whereas *B* shows the largest variance in the left set of graphs, the removal of one point makes both histograms almost identical in those parts of the image devoid of the darkest pixels.[156] The *R* channel, on the other hand, now becomes the most divergent. The single point that was removed, is commonly referred to as outlier or wildshot observation. Barnett & Lewis (1984) distinguish three sources of outliers that can be encountered:

1. inherent variability: the amount of variability belonging to the population;
2. measurement error: inadequacies in measurement superimpose additional variability on the inherent factor (these authors also consider rounding errors as part of the error of measurement);
3. execution error: due to the imperfect collection of the data (*e.g.* the sampling bias and other effects).

But what to do with outliers in your data sample? Let me give three examples of how to properly deal with outliers.

- Appenzeller (1972), when monitoring the luminous blue variable R 81 in the LMC, obtained some magnitudes of the kind that likely would be removed by any automatic outlier-clipper. Tubbesing *et al.* (2002) also collected such "deviant" data, and identified these magnitude drops with eclipses. By combining Appenzeller's published data with their new observations obtained thirty years later, these authors could improve the binary period by a factor of ten, resulting in $P = 74.566 \pm 0.014$ day. Although variability of this amplitude was unexpected, these were outliers that perfectly belong to the inherent variability of the subject. Figure 6 illustrates the statistical properties of these data in a box-whisker plot for three samples where inherent variability is present, clipped, or removed.

- In a CCD-based light curve of a variable star, we noticed a very strange outlier present in one of the photometric bands at only one single moment. After verifying and rejecting all instrumental explanations, a small programming bug[157] was found in the routine for estimating the instrumental magnitudes of the stars: it so happened that only in one single case did the effect become visible in the data. The effect disappeared after the bug was corrected (Tuvikene 2011).

- In Figure 42, two outliers are manifestly present in the data collected since the year 2000: the two data points not only "deviate markedly" (terminology used by Barnett & Lewis 1984) from the data almost simultaneously obtained by other observers, they also deviate from the model orbital solution curve. Similarly large deviations – positive and negative – also showed up in the past, but since these two data points

[156]Note that for this exercise the tapestry image was cropped to include only the textile picture frame.
[157]Misidentification of a sharp-peaked point spread function as a cosmic-ray pixel.

were taken from the literature, there is no ground for eliminating them. Two other outliers (not shown) were, however, removed because we found out that the authors had not applied heliocentric corrections to the times of observation taken from an online catalog (see Tuvikene *et al.* 2010, and references therein).

Graphical inspection of the data can reveal outliers, but a box plot of the full range of variation of the data is most useful too: outliers show up as surprisingly high maxima or surprisingly low minima. The discussion of these box-whisker plots in Section 4.2 mentions the relevant statistical quantities, *viz.* the lower (or first) quartile $Q1$ that cuts off the lowest 25% of the data (also called the 25th percentile), the second quartile $Q2$ or the median, which cuts the dataset in half (50th percentile), and the upper quartile $Q3$ that cuts off the lowest 75% (75th percentile). The distance from the first to the third quartile is called the interquartile range: $IQR = Q_1 - Q_3$. Hence, the box plot displays the full range of variation, the likely range of variation (the IQR), and a typical value (the median). See also Sterken (2011) for an application of box-whisker plots to large datasets with outliers.

John Tukey gave a rule for handling outliers: they are either $3 \times IQR$ or more above Q_3, or $3 \times IQR$ or more below the first quartile. For suspected outliers, a coefficient 1.5 is applied: observation m is a suspected outlier if $m < Q_1 - 1.5 \times IQR$ or $m > Q_3 + 1.5 \times IQR$ (see also the Table in the caption of Fig. 45). Outliers should be carefully dealt with:

1. by visual inspection of the original (raw) data – for example, CCD frames should be inspected for mishaps;
2. the distribution of the *control data* – *i.e.*, the comparison, calibration and standardization data – should be analyzed with an eye towards the presence of outliers;
3. the software pipeline should be thoroughly verified.

Finally, take very seriously the advice of Landolt (2007):

> BE SUSPICIOUS OF, BE KNOWLEDGEABLE ABOUT, ALGORITHMS WHICH CLIP ERRANT DATA POINTS. LOOK AT ALL DATA INTENSELY. YOU MAY BE THROWING AWAY YOUR NOBEL PRIZE!

By all means, you should never automatically delete an outlying point, and if you decide not to show such a point in a graph, the Figure caption should state that there was one point out of scale. It is sad, but outliers are just too often shoved beneath the rug.

11.5 Dot joining, or "helping" the eye by connecting points

Plotted data points are often joined by lines, either because it is the graphing software's default option, or because the designer decided that the points must be linked, either for visual clarity (like in Fig. 13), or because there is a functional basis for doing so. Dot joining is often applied to mask *missing data*, *i.e.*, when no data were collected (for example, because of daytime interruption in the observing window), or when the data are simply not good.

Note that when datapoints are connected, the graph may give the impression that interpolation in between points is allowed, which is not always the case if the data gathering was not a continuous process. Figure 19 is such an example: the first line segment for spectral type A0 may not be used for interpolating color indices, because there is no information whatsoever about what the impact of the Balmer discontinuity on the color index is. In the same vein, Figure 1 of Paper III shows connected journal impact-factor data, although this bibliometric index is released on a year by year basis only. Figure 14 is an example where a bar chart can also be represented as a line plot: since there are no data in between the bar values, interpolation is not justified, nevertheless the left panel version of the data turned out to be a more convincing presentation than the right panel, a confirmation that dot-joining fills data gaps and fools the eye.

Beware also that the eye has a natural tendency to see patterns that in reality do not exist. A telling example is the famous story of the *canali* observed by Giovanni Schiaparelli during the Mars opposition of 1877, and confirmed by later observers – though not by every one. Evans & Maunder (1903) consequently set up experiments to find out whether the impression of a network of lines could be produced during observations of disks of diameters 8 to 16 cm, seen from distances of 5–11 m by a class of schoolkids. These authors concluded that the eye sums up the details that it cannot resolve, and that canal-like impressions seem to arise by the *"tendency to join together minute dot-like markings"*. Similarly, the mind sees data trends that are not present,[158] and hand-drawn curves that bridge the gaps "to help the eye" – typified by spurious inflection points and excessive convexity – are very convincing ways to tell the reader what is in the mind of the author, and not what the data say (as can be seen in Fig. 10). This is especially true when estimating period or cycle lengths on the basis of sparse and unevenly-spaced data.

11.6 The perception of linearity

A linear least-squares fit is a curve fit that is linear in all coefficients: a relationship of the form $y = a + bx$ (where y and x are the data vectors, a is the intercept and b the slope). A non-linear least-squares fit relates to non-linearity in the coefficients. A second difference between the latter and the former, is that for non-linear fitting, preliminary estimates of the parameters are sometimes needed in order to yield a robust result. If a fitted equation serves to derive a prediction of y from x, the fit is called a *regression of y on x*, but if the fit is meant to derive x from y, it is called a *regression of x on y*.

The linear relationship between two variables is about the simplest kind of relationship, yet it has in practice a surprising degree of complication, see Deeming (1968) who points out some subtleties in the analysis of linear correlation in astronomy. Some graphs suggest a linear relationship between the variables: either weak (because the designer [un]consciously wants to see a linear kinship), or strong (because the graph suggests a firm linear correlation: for example, color index (differential log of irradiance) appears to be a straight line in the reciprocal of wavelength in Figure 19. Some non-linear relationships, however, may be linearized, for example power-law relationships[159] will appear as

[158] The eye also tends to recognize circular patterns in semi-random noise.

[159] Power-law distributions occur in nature, for example sizes of cometary nuclei and occurrences of solar flares follow a power law, as well as growth processes – see the discussion on the accruement of citations in Section 5 and in Figure 8 of Paper III.

a straight line in a log–log diagram, see also Section 7.3.6. The following points must be considered when dealing with such apparently straightforward data presentations:

1. the strength of the linearity is quite often expressed by a rather weak numerical parameter, the correlation coefficient r;[160]
2. x and y may not be error-free, see the discussion on Figure 29 in Paper III;
3. the errors on both variables may not be independent;
4. the errors on both variables may not be normally distributed;
5. in regressing y on x, or x on y, you minimize using, respectively, vertical or horizontal distance, a choice that is important in view of the preceding points (see Murtagh 1990);
6. if one of the variables is an observable, it is very well possible that some of the values are of better quality than others, thus weight factors[161] may have to be considered.

Least-squares fitting is very sensitive to outliers, hence one should only apply routines for robust regression. The effect of outliers, moreover, strongly depends on their position in the x-interval: a bad point in mid-range may not harm very much, but outliers near the extremes of the x-range definitely change the slope and intercept of the fit.

An example of a moderate linear correlation is shown by Lemaître's available data in Figure 3: with these data, the linearity is not much more than a trend (albeit with a totally erroneous slope) that, with hindsight, is a true linear relation. Let us now make the exercise for the data shown in Figure 3: a straightforward linear regression yields a slope of 477 ± 105 km s^{-1}Mpc^{-1} for Lemaître's data, and 454 ± 75 km s^{-1}Mpc^{-1} for Hubble's. Which one is the most accurate (see also the discussion on page 103), and which one is the most precise? The latter value agrees, within the errors, with the value given by Hubble (1929), *viz.* 465 ± 50 km s^{-1}Mpc^{-1}, and with a mean error on the slope of two thirds of the fit to Lemaître's data, a result that looks more precise. With some reservations, though:

- the data used in this exercise were not published by Lemaître, but were reconstructed from the literature (by Duerbeck & Seitter 2001);
- Hubble even excluded some data, moreover he obtained a different solution depending on whether data were binned or not (the two lines in his diagram, see Fig. 3);
- the *Hubble constant* is a constant of nature that took almost a century to settle to its presently-known value, and as such one may say that the precision at every attempt was quite reasonable, but that the accuracy of the result of every endeavor at determining the Hubble constant is very strongly time dependent and technology reliant (see Fig. 44).

[160] $-1 \leq r \leq 1$; r is the ratio of the covariance σ_{xy} and the product $\sigma_x \sigma_y$ of the standard deviations. When one variable increases while the other variable decreases, the correlation is negative.

[161] For example, the reciprocal of the variance, or a factor based on the quality of the data (Tuvikene *et al.* 2010).

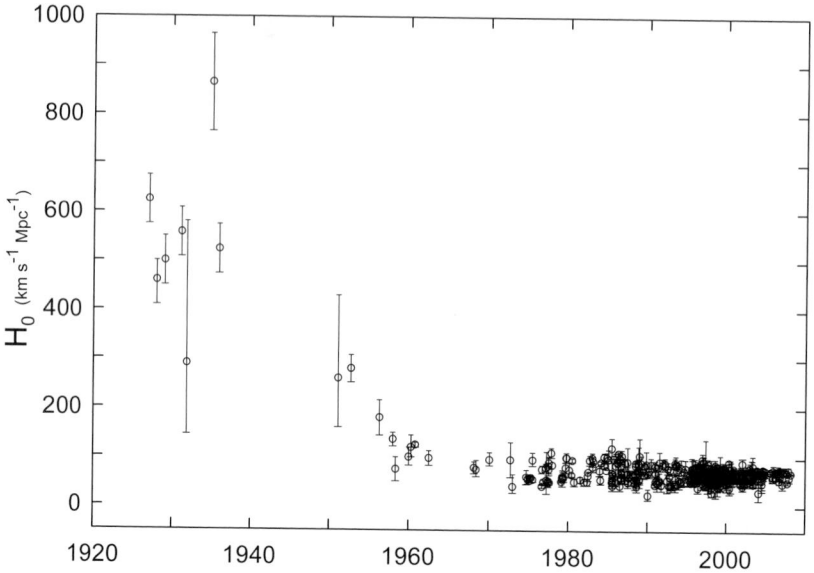

Fig. 44. Evolution of the Hubble constant. Based on data from the *HST Key Project on the Extragalactic Distance Scale*, `http://www.noao.edu/staff/shoko/keyproj.html`.

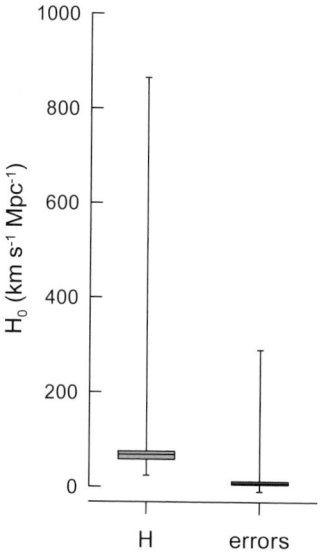

Fig. 45. Box-whisker plot for the Hubble constant and its errors. $\Delta Q = Q_1 - Q_3$ and $k = 1.5$.

	Mean	Median	σ	Q_1	Q_3	$k\Delta Q$
H_0	74.8	68	60.3	25.5	58	75
errors	10.9	8	16.4	10.5	5	12

A straightforward box-whisker plot of the dataset shown in Figure 44, without looking at the graph, may lead the inattentive reader to the conclusion that H_0 is very robustly determined, though it is affected by some outliers in the data and in the associated errors. Figure 45 illustrates this fairly well: data deviating by more than $1.5 \times (Q_1 - Q_3)$ could be flagged as outliers, though in reality they are not. This example clearly shows how dramatic the difference between precision and accuracy can become: consistent observational data have ever led to the seemingly precise but physically inaccurate result that the age of the universe in Hubble's days was much lower than the geological age of our Earth!

A counterexample of a strong suggested linearity that could not stand the test of time is the so-called *Titius–Bode law* of planetary distances.[162] Figure 46 illustrates this relationship for the data that were available in 1766. What is the problem here? Well, the relation

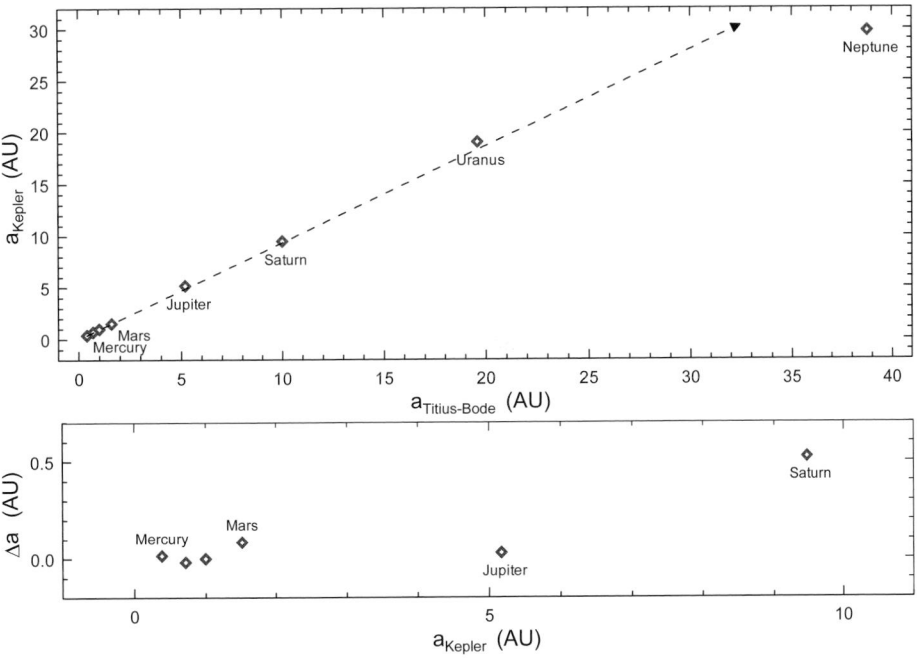

Fig. 46. *Top*: relationship between the semi-major axis a of planetary orbits derived from Kepler's third law (orbital period is the only parameter), and a based on the Titius-Bode law. The dashed line is the least-squares fit for the planets Mercury to Saturn. *Bottom*: differential semi-major axis $\Delta a = a_{\text{Titius-Bode}} - a_{\text{Kepler}}$. The scale ratio of the y axes in both panels is almost 20.

gives a visually perfect fit for the the planets Mercury to Saturn, and this "law" even seemed confirmed when Uranus was discovered by William Herschel in 1781: a linear fit indeed yields a perfect match. But the extrapolation, and the so-called law, completely

[162]The semi-major axis a of each planet in the solar system is given by $a_n = 4 + 3 \times 2^n$ ($n = -\infty, 0, 1, 2, 4, 5$), where a is expressed in units of $\frac{1}{10}$ AU and n is the rank number of the planet (starting with Mercury).

breaks down for Neptune. The problem is a problem of perception in the first place, as illustrated in Figure 46 that plots the differential semi-major axis $\Delta a = a_{\text{Titius–Bode}} - a_{\text{Kepler}}$ for the planets that inspired and supported that law: it is evident that the linearity already was a problem from Saturn on. Note that Figure 46 is a comparison of models, not a comparison of observational data.

When purporting linearity, it is much better to support any claims by evaluating the difference between a quantity and its calculated value (or observed *minus* calculated, see the lower panel of Fig. 46), than to graph the full dynamic range of both involved variables (top panel in Fig. 46). The rule to always analyze residuals (see also the discussion on page 79) is especially true for ephemeris determinations of the form Epoch time = $T_0 + EP$, where T_0, E and P, respectively, are the time of an initial epoch, the cycle number and the period: especially when P is very small, E can become extremely large (see Fig. 42), and an E versus time plot will never reveal a cycle-count error in E, whereas a differential plot of observed *minus* calculated time would immediately reveal such errors. This leads to an often poorly understood truism[163] that in such regressions the independent variable E **is errorless, unless it is wrong** – where the first instance of "error" refers to statistical error, and the last meaning refers to error as mistake.

It often happens that the data suggest correlation or linearity, although there is no reason to assume causality. There are, in effect, two questions to be answered, *viz.*,

1. does the correlation really represent a causative case, and
2. if yes, which variable is the cause?

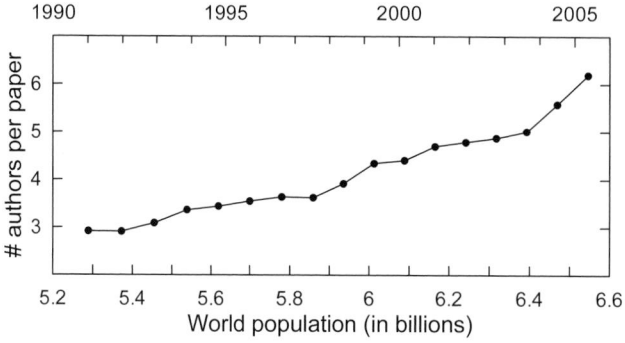

Fig. 47. An example of a nonsense correlation: average number of authors per paper in the non-Letters Sections of A&A and ApJ combined (data shown in Fig. 4 of Paper III) *versus* world population.

The literature is full of so-called *nonsense correlations*, *i.e.*, good-looking correlations that are ascribed to apparently sound causative associations (like total sunspot area and corn price). As an illustrative example, Figure 47 shows the $r = 0.95$ correspondence between the average number of authors per paper in the non-Letters Sections of A&A and ApJ combined (data from Fig. 4 of Paper III) *versus* growth of world population (taken from the *International Data Base* of the U.S. Census Bureau[164]). Even when maintaining

[163] A truism is a claim that is so obvious or self-evident as to be hardly worth mentioning, except as a reminder or as a rhetorical or literary device (Wikipedia).

[164] Population Division `http://www.census.gov/ipc/www/idb/worldpop.php`

that the number of scientists augments when the world population increases, and thus that more people will be expected to share the work, the effect of the increase has a totally different reason.[165] Nonsense correlation is not the same as *spurious correlation*: a relationship suggested by the occasional arrangement of a very small number of data (see Fig. 29 in Paper III).

A special case of nonsense correlation is embedded in the procedure involving *decorrelation*. This is the practice of correcting observational data by so-called "*σ-optimization*", *i.e.*, by best-fitting an equation that combines, for example, air mass, position coordinates on the detector, seeing, sky reading, etc. The coefficients of the fit then provide minute corrections to the results. If you wish to carry out such kind of data fix, then you must verify that:[166]

1) there is an empirical ground for including a particular observable;
2) the improvement, after decorrelation, is really significant;
3) you are sure that the data statistics is not dominated by outliers;
4) the control data (for example the light curves of the constant stars) still behave properly after the decorrelation procedure, and
5) the resulting parameters behave coherently over time, *i.e.*, that the assumption of correlation is valid for **all** data to which this procedure is applied.

11.7 Curve fitting and overshoots

Fitting a smooth curve through the data is done to

1) summarize a large quantity of data;
2) compare sets of data on the basis of a small number of parameters;
3) select a theoretical model;
4) smooth a noisy signal, thus requiring the curve to pass close to (or within measurement error) through each measured point;
5) allow interpolation, thus requiring the curve to pass *exactly* through each point, or
6) allow interpolation of the smoothed signal (*i.e.*, both previous items combined).

The problem is to fit a smooth curve through a series of (noisy) data points. This can be done with polynomial interpolation, though smooth-curve fitting is mostly done through piecewise cubic *spline fitting*. A spline, literally, is a strip or template of wood (or plastics), or a thin strip of rubber used to draw curves. Figure 48 shows an example of a splinewood. This image is included not merely as an illustration of an antiquated graphics tool, but also to illustrate the effect of illumination on recorded color: this is the only picture in this book captured under outdoor lighting.

Such a tool allowed smooth and continuous connection of a series of points, though with "visual" accuracy, *i.e.*, good-looking unconstrained guesswork, often with excessive

[165] The correlation is entirely due to the fact that both variables increase monotonically with time.

[166] For a discussion of these points in context, see Sterken (2010).

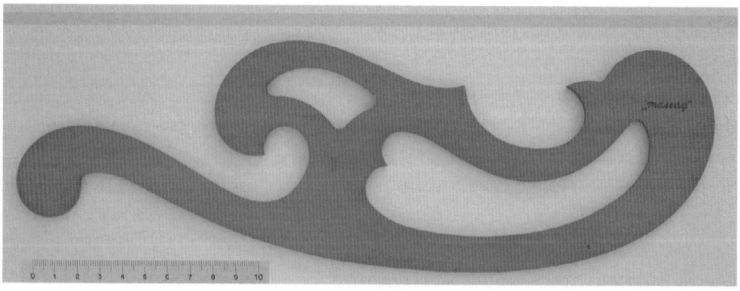

Fig. 48. The splinewood that I used in college in 1964, on a white background, photographed in diffuse daytime sky-illumination. The average *RGB* coordinates of the "white" background are $R = 170.2 \pm 1.5$, $G = 171.6 \pm 1.7$ and $B = 173.8 \pm 1.5$, *i.e.*, a grayish hue. The horizontal band at the top is the background taken from a similar photograph under different sky illumination, yielding average $R = 127.6.2 \pm 1.7$, $G = 157.8 \pm 1.6$ and $B = 203.0 \pm 1.4$, thus a blueish hue (quite close to the lightest × symbol in Fig. 37). The ruler, labeled in 1-cm units, was added to show the scale. This illustration is best viewed in color mode.

curvatures, leading to spurious results. A spline, in our context of computer graphics, is a special function defined, for example, by piecewise polynomial fits (cubic spline or quadratic Bézier[167] curves and surfaces – especially for making vector graphics).

Splines yield smooth lines, but if the derivatives at the ends of the fit interval are not known, the spline fit may turn completely odd. The same happens when gaps in the data series occur. Figure 49 shows an example of a sky brightness measurement as a function of time during the first four hours after twilight, with a fitted spline that shows spurious excursions that will lead any blind interpolation to an erroneous result that is a pure artefact of the applied interpolation method. Such unphysical oscillations – especially when the data are sparse – may be suppressed by adding constraining points nearby the problem areas, or by applying a tension factor that releases the minimum curvature conditions of the fit.[168] Hence, whenever datasets are quantitatively compared on the basis of spline fits, the tension factors should be given along. And, if a smoothing curve is used for analysis, the errors of the fit must be quantified, see Teanby (2007). Spline fitting is very sensitive to outliers, though normal data may also produce outliers in the interpolated dataset, as Figure 49 clearly illustrates. Spline-based extrapolation should be avoided by all means.

11.8 Models and calibrations

The distinction between measurement and model is an aspect that must be preserved at all times. In a not so old paper I read the statement *"The solid line represents a simple sinusoidal fit to the first half of the light curve, and is intended to guide the eye in recognizing*

[167] After the French engineer Pierre Bézier, who pioneered – but did not invent – the concept.

[168] The higher the tension factor, the stiffer the fitted curve will be.

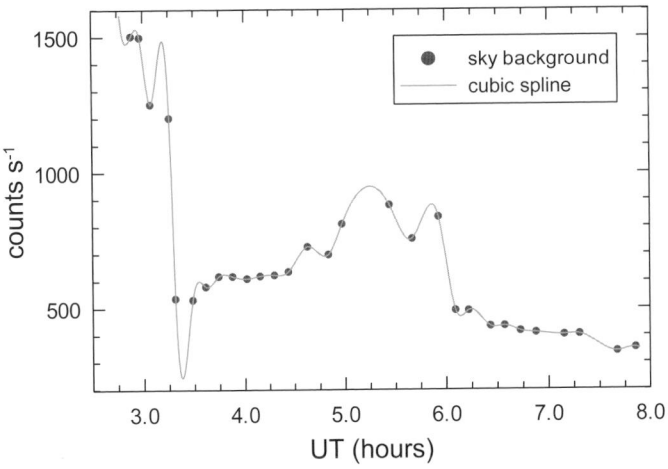

Fig. 49. Example of a cubic spline fit suffering from unphysical oscillations and large overshoots: sky background measurements as a function of time (in detector counts per second).

the asymmetry of the light curve". If a sine curve is added for showing asymmetry only, then the graph will mislead those who do not read the text or the caption. Thus **it is very important to indicate whether a line in a diagram is a fit to the data, or whether it represents a genuine model**. For example, the second-degree polynomial in Figure 8 is a mathematical model that exactly represents the relation between the variables. Likewise, the dashed line in Figure 46 is a fit to data coming from the comparison of two models.

Figures 50 and 51 show another example of differentiation between data and model, in fact between observational data and their calibration in terms of MK spectral type. The photometric color indices of the Geneva photometric system lead to a set of three orthogonal reddening-free parameters X, Y, Z, see Cramer & Maeder (1979), and Cramer (1993) for a definition.[169] Figure 50 shows part of the X, Y plane, where the lines are the calibrations for spectral classes Ia, Iab, Ib, II, III, IV and V. A simple inspection of this diagram gives the impression that deriving spectral class from a set of measured color indices is a straightforward matter that leads to unambiguous and accurate results. Figure 51 shows the same part of the X, Y plane superimposed on the full-range observational X, Y diagram (more than 30 000 stars displayed in Fig. 4 of Cramer 1994). This graph shows that:

1) a substantial part of the X, Y plane is not calibrated – the reason being that the physical role of the indices X and Y changes in the lower part of the diagram, what leads to ambiguity. At more negative Y values, metallicity effects become dominant, and impede photometric spectral classification;

2) X and Y alone can lead to ambivalent results, and

3) discrimination of spectral type very much depends on the X, Y location: especially for stars of types O and early B, it is very hard to derive an accurate spectral type.

This example illustrates, again, the value of graphical support when analyzing data.

[169] Not to confuse with the CIE XYZ color coordinates.

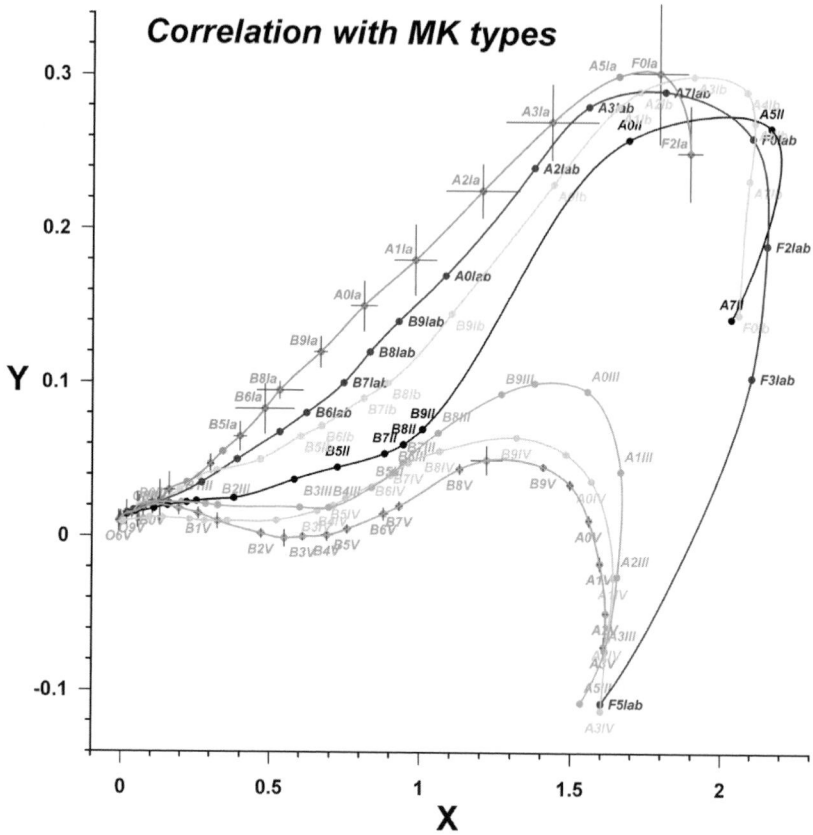

Fig. 50. Schematic diagram of spectral-type calibration in the Geneva photometry X, Y plane. Reproduced from Cramer (2005), with permission.

12 Tables and lists

Tables are used for information that is not suitable for graphical representation, for example observing logs listing exposure times, instrument resolution and scale, observer names, etc. The elements of a Table are often referred to as *cells*.

Tables, just like Figures, are numbered, and their placement follows the same rules as for Figures. It occasionally happens that a Table is presented as a text paragraph, and is not numbered, see the example on page 87.

Lists are used when a small amount of information per item is listed, for example a handful of entries of the form x, y, σ that do not deserve an additional graph. There are two types of lists:

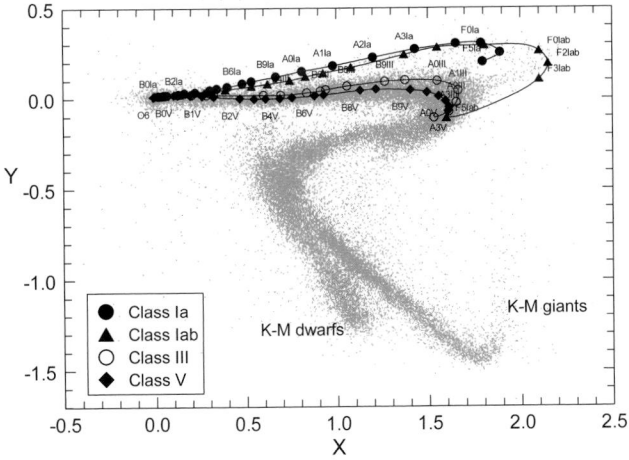

Fig. 51. The X, Y plane shown in Figure 48, superimposed on the full-range observational X, Y diagram taken from Figure 4 of Cramer (1994). Only four spectral calibrations are shown.

1) run-on lists: a handful of items that do not require a specific order, for example in the statement *A&A, AJ, ApJ, and MNRAS are leading astronomy journals*;

2) displayed lists: *e.g.*, bullet lists and numbered lists, with possibly a large number of items. Always prefer neutral marker symbols, including no more than 2 or 3 levels of indentation in 2-column pages, and beware punctuation and capitalization.

Machine-readable Tables are a special case of printed Tables. Some journals offer the possibility to submit data archives for electronic access through an archive or journal website. When importing Tables, you should verify whether

1) missing values are included in between column separators (TAB, OR – or, or ,), and

2) whether there are any empty lines (especially at the end of the file).

Tables often contain lines and rules, and these non-data elements – if needed at all – should follow the same guidelines as in graphics: dense and heavy grids should be avoided at all cost. Horizontal rules are only good for dividing data in topical paragraphs. Always close a Table with a rule (except when there are notes), but a Table with a rule below each line has too much tabular chart junk, and is useless. Tables must be functional, without vertical lines, except when doubling column groups. Color can be a great help for visual clarity, though it does not happen very often (see, as an example, Table 3 in Laurent Cambrésy's paper in Volume 1 of these Proceedings).

The most common typesetting errors in Tables are the column alignment (some authors center every column), the mixing of the math *minus* sign and the dash, the number of significant digits not conforming to the accuracy of the data,[170] and forgetting the leading

[170] See Paper I and the discussion of Figure 18.

ϕ	1874	1875	1876	1877	1878	1879	1880	1881	1882	1883	1884	1885	1886	1887	1888	1889	1890	1891	1892	1893	1894	1895	1896	1897	1898	1899	1900	1901	1902
40																				0.06									
39																													0.04
38																											0.1		0.04
37																													
36																												0.02	
35							0.5													0.1		0.02							
34																	0.9			0.02									
33								0.1									3.6	0.1	0.1			0.1							
32						0.3		0.1									1.5	0.04	0.2										
31							0.03	0.1		0.2								0.9	0.6		0.2								
30				0.5		1.0	0.1	0.1		0.03								1.5	0.3		0.2								
29				0.2		2.2	0.3	11.2	0.2	0.1							0.2	20.5	3.5	0.02		0.03							
28						5.3	1.0	26.9	0.1								0.1	5.5	25.1	0.3	0.1	0.05						0.01	0.04
27						0.4	4.0	20.8	0.04								0.3	8.9	14.9	3.3	0.5	0.1		0.03					0.1
26						0.4	23.8	20.0	0.4	0.01							0.4	17.8	4.3	2.9	1.1	0.1							3.3
25						0.1	10.3	8.8	2.1	0.01						0.1	0.7	20.0	15.4	6.2	0.9	0.7							11.0
24							16.7	19.0	7.0	0.3						0.3	0.5	23.2	2.9	35.1	3.5	3.2	0.1	0.02					2.6
23							20.0	19.0	29.9	2.4	0.05					0.1	5.7	25.8	11.7	7.8	5.0	30.7	0.1						2.8
22	1.2	0.2					36.4	6.6	12.9	.2	0.06	0.1				0.02	10.4	31.4	17	12.6	17.4	25.7							0.4
21	2.4						31.0	34.2	28.4	6.3	0.1	0.2		0.04		0.01	8.1	30.2	31.3	14.2	17.7	11.9	0.1	0.04					0.3
20	0.1					0.3	21.9	31.4	43.2	4.1	0.3	0.4	1.6	0.1			12.6	46.5	14.4	14.2	3.8	16.6	2.2	0.04	0.1				0.8
19	5.0	6.4					17.1	19.1	82.1	6.1	1.3	0.1	1.7	0.1			4.4	38.4	18.1	40.7	11.7	23.4	1.5	0.2					0.04
18	4.2	29.8				0.9	8.7	19.5	8.5	11.8	7.4	2.1	1.3				3.6	32.6	11.2	35.6	21.4	41.5	11.4	0.2	0.3				0.2
17	8.4	4.2				0.1	10.2	31.0	9.7	2.6	19.0	8.8	4.8				0.4	15.9	22.5	45.3	33.1	27.9	12.8	0.03	0.5				0.1
16	1.9	1.6	2.2			0.3	4.5	36.9	33.3	3.4	10.1	5.4	2.0	0.2			0.1	27.0	22.5	13.5	25.1	26.3	28.5	1.2					
15	5.8	3.8	15.3		0.8		8.6	41.2	33.3	9.9	51.6	16.4	0.7	0.02				29.3	44.7	34.1	38.7	41.8	10.5	4.8	20.4	0.2			9.7
14	3.7		2.0		2.1	0.5	11.9	50.1	25.9	19.7	51.5	44.3	0.7	1.4				17.3	43.8	44.4	27.3	32.3	31.1	8.1	3.4	0.2	0.03		2.8
13	8.0		0.1		2.1		7.6	34	23.5	29.7	48	13.6	7.3	9.5	0.2			2.2	37.8	23.4	36.4	36.7	20.7	7.5	3.7		2.0		
12	19.7	3.4	2.8	1.5			15.9	17.2	20.0	27.5	30.9	36.4	5.0	5.6	0.7	0.01		2.4	35.1	35.5	47.4	44.7	19.8	34.4	12.9		0.7		
11	62.1	8.5	4.7	2.1	1.1		2.3	23.6	38.6	84.8	28.5	15.4	1.4	1.3	2.1	0.04		1.2	107.4	32.6	19.6	105.3	3.1	12.1	11.6	0.7	1.3	0.03	
10	19.2	12.6	1.2	1.6	0.7	2.1	4.2	17	17.7	30.4	45.1	44.9	2.7	2.2	7.4	0.2		1.1	58.1	25.3	66.3	44.8	3.4	17.9	14.4		3.5	3.6	6.2
9	4.2	34.7	0.9	14.8	2.0		1.2	0.2	6.5	15.9	30.1	20.3	20.1	3.4	0.7			0.4	37.9	32.7	52.3	25.1	12.5	10.1	1.7	0.1	.2	10.9	1.3
8	12.1	13.4	0.4	2.7	2.6			0.4	8.9	25.2	42.4	13.9	9.4	3.7	0.2			0.02	17.3	31.6	16.8	3.0	4.0	8.7	0.9	0.5	0.8	6.4	
7	39.3	16.7	0.4	1.0	2.2			7.7	10.4	16.5	45.5	21.6	1.9	2.7	0.3	0.9	0.05		3.0	1.6	50.4	0.8	1.6	20.9	9.5	2.1		0.2	
6	28.6	15.2						1.9	1.7	2.8	18.3	8.6	1.7	5.8	0.3	1.5			3.2	0.9	19.8	1.9	11.7	18.2	9.4	17.9	0.1		
5	11.6	1.6			0.6				0.1	15.8	16.4	9.6	7.6	0.4	0.7	1.5	0.02	0.04	1.2	5.4	2.5	7.2	5.9	11.5	13.4	0.1	2.7		
4	3.4	0.2			9.1	0.2				7.7	15.1	11.1	2.7	3.1	0.3			0.3	0.1	5.0	8.1	8.6	0.1	18.7	6.3	0.4	11.0		
3	2.9	0.2	0.6	0.1	1.5				1.0	7.4	9.4	2.2	0.2	2.0	4.2	0.05		0.1		6.5	3.1	2.3	2.2	10.3	0.1	1.0	1.4		
2	25.8		0.1	1.6					3.7	7.4	3.8	4.5	1.1	0.8	1.5					5.0	7.3	4.6	0.5	8.2	1.2	0.2		0.3	
1	5.8	0.2	1.6		0.1				2.4	0.02	3.2	2.5	0.5	1.6	1.0	0.1		0.05		0.4	5.0	0.2	0.04	1.2			0.5	0.4	
0		0.1	0.1	0.2	0.1				0.9	0.02	0.2	0.4	0.9	0.6	0.03	0.1				0.5	0.5	0.1	0.02	1.3					
0			0.5						5.1	0.2	0.7	0.4	4.4	1.0	0.1	0.2				2.1	0.05		0.03	3.1	0.3				
-1									4.9	0.01	0.2	3.5	8.2	1.6	0.4	0.6		0.1		4.4	.2	0.8	0.05	4.0	4.0	0.02	0.03		
-2				3.4						5.7	5.2	10.1	1.1	0.5	1.6			0.1	0.04	2.7	6.1	2.9	0.05	11.3	0.04	1.3	0.01	0.1	
-3				0.4				2.1		0.05	8.9	10.6	9.4	4.7	3.0	0.3				3.6	7.2	8.8	0.3	9.5	0.3	0.1	0.05		
-4	3.1	3.6	3.1	0.7				1.8	1.9	2.5	5.2	22.1	10.8	3.8	8.1	0.5				14.1	19.1	2.7	0.4	11.6	0.5	0.6	0.8	0.2	
-5	23.8	0.9	0.1	0.7			0.1		10.0	9.9	4.2	23.1	27.7	6.4	5.4	6.6	0.2		0.05	35.7	15.7	14.5	0.2	28.9	3	6.3	6.1	0.5	
-6	6.0	6.6	8.3						3.9	17	14.1	14.5	11	3.2	8.5	13.8	1.2		0.3	36.9	11.5	18.9	2.8	60.9	15.5	1.0	8.7	0.3	
-7	36.3	19.7	4.1	1.2	0.1			0.02	4.7	21.0	38.1	12.2	7.2	21.4	9.9	11	2.6		5.2	43.3	8.1	9.7	17.9	65.4	9.6	1.2	11.4		1.8
-8	17.1	23	1.9	23.4	0.1			2.5	4.8	52.3	56.9	21.9	12.0	39.1	11.5	7.3	0.3		1.5	61.8	14.2	18.1	7.4	17.1	32.1	4.0	1.1	0.4	6.9
-9	13.7	2.5	6.7	0.3	0.1		2.2	6.2	5.4	111.5	82.3	17.4	19.6	21.3	5.1	5.9	0.4		7.5	40	14.1	22.8	8.1	26.1	14	16.1	3.4	0.4	
-10	36.7	14.2	4.5	19.9			2.4	14.5	19.2	122.9	59.3	35.9	25.9	10.7	7.3	0.1	0.5	1.6	3.4	39.3	15.1	49.5	30.2	12.3	6.3	21.2	0.6	0.04	
-11	21.6	13.9	11.1	4.8			0.4	8.3	13.3	52.2	55.8	50.8	28.1	4.3	1.1	0.05	0.03	3.5	9.0	64.7	20.7	35	20.9	22.3	14.9	6.0	5.9	0.01	
-12	12.5	5.7	20.6	1.2		0.2	1.0	2.6	20.6	23.5	42.2	32.8	16	7.2	1.9		0.01	3.6	20.7	51.3	81.8	17.5	22.7	11.9	96.5	3.8	8.0		
-13	46.0	3.1	8.8	1.7		0.2	7.5	4.6	14.1	28.5	59.9	40.1	10.2	1.0	4.0			0.9	19.4	48.2	81.2	31	10.8	13.1	49.2	15.1	2.0		
-14	27.2	18.4	12.9	6.1			3.7	7.8	38.1	39.5	51.6	53.3	16.7	1.4	0.8			5.0	36.8	35.2	56.9	26.8	20.4	2.9	14.1	7.5	0.1		
-15	21.2	4.9	6.2	2.4			6.4	18.9	16.8	38.3	29.7	44.2	28.8	0.8	1.2			12.8	17.1	25.7	118	32.4	53.6	3.1	3.4	4.9			0.01
-16	27.8	0.8					12.3	16	8.5	61.9	19.9	43.5	9.7	0.5	0.1			6.7	34.1	46.4	31.5	39.6	34.7	4.7	1.4	0.1			
-17	14.0	1.2	0.6			0.7	33.6	15.4	26.8	28.8	21.2	12.5	19	0.9	0.1			7.4	5.5	60.7	27.8	24.3	12.1	1.2	0.2				
-18	41.0						15.9	9.3	44.5	23.9	17.6	14.7	19.4	2.1	0.01		0.6	11.8	33.1	74.9	28.6	21.8	10.5	0.04	0.5			0.1	0.2
-19	6.8			0.5		0.2	18.4	30.0	25.2	15.7	7.2	28.4	7.4	1.4		3.7	1.3	23.1	44.2	43	46.9	8.9	12.7	8.9	0.2			0.1	0.01
-20	7.3					5.0	17.4	31.2	29.2	3.9	4.2	12.9	1.7	0.9		11	4.2	32.7	26.6	62.9	19.6	2.8	34.1	1.2	0.03			2.4	1.8
-21						2.1	3.5	5.0	21.1	8.1	3.0	11.9	0.8	0.01		3.8	3.8	14.2	18.0	37.5	4.6	1.9	29.8				0.02	2.1	10.0
-22						7.7	1.6	4.6	74.2	66.3	0.4	14.7	0.1			1.5	2.4	5.2	33.1	50.4	10.8	1.9	7.8	0.1			0.03	0.04	
-23						11.2	3.3	4.1	37.0	31.3	2.1	1.0	0.2			0.7	7.5	10.7	36.8	5.1	1.0	2.7	2.3					0.02	
-24						0.5	6.2	1.5	10.4	23.8	16.2	0.6	0.06			0.4	8.4	5.2	36.4	9.5	0.2	9.3							
-25						1.7	7.7	13.9	0.9	4.7	6.5	0.1				0.1	5.2	5.4	10.8	3.8	0.9	2.5							0.01
-26						3.0	1.5	4.6	0.5	17.5	3.7		0.04			1.3	1.6	12.4	12.9	3.2	8.9	0.2							0.01
-27						1.8	0.5	5.2	8.5	3.0	2.0					2.4	3.5	3.5	11.2	10.8	28.9	1.2							
-28						0.04	0.9	4.1	40.1	3.8	18.2		0.01			0.03	1.4	2.2	70.6	13.2	6.4	0.6		0.2					
-29							5.1	2.8	38.2	2.9	0.06					0.02	0.9	0.9	40.9	5.1	2.2	0.02							
-30							0.1	5.2	27.8		0.2						0.3	0.3	17.1	3.0	3.2								
-31							0.2	2.4									0.1	0.3	29.2	1.5	2.1	0.01							
-32							1.5	2.0									0.1	0.2	15.8	0.5	34.8								
-33							0.05										0.1	0.3	0.4	0.1	9.7								0.1
-34							1.6										0.3	0.2	0.3	0.4		0.01							
-35							0.2	1.2									0.1	0.1	0.01	0.6							0.01		
-36							2.5	1.3										0.1											
-37							0.7											0.02											
-38							1.8																	0.01					
-39							0.6											0.02											
-40																0.03		0.02											

Fig. 52. Data from Maunder's "anonymous" 1903 paper (reformatted through typesetting). The leftmost column lists heliocentric latitude ϕ, the column headers are years. Compare to Figure 11.

zero before a decimal (*i.e.*, one should write 0.1000, not .1000 nor 0.1). Another nuisance encountered in Tables, but also in Figure labels and legends, is the mixing of the decimal dot and the decimal comma. Figure 52 is a reproduction of the most important Table of Maunder's "anonymous"1903 paper, and illustrates the omission of 0-values, the alignment on the decimal dot, and the inconsistent use of the number of decimals (one decimal, except when the data entry is smaller than 0.1). This Table, however, is one of the most beautiful illustrations that a Table is also a graphical element: indeed, the butterfly pattern jumps out without even the need for a plot!

13 What to avoid at all price

The following list summarizes the most common conceptual errors and mistakes encountered in graphics. All these examples lead to unwanted noise in the communication system as illustrated in Figure 2.

- using software that you have not mastered, and letting its wizard take over;

- referring to Figures and Tables that are not included in the text;
- including Figures and Tables to which you do not refer in the text;
- placing Figures and Tables out of order, or too far away from the discussion;
- thin lines;
- unlabeled axes;
- too small text labels;
- using a medley[171] of fonts;
- errors in textual elements, to be avoided by copy & paste instead of retyping;
- wasting valuable space, especially white space around Figures;
- poor contrast with background, resulting in invisible labels;
- mixing 1- and 2-column graphics on the same page;
- destroying artwork by sloppy digitization;
- undocumented image enhancement that manipulates the image towards an aesthetically pleasing result at the cost of data fidelity;
- bad multipliers on axes;
- inconsistent design of graphics in a multi-authored paper;
- using all capitals in axis titles and legends;
- omission of units of measure;
- ticks interfering with the data;
- mixing decimal dots and commas in graph labels and in tabular entries;
- fake perspectives;
- omission of axes;
- using color only for data separation (a fatal error in research-grant applications);
- too dense axis labels;
- varying zoom percentages;
- needless resampling of images to fit the size planned for print;
- image rotations that involve resampling;
- 3-D graphics where the extra dimension is not needed, and
- bending the rules of statistics to prove your point.

[171] A medley is a potpourri or collection of sorts of everything.

14 Summary

Before submitting your paper to a journal or to an Editor, the PDF must be 100% portable. Portability can be checked with a so-called PDF *preflight*, the process of detecting problems that may impede error-free printing. ACROBAT, unfortunately, provides full preflight functionality only for its high-end professional products. Preflight output may easily amount to more than a hundred pages for a book like this one. But some straightforward controls can be easily performed:

- check the color pages, and make sure that the color tones and resolutions are OK;
- verify that color illustrations can be read by everyone;
- examine the text proof;
- review the entire manuscript page by page to the end;
- check page numbering to make sure no pages are missing;
- pay special attention to each image: check for shading, missing text and symbols, extra artifacts in the image, background, etc., and
- carefully look at each page.

Communication is the act of successfully conveying information to another reader, but also to a referee, to the Editor, and to the Publisher and Printer. Getting acquainted with the technical aspects and the design facets of artwork is a slow, long and tedious process. But reading graphs also needs training, because the viewer tends to see in them mainly those things which she or he knows in the first place. As McLuhan (1962) puts it

> MY POINT IS THAT I THINK WE'VE GOT TO BE VERY WARY OF PICTURES; THEY CAN BE INTERPRETED IN THE LIGHT OF YOUR EXPERIENCE.

McLuhan, furthermore, explains why non-literate societies cannot see films or photos without much training, and calls for some sort of process of education in reading communication media.[172] Education in reading, and training in making good graphics, can make ideas easier to understand. A poorly designed graph, to the contrary, may render the impression that the underlying science is bad: **a graph can never turn bad science into good science, nor upgrade bad theories into good ones**. Tukey (1977) observed that

> THE GREATEST VALUE OF A PICTURE IS WHEN IT FORCES US TO NOTICE WHAT WE NEVER EXPECTED TO SEE.

Paper III, in its Sections on bibliometrics, includes several graphs that forced me to notice what I never had expected to see.

[172] McLuhan (1964) describes medium as "any extension of ourselves".

I thank Mrs. Lena Holst and Mr. Mikael Lammgård of the *Röhsska Museet of Fashion, Design and Decorative Arts*, and Dr. Per Dahlström (Göteborg, Sweden), for most efficient help in providing detailed information on the tapestry reproduced on the cover of this book.

I express my gratitude to Mrs. Sophie Hosotte and to Dr. Agnès Henri (EDP) for their most adequate communication and support during the composition phase of the front and back cover of this book.

I thank Dr. David Hathaway for a high-resolution version of Figure 12, Drs. Ulrich Bastian and Tilmann Althaus for their help in acquiring Figure 21, Dr. Thierry Chanut for kindly providing the data used in Figure 41, and Dr. N. Cramer for providing Figure 50. Dr. Hilmar W. Duerbeck, Dr. F. Richard Stephenson, and Dr. Koen Brosens were very helpful in discussing diverse aspects of the cover illustration.

References

Allen, C.G., 1961, The most probable choice between several discrepant observations and the formation therefrom of the most likely induction, Biometrika, 48, 1

Appenzeller, I., 1972, Light Variations of High Luminosity O and B Stars in the Large Magellanic Cloud, PASJ, 24, 483

Barnett, V., & Lewis, T., 1984, in Outliers in Statistical Data (John Wiley and Sons), 2nd edition, p. 4

Belton, M.J.S., Thomas, P., Veverka, J., *et al.*, 2007, The internal structure of Jupiter family cometary nuclei from Deep Impact observations: The "talps" or "layered pile" model, Icarus, 187, 332

Brosens, K., 2000, Astronomia, in Wereldwijs. Wetenschappers rond Keizer Karel (Leuven, Stedelijk Museum Van der Kelen–Mertens, Leuven), p. 259

Butler, J.C., 1979, Trends in ternary petrologic variation diagrams – fact or fantasy?, Amer. Mineral., 64, 1115

Chanut, T., Winter, O.C., & Tsuchida, M., 2008, Nebular gas drag and co-orbital system dynamics, A&A, 481, 519

Cole, B.L., 2004, The handicap of abnormal colour vision, Clinic. Exper. Optom., 87, 258

Cramer, N., 1993, Intrinsic colours of O, B and early A-type stars in the Geneva system, A&A, 269, 457

Cramer, N., 1994, in The Impact of Long-Term Monitoring on Variable Star Research: Astrophysics, Instrumentation, Data Handling, Archiving (Dordrecht: Kluwer), ed. C. Sterken & M. de Groot. NATO Adv. Sci. Inst. Ser. C, 436, 405

Cramer, N., 2005, Aspects of Geneva Photometry, Part 3: Doing physics with colours, Orion, 326, 4

Cramer, N., & Maeder, A., 1979, Luminosity and $T_{\rm eff}$ determinations for B-type stars, A&A, 78, 305

Dalton, J., 1798, Extraordinary facts relating to the vision of colours, Memoirs of the Literary and Philosophical Society of Manchester, 5, 28

Deeming, T.J., 1968, The analysis of linear correlation in astronomy, Vist. Astron., 10, 125

Delambre, J., & Méchain, P., 1806, Base du Système Métrique Décimal, Discours Préliminaire, p. 158

Delmarcel, G., 1999, Flemish Tapestry (London: Thames & Hudson Ltd.)

Delmarcel, G., & Brosens, K., 2009, personal communication

Duerbeck, H.W., & Seitter, W.C., 2001, In Hubble's shadow: early research on the expansion of the universe, in Homage to Miklós Konkoly Thege (1842–1916). 100 Years of Observational Astronomy and Astrophysics., ed. C. Sterken & J.B. Hearnshaw, J. Astron. Data, 7, 7

Eddington, A.S., 1931, A Homogeneous Universe of Constant Mass and Increasing Radius Accounting for the Radial Velocity of Extra-galactic Nebulae, MNRAS, 91, 483

Evans, J.E., & Maunder, E.W., 1903, Experiments as to the actuality of the "Canals" observed on Mars, MNRAS, 63, 488

Fisher, R.A., 1925, Statistical methods for research workers (Edinburgh: Oliver and Boyd)

Fu, J.N., & Sterken, C., 2003, Long-term variability of the SX Phoenicis star CY Aquarii, A&A, 405, 685

Funkhouser, H.G., 1937, Historical Development of the Graphical Representation of Statistical Data, Osiris, 3, 269

Ho, P.Y., 1962, Ancient and mediaeval observations of comets and novae in Chinese sources, Vist. Astron., 5, 127

Hoaglin, D.C., Mosteller, F., & Tukey, J.W., 1983, in Understanding robust and exploratory data analysis (New York: Wiley), p. 3

Hermann, L., 1870, Eine Erscheinung simultanen Contrastes, Pflüg. Arch. Gesamte Physiol., 3, 13

Holweger, H., & Oertel, K.B., 1971, Influence of Hyperfine Structure on the Solar Cobalt Abundance, A&A, 10, 434

Hosotte, S., 2009, personal communication

Hubble, E., 1929, A Relation between Distance and Radial Velocity among Extra-Galactic Nebulae, Proc. Nation. Acad. Sci. USA, 15, 168

Hungerford, M.W. (née M. Wolfe Hamilton) 1878, Molly Bawn (New York: Hurst and Company Publishers)

Kodak Handbook for the professional photographer, 1979, 2nd edition

Kraus, S., Weigelt, G., Balega, Y.Y., *et al.*, 2009, Tracing the young massive high-eccentricity binary system θ^1 Orionis C through periastron passage, A&A, 497, 195

Kurtz, D.W., 1983, The High Frequency Limit to Fourier Analysis. A Reminder of the Nyquist Frequency, IBVS, 2285

Landolt, A.U., 2007, Standardization in the Classical *UBVRI* Photometric System, in The Future of Photometric, Spectrophotometric and Polarimetric Standardization, ASP Conf. Ser., ed. C. Sterken, 364, 27

Lemaître, G., 1927, Un Univers homogène de masse constante et de rayon croissant, rendant compte de la vitesse radiale des nébuleuses extra-galactiques. Annal. Soc. Scient. Bruxelles, XLVII, 49

Littmann, M., 1998, The Heavens on Fire (Cambridge: Cambridge University Press)

Lockyer, W.J.S., 1903, Sun-spot variation in latitude, 1861–1902, MNRAS, 64A, 5

Maunder, E.W., 1904, Note on the distribution of sun-spots in heliographic latitude, 1874–1902, MNRAS, 64, 747

Maxwell, J.C., 1855, Experiments on colour, as perceived by the eye, with remarks on colour-blindness, Trans. Royal Soc. Edinburgh, 21, part II

Maxwell, J.C., 1860, On the Theory of Compound Colours, and the Relations of the Colours of the Spectrum, Trans. Royal Soc. London, 150, 57

McLuhan, M., 1962, The Gutenberg Galaxy: The Making of Typographic Man (University of Toronto Press)

McLuhan, M., 1964, Understanding Media (Routledge Classics)

Montgomery, S.L., 1999, The Moon and the western Imagination (The University of Arizona Press)

Mortimer, C., 1731, An Account of Mr. James Christopher Le Blon's Principles of Printing, in Imitation of Painting, and of Weaving Tapestry, in the Same Manner as Brocades, Phil. Trans. R. A. S., 37, 101

Murtagh, F., 1990, in Errors, Bias and Uncertainties in Astronomy, ed. C. Jaschek & F. Murtagh (Cambridge: Cambridge University Press), p. 385

Parkhurst, C., & Feller, R.L., 2007, Who invented the color wheel?, Color Res. Appl., 7, 217

Pelkonen, V.-M., Juvela, M., & Padoan, P., 2009, Predictions of polarized dust emission from interstellar clouds: spatial variations in the efficiency of radiative torque alignment, A&A, 502, 833

Rada, W.S., & Stephenson, F.R., 1992, A Catalogue of Meteor Showers in Mediaeval Arab Chronicles, QJRAS, 33, 5

Rozelot, J.P., 1969, Photography of the U.V. Lines in the Solar Corona, A&A, 2, 122

Shannon, C.E., 1948, A Mathematical Theory of Communication, Bell Sys. Techn. J., 27, 379–423, 623–656, available at `http://portal.acm.org/`

Shannon, C.E., 1949, Communication in the presence of noise, Proc. Inst. Radio Eng., 37, 10 (Reprinted Proc. IEEE, Vol. 86, No. 2, Feb. 1998)

Shannon, C.E., & Weaver, W., 1949, The Mathematical Theory of Communication (Univ. of Illinois Press)

Shore, S.N., 2009, Modern astronomical surveys and the development of informatics, A&A, 500, 491

Stebbins, J., & Whitford, A.J., 1945, Six-color photometry of stars. III. The colors of 238 stars of different spectral types, ApJ, 102, 318

Steer, W.A., 2004, Eye sensitivity, `http://www.techmind.org/colour/`

Stephenson, F.R., 2010, private communication

Sterken, C., 1992, On the future of existing photometric systems, Vist. Astron., 35, 139

Sterken, C., 2000, Mode Identification through Photometry: Full Colours?, in Delta Scuti and Related Stars, Reference Handbook and Proceedings of the 6th Vienna Workshop in Astrophysics", ed. M. Breger & M. Montgomery. San Francisco, ASP Conf. Ser., Vol. 210, p. 99

Sterken, C., 2009, Over Wetenschapshistorisch Onderwijs, ed. C. Sterken & D. Segers (Brussels: Vrije Universiteit Brussel, Belgium)

Sterken, C., 2010, Photometric Standardisation in the Variability Domain, in Variable Stars, the Galactic Halo and Galaxy Formation, ed. C. Sterken, N. Samus & L. Szabados, Moscow: Sternberg Astronomical Institute of Moscow University, p. 125

Sterken, C., 2011, Just one new measurement of the B[e] supergiant Hen-S22, IBVS, 6000

Sterken, C., Milone, E.F., & Young, A.T., 2011, Photometric Precision and Accuracy, in Astronomical Photometry: Past, Present, and Future (Springer New York), p. 1

Teanby, N.A., 2007, Constrained Smoothing of Noisy Data Using Splines in Tension, Math. Geol., 39, 419

Tubbesing, S., Kaufer, A., Stahl, O., *et al.*, 2002, The eclipsing hypergiant R 81 (B2.5Ia-O) in the Large Magellanic Cloud. System properties from spectroscopic and photometric monitoring, A&A, 389, 931

Tufte, E.R., 1983, The Visual Display of Quantitative Information (Cheshire, CT: Graphics Press)

Tufte, E.R., 1990, Envisioning Information (Cheshire, CT: Graphics Press)

Tufte, E.R., 1997, Visual Explanations (Cheshire, CT: Graphics Press)

Tukey, J.W., 1977, Exploratory Data Analysis (Boston: Addison-Wesley)

Tuvikene, T., 2011, Ph.D. Thesis, Vrije Universiteit Brussel

Tuvikene, T., Sterken, C., Brogt, E., *et al.*, 2010, New times of maximum of CY Aquarii, J. Astron. Data, 16, 2

Usoskin, I.G., Solanki, S.K., & Kovaltsov, G.A., 2007, Grand minima and maxima of solar activity: new observational constraints, A&A, 471, 301

van Genderen, A.M., & Sterken, C., 2007, Orbital Effects on the Light Curves of η Car, BP Cru, and other Eccentric Binaries, IBVS, 5782

Wing, R.F., 2011, in Astronomical Photometry: Past, Present, and Future, ed. E.F. Milone & C. Sterken (Springer Verlag, New York), p. 143

Young, T., 1802, The Bakerian Lecture: On the Theory of Light and Colours, Phil. Trans. R. Soc. Lond., 92, 12–48

Young, T., 1845, A Course of Lectures on Natural Philosophy and the Mechanical Arts (London: Taylor & Walton)

SELECTED COLOR GRAPHICS

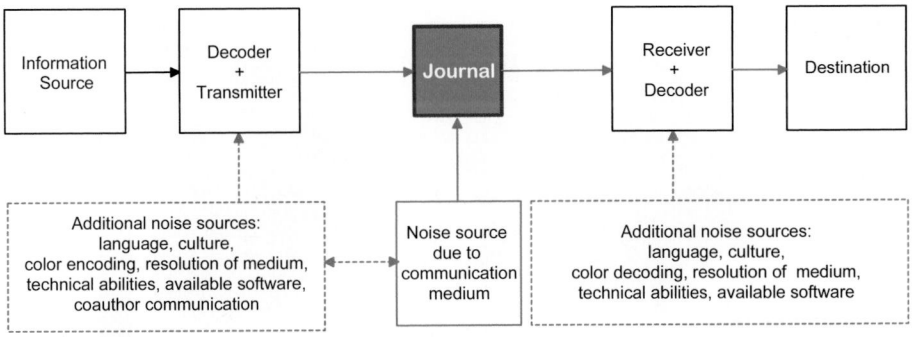

Fig. 2. Schematic diagram of an "extended Shannon" transmission model of communication.

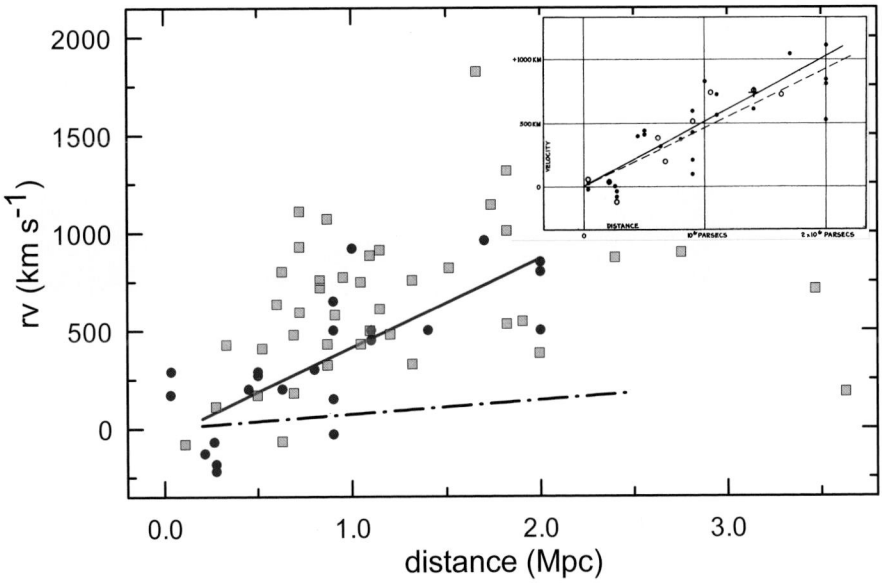

Fig. 3. Relation between observed distance and radial velocity of Hubble's "nebulae" (•: data from Hubble 1929; ■: Lemaître's available data, compiled by Duerbeck & Seitter 2001). The full line is the linear fit to Hubble's data, the dash-dotted line indicates the modern relation. The inset shows Hubble's original diagram (the dashed line in Hubble's drawing is just another way Hubble fitted his data). Source: Sterken (2009).

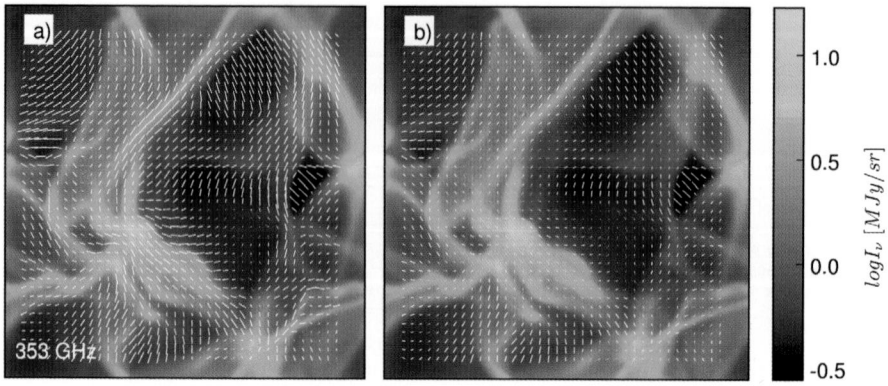

Fig. 4. Simulated polarized dust emission: two models. The scatter plot symbols represent polarization vectors, the colors are log intensity. Source: Pelkonen *et al.* (2009).

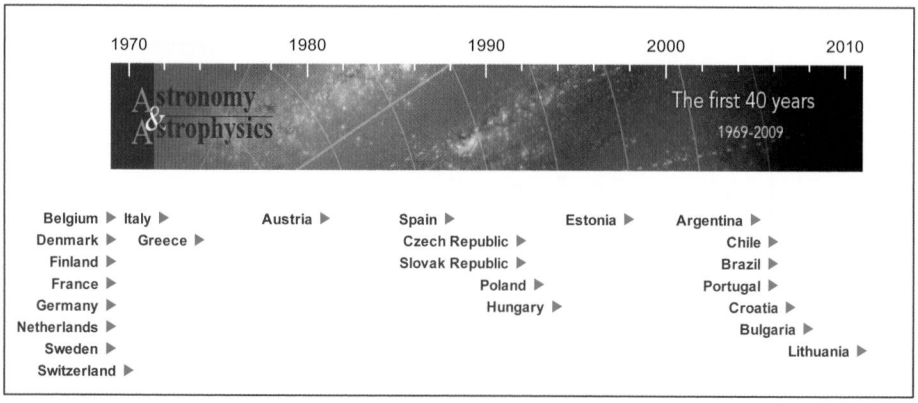

Fig. 5. A&A country membership timeline. The light- and dark-blue tints of labels and symbols are the colors used on the journal's cover (as listed in Table 7).

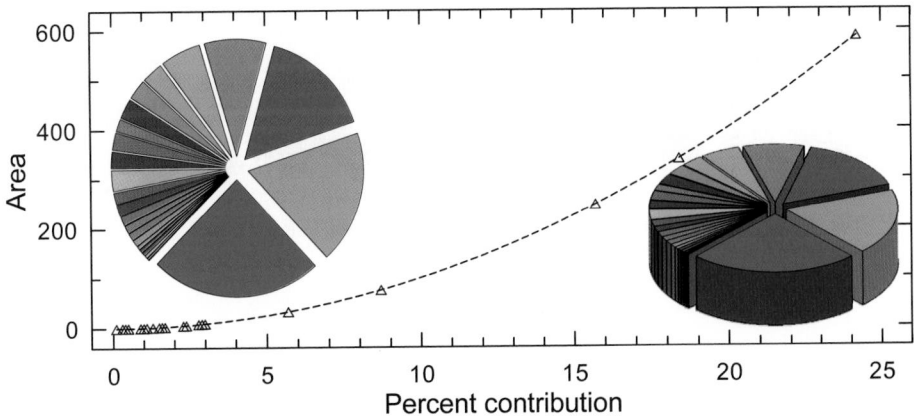

Fig. 8. Submissions from A&A sponsoring countries in 2010 in two pie chart graphs. The line is the parabolic fit of percentage contributions *versus* area for the left pie chart.

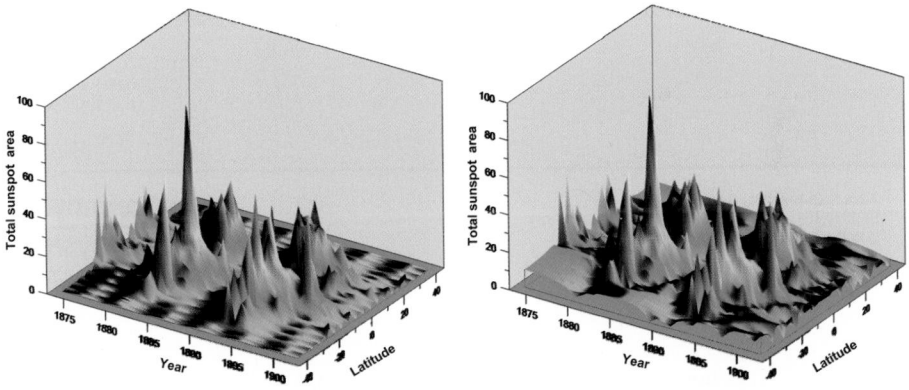

Fig. 9. Sunspot data from Maunder (1904). *Left*: the missing values from Maunder's Table were included as zeros. *Right*: based on the original Table without replacing the missing values with 0.

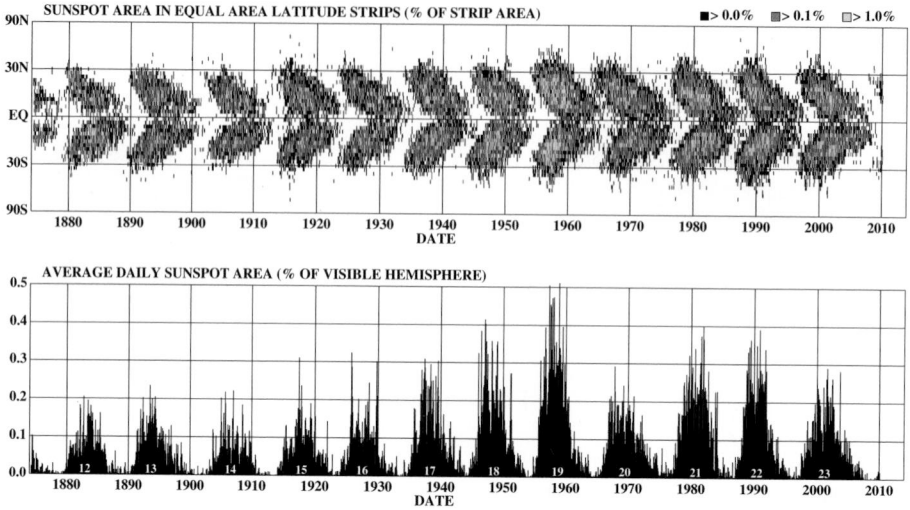

Fig. 12. Daily sunspot area averaged over individual solar rotations. *Top*: sunspot area in equal-area latitude strips; *bottom*: average daily sunspot area (in units of 0.001 visible solar hemisphere). The white labels 12–23 in the lower diagram indicate the solar-cycle number. Source: David Hathaway, NASA/MSFC.

A&A 500, 491–492 (2009)
DOI: 10.1051/0004-6361/200912200
© ESO 2009

Astronomy & Astrophysics
Special issue

COMMENTARY ON: BERTIN E. AND ARNOUTS S., 1996, A&AS, 117, 393

Modern astronomical surveys and the development of informatics

S. N. Shore

Dipartimento di Fisica "Enrico Fermi", Università di Pisa and INFN-Sezione di Pisa, largo B. Pontecorvo 3, Pisa 56127, Italy
e-mail: shore@df.unipi.it

Fig. 17. Example of A&A typefaces and fonts.

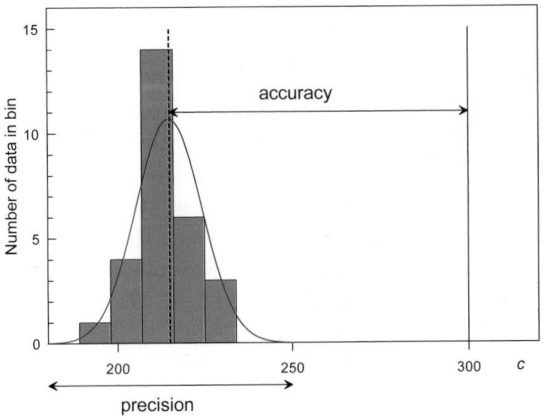

Fig. 18. Difference between accuracy and precision. The bell curve is a Gaussian fit (to a non-normally distributed sample). The solid vertical line corresponds to the value of the natural constant c, the dashed vertical line corresponds to 215 000 km s^{-1} resulting from the Gaussian fit. The unit on the x-axis is 1000 km s^{-1}, tick marks and labels on duplicate axes were omitted on purpose.

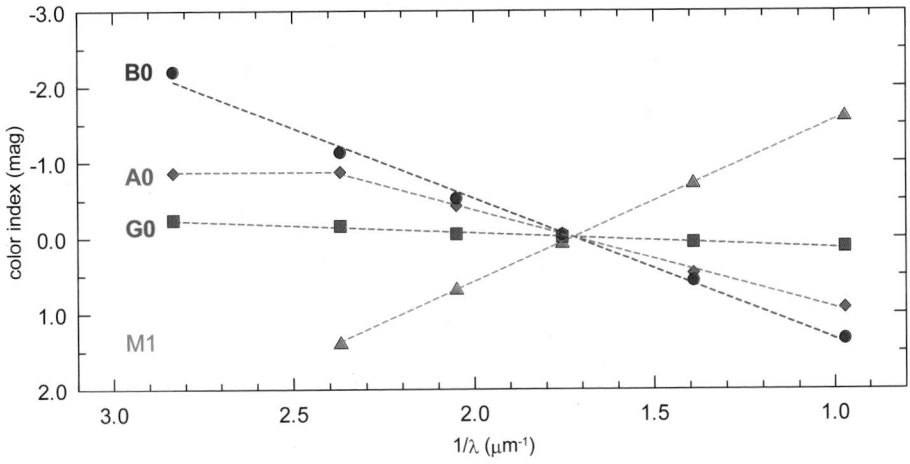

Fig. 19. Example of non-linear axis scales. Observed color index (logarithm of the ratio of two stellar irradiances) for main-sequence stars of different spectral type, as a function of $1/\lambda$. The dashed lines are linear fits that visualize black-body spectral energy distributions (for spectral type A0 the fit is constrained to $1/\lambda < 2.37$). Stebbins & Whitford (1945) list no data for M1 beyond 422 nm. Source: Sterken *et al.* (2011).

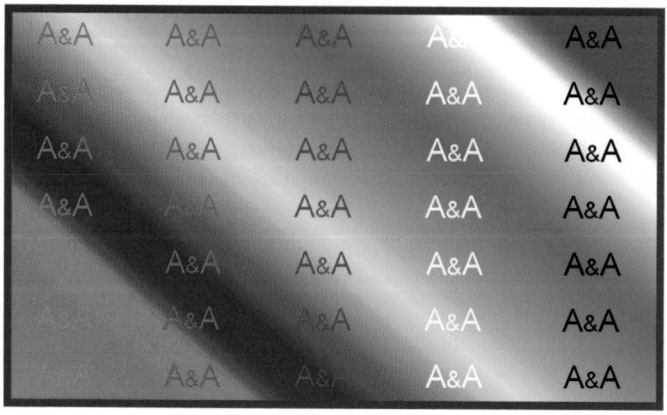

Fig. 20. Example of background contrast problems using, from left to right, *RGB*, white, and black labels. The typeface is HELVETICA, the font size 14 pt. The reduction factor is 0.365, what results in 8.6-pt printed characters.

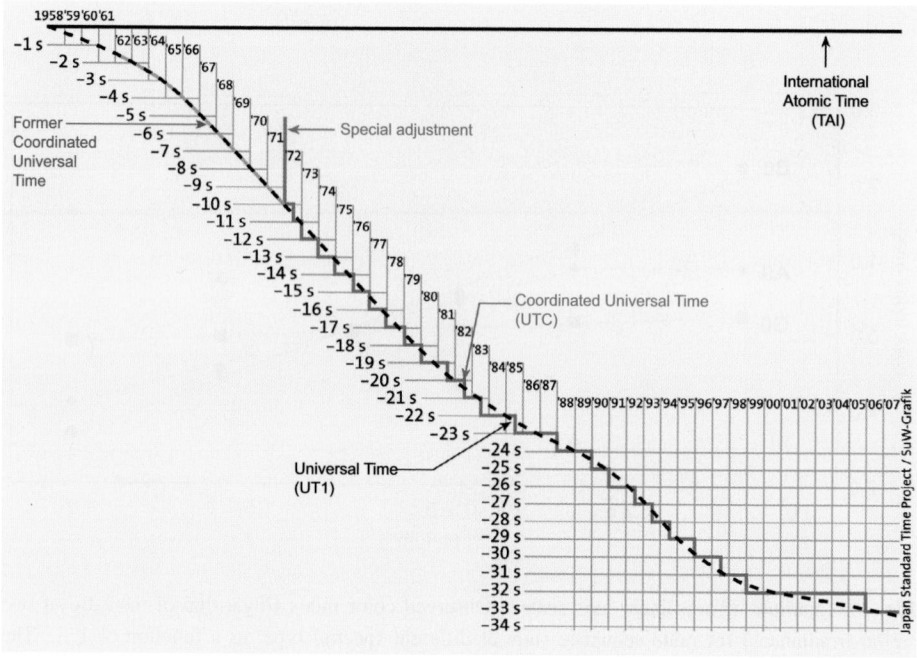

Fig. 21. Difference between uniform International Atomic Time (TAI) and Coordinated Universal Time (UTC), respectively UT1. The dashed curve 1958–1972 is the former Coordinated Universal Time. On January 1, 1972 the difference between TAI and UTC was set to 10 seconds. This graph is a timeline. Adapted from *Sterne und Weltraum*, October 2009, with permission (the only adaptation is a slightly different color rendering for enhancing contrast).

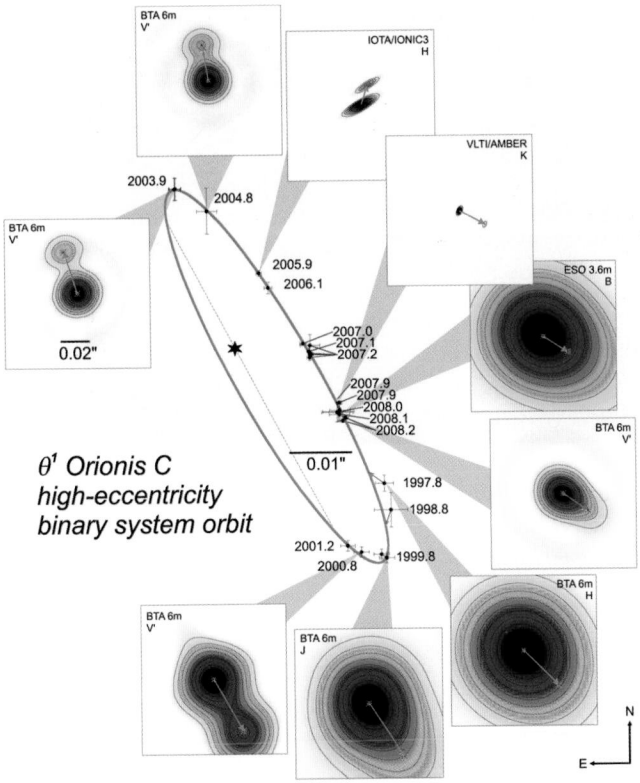

Fig. 23. Example of proper combination of data and decorative graphics: the dynamical orbit of a young massive binary system. Based on Kraus *et al.* (2009).

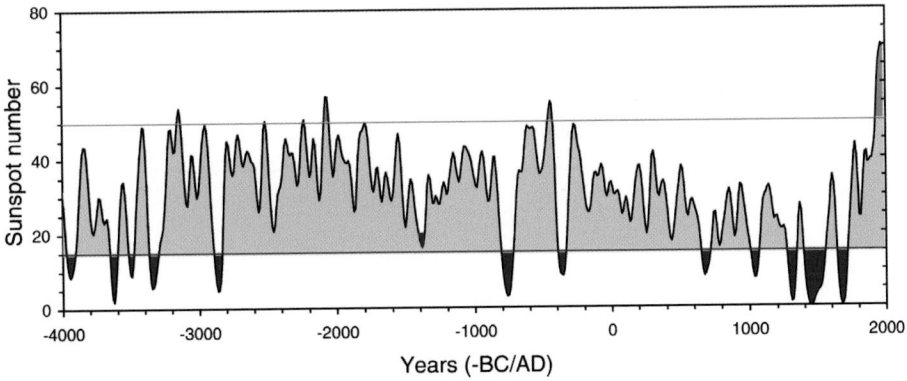

Fig. 25. Reconstruction of decadal average sunspot numbers through part of the Holocene. The x-axis label is explained on page 106. Source: Usoskin *et al.* (2007).

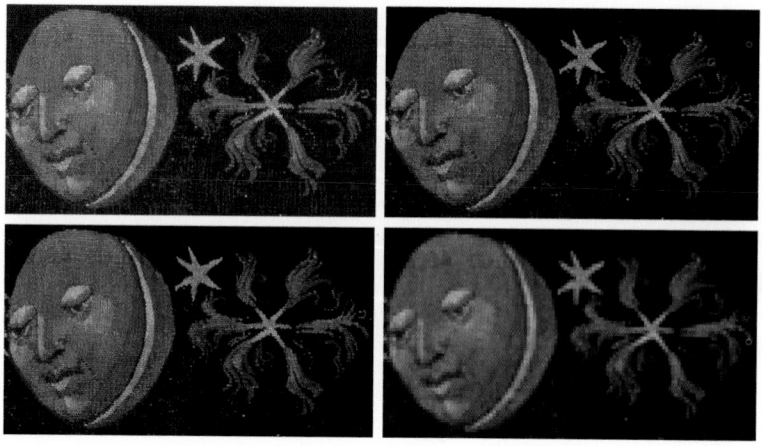

Fig. 27. *Top left:* detail of cover illustration at 600 dpi. *Top right:* same at 100 dpi. *Bottom left:* detail of cover illustration at 300 dpi. *Bottom right:* same at 72 dpi. The lowest-resolution images clearly show ragged outlines and also *pixelation*, *i.e.*, individual image pixels becoming visible.

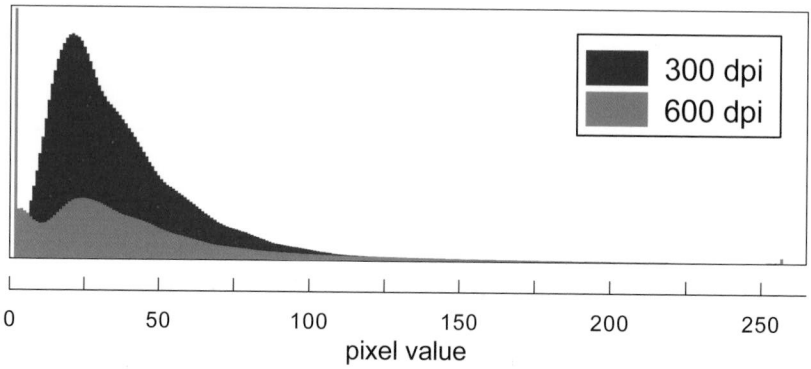

Fig. 28. Normalized histograms of the cover image, based on scans at 300 and 600 dpi resolutions.

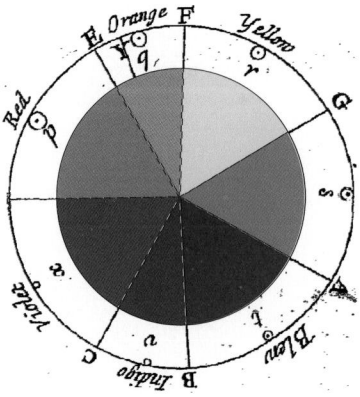

Fig. 29. Newton's color circle (Fig. 11 from *Opticks* 1704, Book I, Part II, Plate III), displaying seven colors selected by Isaac Newton. Probably the first graphical record of the concept of circular color systems. Newton's basic drawing is overplotted here with the 7 *roygbiv* colors: red, orange, yellow, green, blue, indigo and violet.

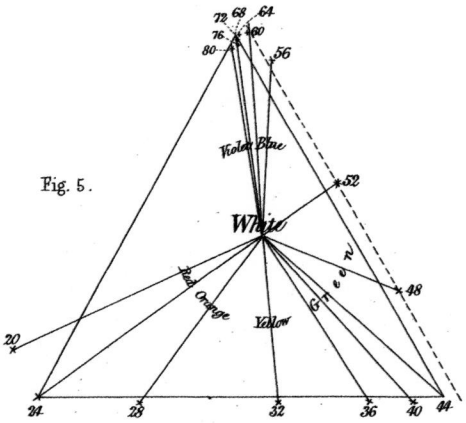

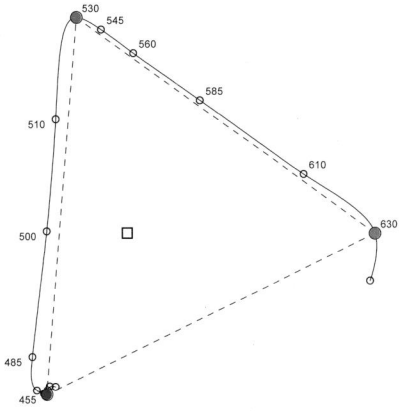

Fig. 30. *Left:* Figure 5 from Maxwell (1860). Channels 24, 44 and 68 correspond, respectively, to **his** selection of red, green and blue primaries. *Right:* rendering of Maxwell's triangular graph, rotated by 145° for comparison with the chromaticity diagram in Figure 33. The central point (□) is the position of "white" as derived from twenty observations. The numbers along the curve are wavelengths of pure spectral hues (in nm), and correspond to Maxwell's channels numbered 20–80.

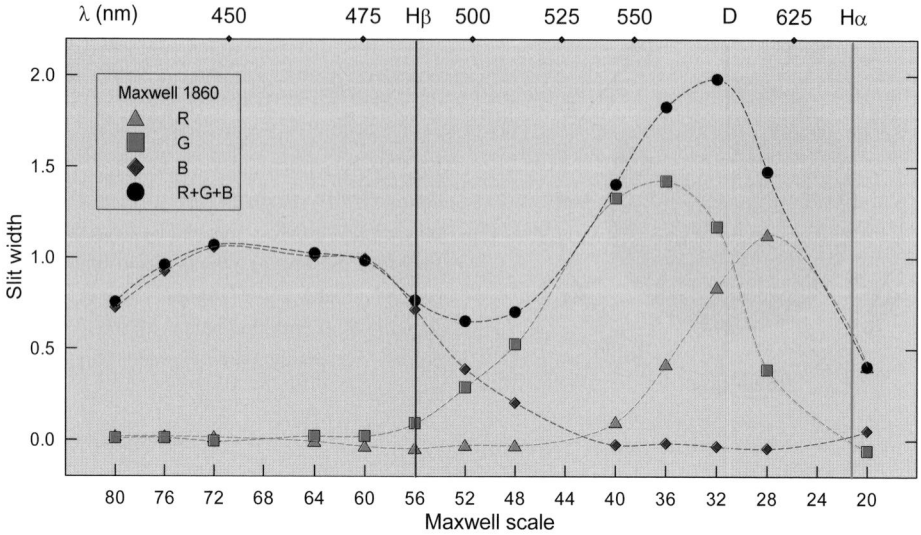

Fig. 31. *RGB* spectral response curves for James Maxwell's own eyes. Dashed lines are spline curve fits. The vertical line near scale value 31 is the doublet of the sodium spectrum near 589 nm (rendered in yellow and thus almost invisible on a grayscale print). Hα and Hβ (Fraunhofer C and F, respectively), are also indicated. The linear scales of the x and y axes are in laboratory units. The duplicate x-axis is expressed in wavelength on an approximate (nonlinear) scale, with small ♦ symbols instead of tick marks.

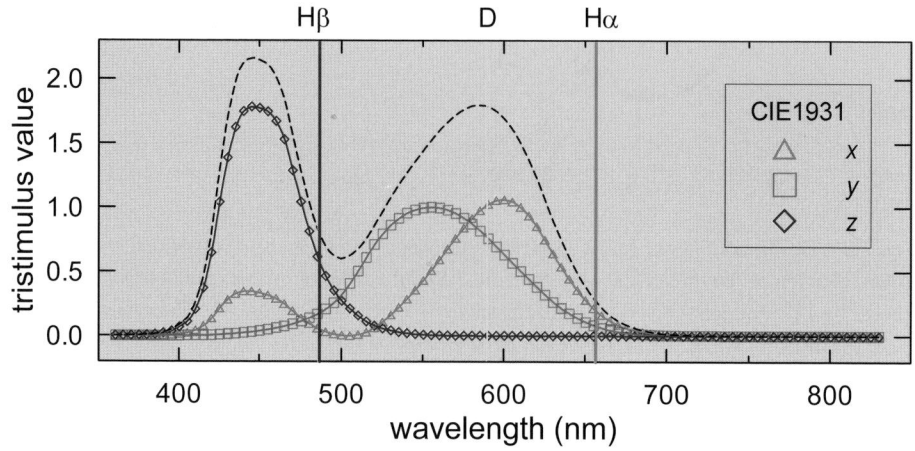

Fig. 32. CIE xyz standard observer functions, and their sum (dashed line). The top x-axis gives the approximate positions of the spectral lines shown in Figure 31. Both wavelength scales are linear.

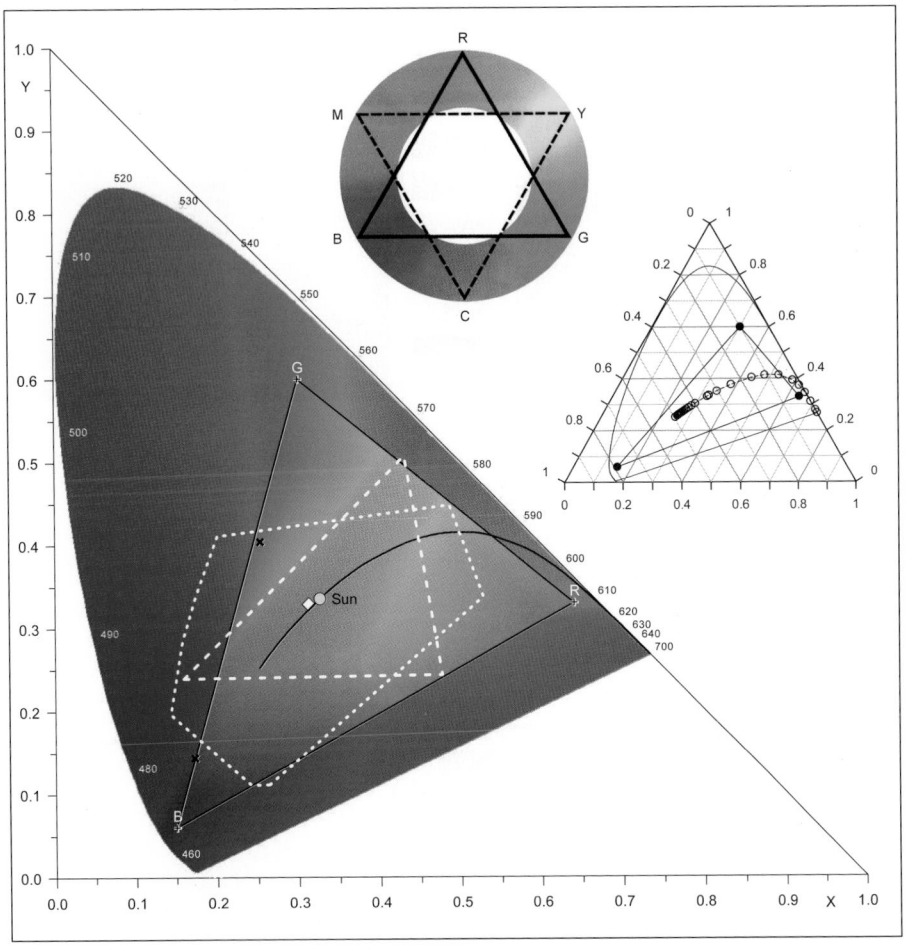

Fig. 33. CIE *xy* chromaticity diagram for the average person, with the *RGB* gamut triangle (typical color CRT monitor) and approximate *CMY* gamut (for color printing). The five-sided polygon is the gamut of a generic *CMYK* color space. The curved black line is the locus of black bodies (500–25 000 K). The pure (fully saturated) colors lie on the ridge of the horse-shoe curve. The position of the Sun is indicated, as well as the white point for a CRT monitor (◇: point of equal energy corresponding to the CIE standard for white light). × represent both A&A blue colors in *RGB* (coordinates from Table 7). The triangular inset shows some of the same data elements in a classic ternary diagram triangle, with some positions of black-body radiators (○) in the stellar-temperature range up to 25 000 K. The colored circle is a rendering of the color wheel representing the visible light spectrum with the positions of the *RGB* and *CMY* colors (Source: Kodak Handbook 1979). Primary colors are separated by 120°, and complementary colors are located directly opposite each other. The colors inside the horseshoe area were adapted from Steer (2004), with permission.

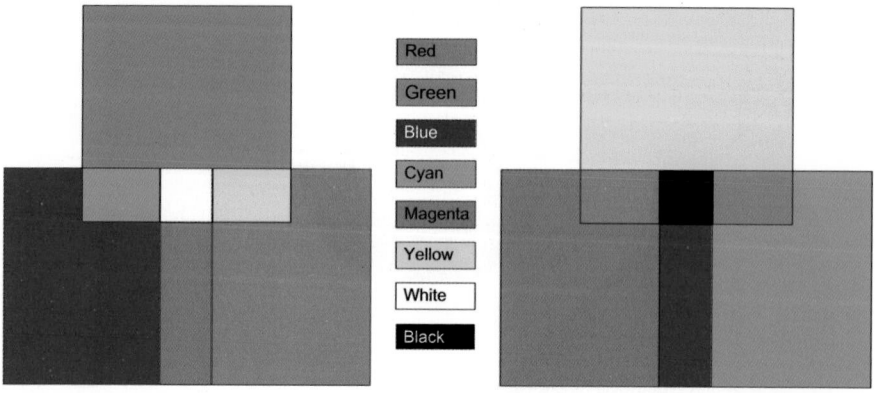

Fig. 34. Primary additive *RGB* (*left*) and subtractive *CMY* (*right*) colors. One graph follows from the other by permutation of colors and swapping white to black.

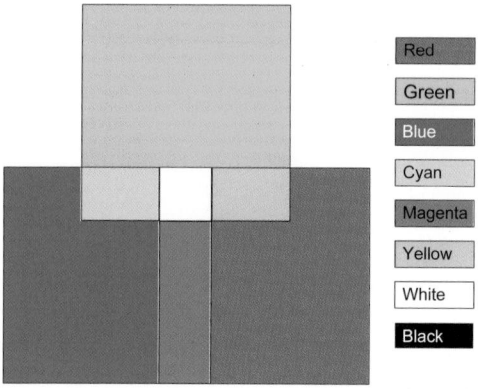

Fig. 35. Simulated *RGB* colors as perceived by a viewer suffering from protanopia. Black and white were untouched, as well as the labels in the legend.

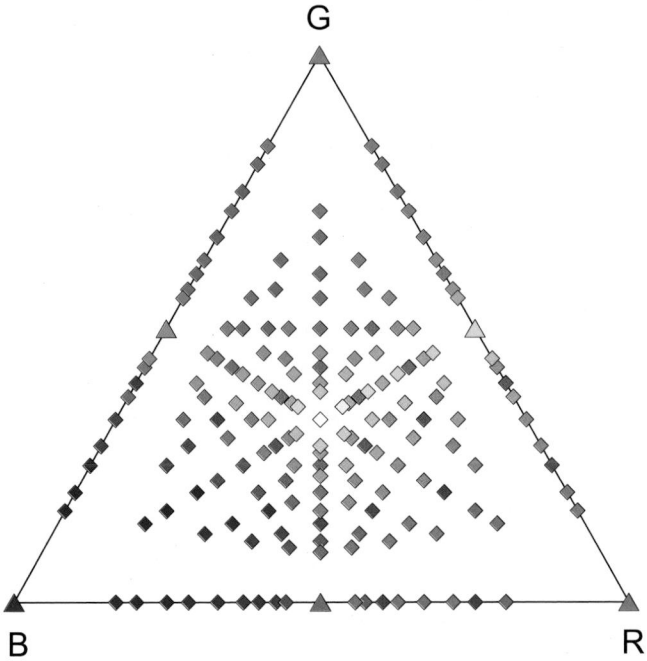

Fig. 36. *Top*: WEBSAFE palette of 256 regularly-spaced colors. The △ symbols for the *RGB* and *CMY* primaries are larger than the ◇ for the composite colors. Black, and the shades of gray (20%, 40%, 60%, 80%), are not shown, since they coincide with the white point with coordinates (1, 1, 1). *Bottom*: simulation of WEBSAFE colors as perceived by a deuteranope, accompanied by Dalton's *verbatim* description of his own color gamut.

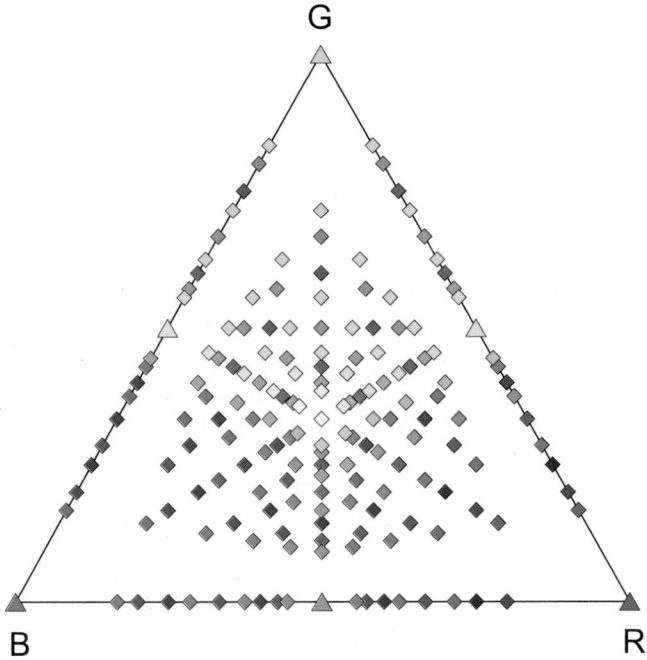

THAT PART OF THE IMAGE WHICH OTHERS CALL RED, APPEARS TO ME LITTLE MORE THAN A SHADE OR DEFECT OF LIGHT; AFTER THAT THE ORANGE, YELLOW AND GREEN SEEM *one* COLOUR, WHICH DESCENDS PRETTY UNIFORMLY FROM AN INTENSE TO A RARE YELLOW, MAKING WHAT I SHOULD CALL DIFFERENT SHADES OF YELLOW (DALTON 1798).

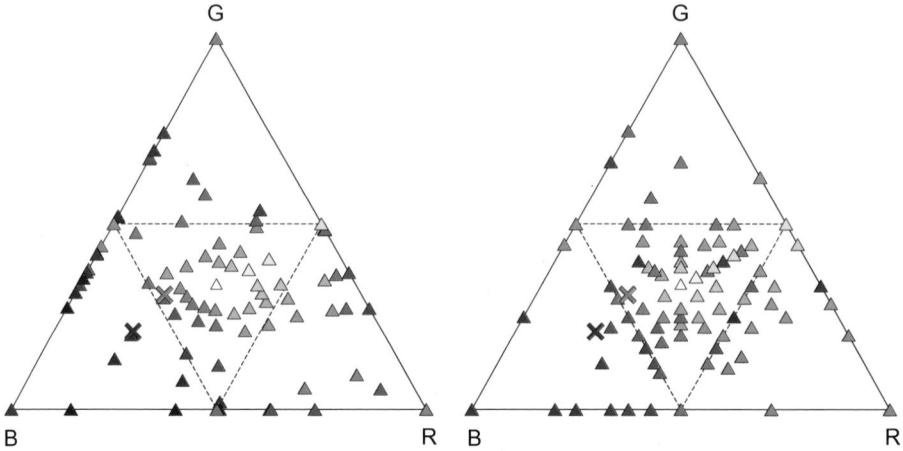

Fig. 37. Two standard color palettes. *Left*: PHOTOSHOP, *right*: GRAPHER. The × signs indicate the positions of the "A&A blues". Axis tick marks and labels have been omitted. Each apex of the triangle has the corresponding tristimulus value 1 with the two other values 0.

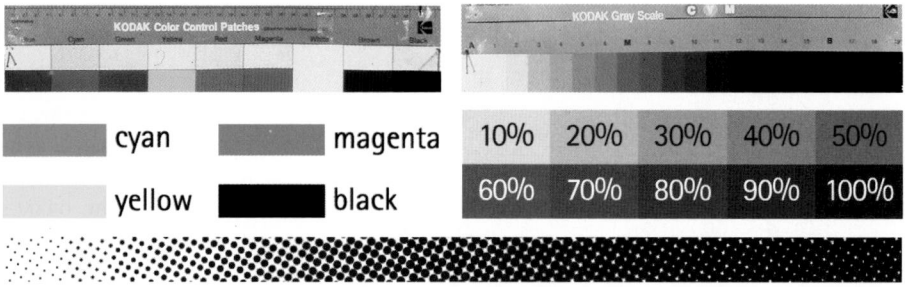

Fig. 38. *Top left:* Kodak control color patches for the cover illustration. *Top right:* Kodak grayscale strip. *Middle*: *CMYK* patches and gray scale used for printing the cover (both middle images are also reproduced in full color on the back cover of this book). Note that the labels for densities 10–50% are dark on a relatively bright background, and *vice versa* for the labels 50–100%. *Bottom*: halftone image of the Kodak grayscale strip, maximum dot size 7 pixels, screen angle 120°.

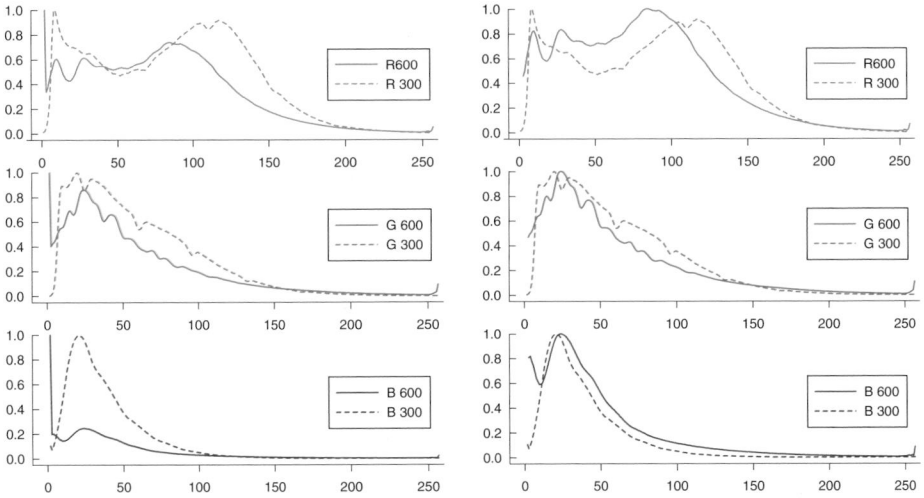

Fig. 43. *RGB* histograms of the cover image based on 300 and 600 dpi scans. *Left*: unedited histogram data. *Right*: after removal of an outlier at pixel value 0 in all channels.

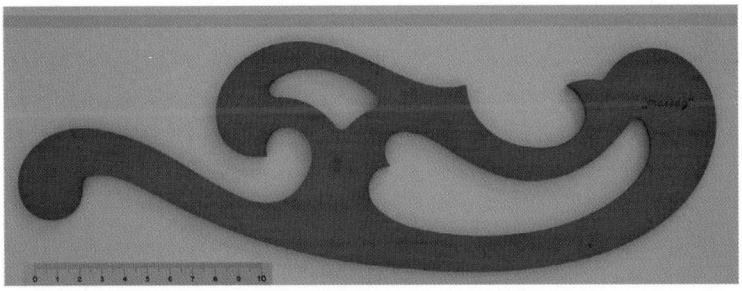

Fig. 48. The splinewood that I used in college in 1964, on a white background, photographed in diffuse daytime sky-illumination. The average *RGB* coordinates of the "white" background are $R = 170.2 \pm 1.5$, $G = 171.6 \pm 1.7$ and $B = 173.8 \pm 1.5$, *i.e.*, a grayish hue. The horizontal band at the top is the background taken from a similar photograph under different sky illumination, yielding average $R = 127.6.2 \pm 1.7$, $G = 157.8 \pm 1.6$ and $B = 203.0 \pm 1.4$, thus a blueish hue (quite close to the lightest $\times$ symbol in Fig. 37). The ruler, labeled in 1-cm units, was added to show the scale. This illustration is best viewed in color mode.

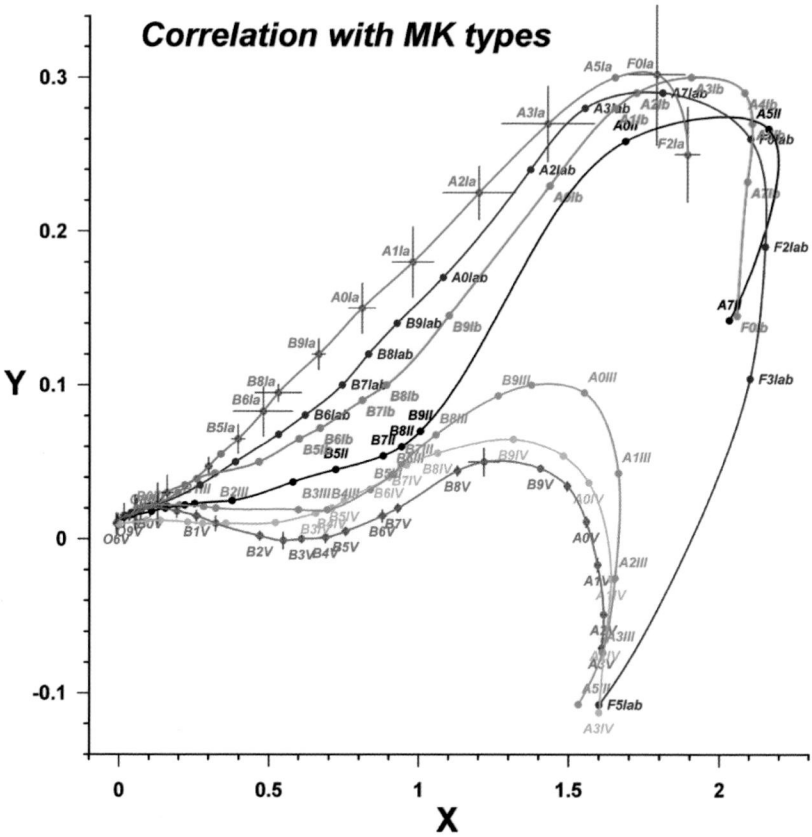

Fig. 50. Schematic diagram of spectral-type calibration in the Geneva photometry X, Y plane. Reproduced from Cramer (2005), with permission.

Scientific Writing for Young Astronomers
C. Sterken (ed)
EAS Publications Series, **50** *(2011) 173–282*
www.eas-journal.org

WRITING A SCIENTIFIC PAPER III. ETHICAL ASPECTS

C. Sterken [1]

Abstract. The main theme of this paper is truthful communication of scientific results. Therefore, concepts of truth, error, quality and value are elaborated.

The following bibliometric parameters are explained: the journal impact factor, the journal cited half-life, and the journal immediacy index, as well as paper counts, citation rates, citation index and the Hirsch index. These bibliometric indices and indicators are illustrated with examples derived from bibliometric analyses of the astronomical literature.

Scientific misconduct in the broadest sense is discussed by category: researcher misconduct, author misconduct, referee and grant-reviewer misconduct. But also publisher misconduct, editorial misconduct and mismanagement, and research supervisor misbehavior are dealt with. The overall signatures of scientific misconduct are focused on, as well as the causes and the cures. This is followed by a Section devoted to whistleblowing.

The biases of bibliometric indices, and the use and abuse of bibliometrics are illustrated. Moreover, suggestions for remediating the present defective system of bibliometric measurement and evaluation are worked out. Finally, the hopes and concerns of our students – either expressed during or after the lectures, or through subsequent private contacts – are passed on.

1 Introduction

Scientific research implies, first of all, solid codes of moral conduct, in addition to responsible attitudes of every research worker. Scientific publishing, in addition, involves ethical issues that affect Editors, Publishers, referees and readers. Many people – even scientists – believe in the myth that research workers are always objective and impartial when acquiring, handling and analyzing data, and that they are sincere and honest when interpreting and reporting the results. But in reality, there is also passion and subjectivity involved, as well as self-deceit and, unfortunately, cheating and stealing as well.

[1] Vrije Universiteit Brussel, Pleinlaan 2, 1050 Brussels, Belgium

©The authors, published by EAS, EDP Sciences, 2011
DOI: 10.1051/eas/1150003

Writing about dishonesty and deceit is most delicate and difficult, especially when it relates to upfront experience in your own research field. I, therefore, commence with one of the very first systematic descriptions of failures to adhere to moral principles, written by Charles Babbage, a British mathematician who laid the design foundations of the mechanical computer.[1] Babbage's vision that a mechanical computer would prove to be of extraordinary value for astronomical calculations – and why he presented his machine to astronomers in the first place – has more than come true. In 1830, in a paper entitled *Reflections on the Decline of Science in England, and on Some of its Causes*, he discusses – among other items – the problem of dishonesty in science, and he lists several types of cheating: fraud and hoaxing, in addition to forging, trimming and cooking of data. He makes these facts known because he trusts that doing so *"may deter future offenders"*. Only, a lot more goes wrong in scientific research and in reporting its results, than what is included in Babbage's scheme, and this wider spectrum of wrongful conduct is now named *scientific misconduct*.

Babbage (1830), moreover, deals with other types of misbehavior of persons pursuing science, such as: excessive expenses for printing with no sort of regard to economy, the system of management (and accountability) of learned societies, the vesting of multiple functions in one person, the results of counting papers, and even the "culpable silence" of colleagues who *"have spoken on one side of the question, and have voted on the contrary"*. He also includes one Section entitled *Of Minute Precision*, and a subsequent Section *On the Art of Observing*, and he bestows the highest praise on the Royal Society for the extreme regularity with which the Volumes of its *Transactions* appear, for the shortness of the time between reception and publication date, and for the rarity of any alterations made previously to the printing – exactly the way a modern journal is expected to appear on a regular basis, fast, and reliable. In short, Babbage's paper deals with research, with publishing, with truth, with error, and with valuation, and I use this paper as a key text throughout this chapter.

The next three Sections delineate some basic aspects of *truth*, *error*, and *value*. Then I attempt to give a definition of scientific misconduct. This is followed by Sections dealing with the misconduct problem by category of actors (referees, editors, authors, etc.). The succeeding Section deals with the probable origins and causes of the misconduct problems, and is followed by a Section describing remedies. This paper addresses aspects that are partially dealt with in Papers I and II, and also in Uta Grothkopf's and Steve Shore's papers in Part 1 of these Proceedings.

2 On truth

Science is a system for the production of truth, and in Paper I it was stated that I proceed "on the premise that the reader is strongly committed to the core business of science: *verity* and *truth*, even more so to the *truthful communication* of scientific findings. Bronowski (2008) even sees truth as the drive at the center of science.

[1] In 1822 he read a paper to the Royal Astronomical Society *Note on the Application of Machinery to the Computation of Astronomical and Mathematical Tables.*

The famous question *"what is truth?"* was uttered by Pontius Pilate, who received (and perhaps expected) no answer.[2] The answer, of course, depends on the context. In scientific context, truth is *empirical truth*, which means **correspondence with reality**,[3] *i.e.*, knowledge that passed the test of verification to which all scientific data, theories and conclusions will eventually be subjected to. Truth in science also means *objective truth*, *i.e.*, coherence in facts, ideas and theories. Truth in science also means **to be true in the communication of results**, *i.e.*, *speaking the truth*.[4]

Objectivity, though, does not necessarily lead to **certainty**: a coherent set of data can very well lead to incorrect views, hence models can be accepted as truth only until they are proven wrong. Moreover, as Feyerabend (1987) points out,

> ... THE FACT THAT A MODEL WORKS DOES NOT BY ITSELF SHOW THAT REALITY IS STRUCTURED LIKE THE MODEL,

as is demonstrated in Paper II with the Titius-Bode law of planetary distances. Truth in science, furthermore, excludes anything that the scientist *wants* to be true, or *believes* to be true, or is coerced to *accept* to be true.

Medawar (1990) does not refer to *absolute truth*, but to the principle that a scientific statement (or theory, or conclusion)

> WILL BE JUDGED TRUE IF IT IS ALL OF A PIECE, HANGS TOGETHER, DOESN'T CONTRADICT ITSELF, LEAVES NO LOOSE ENDS, AND CAN COPE WITH THE UNEXPECTED.

But be very prudent: Hetherington (1972), on van Maanen's "discovery" of internal motions in spiral nebulae[5] draws a moral: *"van Maanen's results had neatly fitted into a system in which each part verified the other, a system of scientific theories and observations in which the correctness of each part was judged on the basis of its agreement with the whole."* He calls this *"a vicious circle operating unchecked"*, and adds that the criterion of self-consistency, which had initially admitted van Maanen's (1916) internal motions into the domain of scientific truth, eventually demanded their removal. The lesson taught by the van Maanen case is that **some results accepted by the scientific community at large, and published in leading (or peer-reviewed, as we call them today) journals, may indeed be far away from truth**.

But what is the use of questioning the nature of truth in a paper about scientific writing? Well, scientific research involves imagination – and science without imagination is sterile. But inventive imagination is inutile if it is not moderated by a critical process that includes replication and validation. So, if a scientific result is "true", what then about scientific writing? Franklin (1984) points out that

> IF A LARGE MAJORITY OF SCIENTISTS DO NOT SUBSCRIBE TO THE HONEST REPORTING OF EXPERIMENTAL RESULTS, THEN SCIENCE MAY BE FACED WITH AN INSOLUBLE PROBLEM.

---

[2] *Pilate saith unto him, What is truth?* (Gospel of John 18:38, King James Version of Bible).

[3] Definition from Medawar (1990).

[4] In astronomy, the adjective *true* is also used to denote the difference between a calculated and an apparent (*i.e.*, observed) quantity, such as, for example, a zenith angle.

[5] Historically, the name spiral nebula was used for objects such as the Andromeda galaxy.

Paper II referred to the truism that one very specific observable can be **errorless, unless it is wrong**; in the same vein, you can say that some scientific result can be true, while its reporting can very well be a lie – simply because the form in which the result is communicated to a journal very often misrepresents the process of thought that led to the scientific outcome that is described. Some of this is because of the IMRAD structure (see Sect. 9 in Paper I) used in scientific reports, and part of this situation is the consequence of the attitude that scientists tend to embed their experimental work in a theoretical or empirical paradigm.

Popper (1996) lists three ethical principles that form the basis of discussions around the search for truth, *viz.*,

1. **fallibility**: perhaps I am wrong and perhaps you are right – but we could easily both be wrong;
2. **rational discussion**: we want to try, as impersonally as possible, to weigh up our reasons for and against a theory, and
3. **approximation to the truth**: we can nearly always come closer to the truth in a discussion that avoids personal attacks.

Popper underlines that these ethical principles imply tolerance:

> IF I HOPE TO LEARN FROM YOU, AND IF I WANT TO LEARN IN THE INTEREST OF TRUTH, THEN I HAVE NOT ONLY TO TOLERATE YOU BUT ALSO TO RECOGNIZE YOU AS A POTENTIAL EQUAL.

I consider point 3, and the subsequent remark on tolerance, very important in the context of refereeing, see Section 8.

3 On error

To err is human: error is a normal part of human fallibility, and scientific error is an absolute normal component of any research activity. Error is sometimes described as *negative knowledge* that – even when detected – affects work in the sense that it interrupts and even reverses the workflow. In addition, error – in the statistical, in the experimental, and in the human sense – is just unavoidable. The term *error*, though, can have many different context-dependent meanings, see Table 1 of Paper I for several alternatives: uncertainty, mistake, defect, flaw, blunder, etc.. Error is everywhere, in whatever we observe, calculate, investigate, and write. But some mistakes are avoidable: for example, committing the same error again on purpose (because it makes your life easier, see Sect. 7.7), or omitting deep proofreading before submitting a paper for publication. Such carelessness should result in random errors in most – but not in all – circumstances. The following list explains a few types of error:

- error: a deviation from accuracy or correctness, for example, erroneous data;
- mistake: an error caused by a fault, like a misjudged conclusion;
- blunder: a very bad mistake, like publishing the same graph to document totally different datasets – the kind of mistake resulting from working under pressure;

- error caused by poor knowledge of the theory of the measuring apparatus;
- statistical error: the difference between an estimated value and the true value, see the discussion on precision and accuracy in Papers I and II;
- observational error: the difference between a measured value and its true or calculated value, see the discussion on $O - C$ in Papers I and II;
- rounding error: resulting from rounding off to a number of decimals, but also the difference between the calculated approximation of a number and its exact value;
- communication error: often due to lack of proofreading, or because of too many last-minute changes in a manuscript;
- procedural error: for example, extrapolation, sloppy transformations or calibration, or the application of inappropriate statistical models;
- documentation error: when a description falls short of allowing close replication of the followed procedures;
- software error, particularly in black-box[6] software and data-reduction pipelines, including the plug & play approach (or lack of training in calibrated measurement);
- sampling error: caused by non-continuous measurement, for example due to the observing window;
- numerical magnification error: extrapolation from a very limited amount of data;
- personal error: for example, personal bias and lack of scientific objectivity;
- conceptual error: for example, equating correlation and causation;
- fruitful error: the kind of mistake or blunder that occasionally leads to a discovery;
- error of the system of science, *viz.*, all varieties of scientific misconduct;
- interpretation error, including "being right for the wrong reason", and
- fooling yourself – and others.

Some of the errors listed above belong to the category of *fooling yourself*, in the sense of Richard Feynman's statement in his 1974 Caltech commencement address[7] on the topic of *cargo cult science*, *i.e.*, soft science that has the semblance of science, but is only *pseudoscience* because it lacks scientific integrity and honesty:

> THE FIRST PRINCIPLE IS THAT YOU MUST NOT FOOL YOURSELF – AND YOU ARE THE EASIEST PERSON TO FOOL.

Feynman (1974) illustrated the analogy between certain fields of research in the sciences, and cargo cults.[8] What he implied is that such mimicry leads to a type of "science" – of a form that *looks* perfect, but yields a product that just does not work. One example of cargo cult science in astronomy is the kind of scientific analysis pictured in Figure 42 of

[6] A black box is a device or a system of devices of which only input and output characteristics are accessible, without any information about its internal mechanism.

[7] A commencement address is a speech given to graduating students.

[8] *Cargo cult* is the practice of some tribal societies to imitate technologically advanced products by functionally irrelevant goods of their own creation.

Paper II that, quite often, leads to totally spurious results and unfounded claims, simply because some kind of game mechanism is applied to deficient and non-cogent sets of data (see Sterken 2005 for a deeper discussion).

Feynman consequently stresses that all details that could throw doubt on your interpretation must be given – if you know them – and *"if you know anything at all wrong, or possibly wrong"*, you must explain it.[9] He furthermore states that you – the author –

> MUST TRY TO GIVE ALL THE INFORMATION TO HELP OTHERS TO JUDGE THE VALUE OF YOUR CONTRIBUTION; NOT JUST THE INFORMATION THAT LEADS TO JUDGMENT IN ONE PARTICULAR DIRECTION OR ANOTHER.

His "others" includes authors, referees and editors. Thus follows Feynman's second principle, which expands from not fooling the individual person to not fooling the scientific community at large:

> AFTER YOU'VE NOT FOOLED YOURSELF, IT'S EASY NOT TO FOOL OTHER SCIENTISTS. YOU JUST HAVE TO BE HONEST IN A CONVENTIONAL WAY AFTER THAT.

Langmuir (1954), in fact, had already dubbed such deviation from the principles of the scientific method as *pathological science*, *i.e.*, research that hangs on long after it was demonstrated to be false, or as he describes it as *"the science of things that aren't so."* He lists the symptoms of pathological[10] science as:

1. the maximum effect that is observed is produced by a causative agent of barely detectable intensity, and the magnitude of the effect is substantially independent of the intensity of the cause;

2. the effect is of a magnitude that remains close to the limit of detectability, or the results are of very low statistical significance;

3. claims of great accuracy are made;

4. fantastic theories, contrary to experience, are proposed;

5. criticisms are met by *ad hoc*[11] excuses thought up on the spur of the moment, and

6. the ratio of supporters to critics rises to ~50% and then falls gradually to oblivion.

Popper (1996) rightly points out that

> THE MAINTENANCE OF A SELF-CRITICAL ATTITUDE AND OF PERSONAL INTEGRITY THUS BECOMES A MATTER OF DUTY.

---

[9]This is one aspect in which scientific practice differs from the practice of law: whereas the scientist's duty is to play devil's advocate, the lawyer will present to the judge only those (true or false) arguments that support his or her own truth, fit the defense, and secure an optimal legal outcome. Part of Feynman's quote is also given in Section 9.8 of Paper I.

[10]Not to confuse with *pathology*: the study and diagnosis of disease.

[11]*Ad hoc*: Latin phrase that means *for this*, *i.e.*, only applicable for this case.

This very specific aspect of scientific writing is not so self-evident, because this aspect is not necessarily rooted in the native culture of the author, nor is it often included in the PhD training of scientists. Kale, senior clinical Editor at the *British Medical Journal*, cites a Chinese medical colleague who confirms this: *"We are not used to this style of writing, where you present both the advantages and disadvantages of a policy and its successes and failures"*. But Feynman's statement, and Kale's (2006) quote, exactly say what we are so used to doing in physics and astronomy: present advantages **and** disadvantages or, more formally, express that principle in the format

$$\boxed{\textit{result} \pm \textit{error.}}$$

Such form of communication is far from natural, and must be acquired, and this learning process can be quite agonizing to the beginner.

In short, the above discussion relates to three aspects that are felt as somewhat painful by the novice scientist, *viz.*:

1. what you have written must first be thoroughly checked by yourself;
2. what you have written must be checked by others (proofreaders, referees, editors);
3. what you have published will, for sure, be taken apart and verified by others.

Popper (1996) points out that it is impossible to avoid all mistakes, but that we should avoid them whenever possible.[12] In attempting to do so, criticism by others is a necessity. Watson (2007) emphasizes this very clearly:

> NOTHING CAN REPLACE THE COMPANY OF OTHERS WHO HAVE THE BACKGROUND TO CATCH ERRORS IN YOUR REASONING OR PROVIDE FACTS THAT MAY EITHER PROVE OR DISPROVE YOUR ARGUMENT OF THE MOMENT. AND THE SHARPER THOSE AROUND YOU, THE SHARPER YOU WILL BECOME.

4 On quality and value

Quality. The quality of a scientific work refers to something inherent to the work itself. Sechrest (1987) distinguishes three problems that relate to the quality of scientific findings, and I add a fourth problem to his list:

1. quality of our data (individual numbers, but also sums, means and indices – that are corrupted by bias, carelessness, undetected error, or incompetence);
2. quality of our conclusions, including the risk of spoiling good data during analysis;
3. quality of our interpretations, including the failure to detect limitations in the process of data collection, and
4. **quality of the scientific reporting**.

[12] See also the epilog on page 274.

Value. The value of a scientific work (and of a scientist) refers to how the research (and the researcher) are **evaluated**, and how the result of this assessment is perceived by the society (*i.e.*, the colleagues, the peers,[13] and last but not least, the scientific community).

The meaning of scientific quality and value – like the meaning of truth and the meaning of error – is context-dependent and relative, especially since we talk about value of an abstract object. Relative value is subjective, and depends on individual and cultural views; absolute value, on the other hand, strongly depends on the way how that value is measured or assessed,[14] although most such measurements are relative – even biased (see Sect. 18).

The prime question to ask is: can scientific quality and value be *measured* in the first place? There are two approaches to the measurement of quality, *viz.*,

1. by way of the peer-review process: refereeing of papers or reviewing grant applications by a referee or by one or more experts, and
2. via computer-generated science parameters or indicators used to gauge the quality of journals (*e.g.*, the impact factor), and the quality of the individual (*e.g.*, citation rates).

Unfortunately, peer-review committees cover a much wider field of knowledge than the specialized cases they are to pass judgment on, so more often than not these commissions fall back on easily available indicators such as journal reputation[15] and rough citation counts for evaluating an individual, a research proposal, or a research team. Traditional peer assessment is now more and more replaced by just taking a performance indicator from a database as an absolute dictum.

But what is "valuable science", how do we value science and, consequently, how do we value scientists? In fact, exactly the same question is posed on art and artists: artwork certainly possesses inherent value, such as beauty, originality and expression, but the value of art is nowadays measured primarily by its market value as expressed in monetary terms.

Most scientific papers bring the authors no monetary profit (except for royalties, which usually are peanuts when taking into account the large amount of work involved in producing scientific monographs). The scientific author presents the product to the community, which can enjoy it for free through *Open Access*,[16] or through purchase (often online payment that benefits publisher or copyright holder).

Kennedy (1997) views publication as the fundamental currency of the scientist, and just like currencies are organized in a monetary system, scientific papers are also ranked in a *value system*. A value system is a set of consistent value norms (standards) and measures used for the purpose of attributing a value to a concept. Consistency means that

[13] A peer is a person who is equal to another in abilities, qualifications, and knowledge background.

[14] Measurement refers to obtaining a number or index through counts or calculations, assessment refers to evaluation of the measured quantity by a deeper analysis.

[15] For example "a paper in *Nature*".

[16] See Dewatripont *et al.* (2006), Grothkopf & Erdmann (2008), and also Grothkopf's paper in Vol. 1 of these Proceedings.

1. assignments of value do not contradict each other;
2. assessments of value are stable in time, and
3. different parties (authors, publishers, administrators) rank those values equally.

Value tends to result in reward: Goodstein (2010) defines the "Reward System" of science as *"the set of all means by which scientists express their esteem"*. This system possesses the following general qualitative properties:

- the system is set up to reward people for proving that something is right – consequently, a negative result will less likely make it into print than a positive discovery;
- evaluation is mainly outcome-conscious, and not procedure-conscious: a paper with results is held in higher esteem than one dealing with procedures[17] to get these results;
- therefore, experimental results are either underevaluated, or considered as being unproblematical (Franklin 1989), and
- whereas accurate data are highly valued, papers with collections of data are held in lower esteem (see also the discussion on Fig. 21).

Feynman, however, defends the principle that negative results should also be published:

> IF YOU'VE MADE UP YOUR MIND TO TEST A THEORY, OR YOU WANT TO EXPLAIN SOME IDEA, YOU SHOULD ALWAYS DECIDE TO PUBLISH IT WHICHEVER WAY IT COMES OUT

– that is, if we would only publish results of a certain kind, we would get a biased view – therefore, "negative" results are *valuable* as well.[18]

Feynman then proceeds by demonstrating the need for *validating* scientific results, *i.e.*, repeating experiments of other scientists **under the same conditions** to see if the same result will **always** be obtained. And only when the same result is obtained, you proceed with experiments under changed conditions, and only then you may know that the real difference was the thing you thought you had under control.[19] Some scientists continue to run some experiments, he says, "in the same old way", with total neglect of dominant systematic errors, still producing many papers, while the relevant technical papers are not even quoted – most likely not even read. This, exactly, is *cargo cult science*.

Remember, though, that validation is not the *nihil obstat*[20] nor the *imprimatur*: a paper accepted by an Editor on the basis of a referee's recommendation has been verified, but not necessarily validated (see Agassi 1981, and also the van Maanen case on page 175).

---

[17] For example, standardization and calibration procedures – see Sterken (2011) for an example of a procedure-oriented paper.

[18] This statement somewhat contradicts Goodstein's first point.

[19] This leads him to describe a situation with an animal-behavior experiment that very well shows the need to know the experimental equipment in all its aspects, so to avoid error by poor knowledge of the theory of the measuring apparatus, see page 177.

[20] *nihil obstat*: Latin for nothing stands in the way; *imprimatur*: consent to print.

5 On bibliometry and bibliometrics

The scientific Reward System is based on *bibliometrics*, the analysis of citations of journal articles. Bibliometrics comprises statistical methods used to measure specific parameters of scientific publications, and includes citation analysis and impact analysis.[21] Bibliometrics is used to rate researchers, institutes, publishers, and journals. Evaluation of scientific knowledge, scientific knowledge carriers and scientific knowledge producers, in principle requires – just like any other measurement in the physical world – **right measurement procedures, sets of truthful data including correct error estimates, and correct interpretations**. In other words, the dimensions of *truth, error* and *value*, as described in the previous Sections, must also prevail in all bibliometric indicators used to rate scientists and all aspects of their scientific work.

5.1 The measure of journals

The three most current bibliometric journal indices are the journal impact factor (IF), the cited half-life, and the immediacy index. These indices are calculated each year by the *Institute for Scientific Information* (ISI)[22] **for those journals that ISI indexes**, and the results are released annually in the *Journal Citation Reports*. When reading Sections 5.1.1–5.1.3, you should never forget that these indices are

1. just an **average** for **all** the papers in a journal for a **specific year**;
2. never statistically representative of **any individual paper**,[23] and
3. determined by technicalities **unrelated to the scientific quality of the articles**.

5.1.1 The impact factor

The original idea of the impact factor came from Eugene Garfield, who in the 1950s designed a grading system for journals, to be used mainly by librarians. The impact factor is a scientometric indicator that measures the typical **citation rate** of a journal, and is calculated for year i with the formula

$$\mathbf{IF}_{\mathbf{year}_i} = \frac{\textbf{NUMBER OF CITATIONS IN YEAR}_i \textbf{ TO PAPERS PUBLISHED IN } (\textbf{YEAR}_{i-1} + \textbf{YEAR}_{i-2})}{\textbf{NUMBER OF CITABLE PAPERS IN } (\textbf{YEAR}_{i-1} + \textbf{YEAR}_{i-2})}$$

where the numerator is the number of all citations received by a journal, and the denominator is the number of *citable* items: research papers and review articles are citable, whereas book and software reviews, editorials, abstracts of meetings and errata, are non-citable items. But there is a catch here: the non-citable items do not go into the denominator, but if those uncounted items get cited in the literature, ISI still adds those citations into the numerator, thereby increasing the journal's impact factor (Monastersky 2005).

[21] `http://en.wikipedia.org/wiki/Bibliometrics`

[22] ISI was created by Eugene Garfield in 1960, and is now owned by THOMSON REUTERS.

[23] As explained in more detail in Section 18.

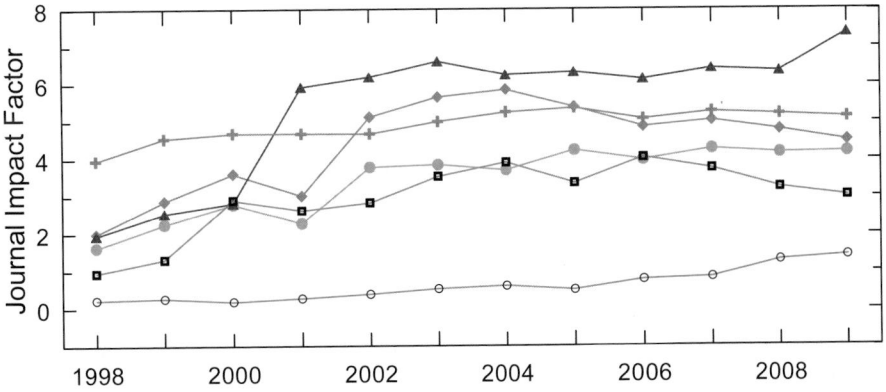

Fig. 1. Impact factor for six wide-field astronomy journals.

Figure 1 illustrates the trend of the impact factor of six wide-field (*i.e.*, not topic-specialized) astronomy journals: the trend is coherent and increases with time, the pattern is quite stable over the last five years, and a strong upturn appeared in 2001–2003. A retrospective analysis of trends for seven medical journals (1994–2005) by Chew *et al.* (2007) reveals a similar upturn since 2001–2002, though with higher percentage change in impact factor than in astronomy. The stronger trends in medical science may be ascribed to the Editors' approach of actively "courting researchers", as one Editor put it: *"Our IF increased because I hustled for key papers – I talked to people I'd known for years and who hadn't previously submitted to [our journal]. . . "*.

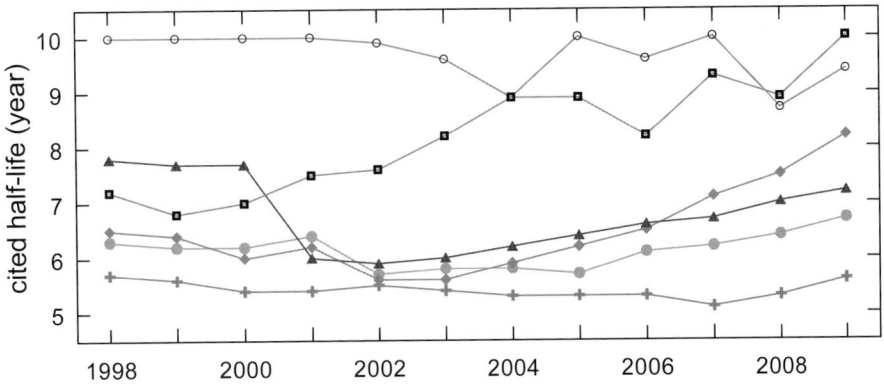

Fig. 2. Cited half-life for six wide-field astronomy journals. The ordinate values of 10 correspond to a cited half-life of 10 or more.

5.1.2 The cited half-life

The cited half-life of a journal is a measure of the rate of decline of the citation curve: it is the time interval that contains half of the citations – *i.e.*, the number of years for citations to drop to 50%. Figure 2 shows the evolution of this index for the same six journals as displayed in Figure 1. The cited half-life does not necessarily correlate with the impact factor: low-IF journals have notably longer cited half-lives.

5.1.3 The immediacy index

The immediacy index of a journal is a measure of how often very recent articles are cited, *i.e.*, how rapidly the average paper is cited in the literature, and is calculated with

$$\text{IMMEDIACY INDEX}_{\text{year}_i} = \frac{\text{NUMBER OF CITATIONS IN YEAR}_i}{\text{NUMBER OF CITABLE PAPERS IN YEAR}_i}$$

Figure 3 shows the immediacy index for the same six journals discussed previously. The immediacy index seems to mimic the impact factor quite well.

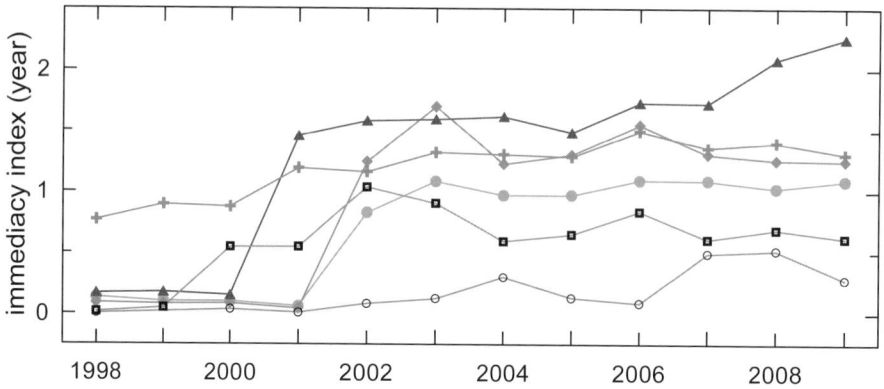

Fig. 3. Immediacy index for the same six wide-field astronomy journals of Figure 1.

Figures 1–3 clearly show that since 2002 the patterns have become quite consistent, with some fluctuations. Smaller players and discipline-specialized journals with a low number of citable papers (not shown) display more erratic behavior.

5.2 The measure of scientists

5.2.1 Paper counts

A simple walk along library shelves demonstrates the explosive growth of the volume of scientific literature: the total information load has a doubling time of a decade or less.

Kennedy (1997), more than ten years ago, described the situation as having reached a near-crisis situation that results from **the contemporary fixation on quantity of publication**. Broad (1981) already argued that the phenomenon of paper inflation stems not from an increase in productivity, but rather from changes in the way people publish. This is very well illustrated by Figure 4, which shows statistics of the non-Letters Sections of A&A and ApJ combined, *viz.*, the average number of authors per paper, and the average number of pages divided by the average number of authors.

The reasons for paper inflation can be summarized as:

- the unavoidable consequence of interdisciplinarity, involving many researchers;
- premature publication of unfinished studies, and
- proliferation of scientific workshops and conferences.

Papers also tend to become shorter over the years for diverse legitimate reasons, such as: limited space in journals, high page charges, inclusion of online material, the pressure on authors to publish in slices, the fact that large-facility data become public domain after several months, and because over the last decade, the page format in some journals allowed the information content to grow.

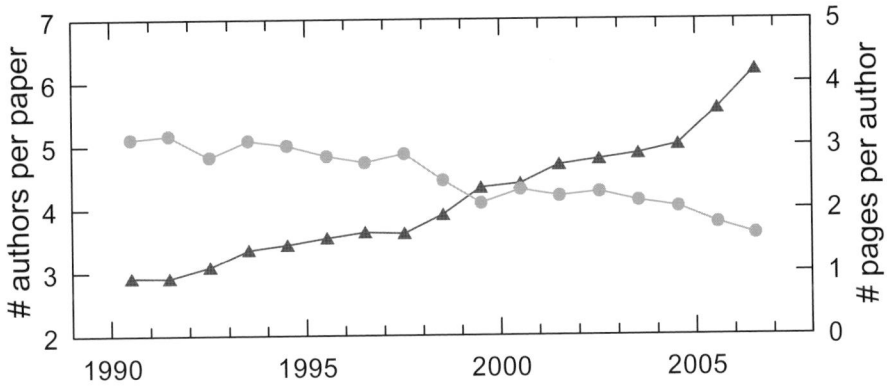

Fig. 4. Statistics of the non-Letters Sections of A&A and ApJ combined. ▲: average number of authors per paper; •: average number of pages divided by the average number of authors. Based on Figures 14 and 16 of Burton (2007).

The first and foremost victims of paper inflation are the students, who have to digest the massive literature, but also the supervisors and administrators, who have no time nor sophisticated means to analyze the quality of these writings. Figure 5 shows, as an example, the cumulative number of publications of four very productive stellar astronomers, who work in closely related research fields. The number of papers published in journals listed by ISI, and the number of entries as given by ADS are shown. Most of this non-ISI article production also includes conference papers, though the ADS compilation also lists conference proceedings books, conference abstracts, monographs, book reviews,

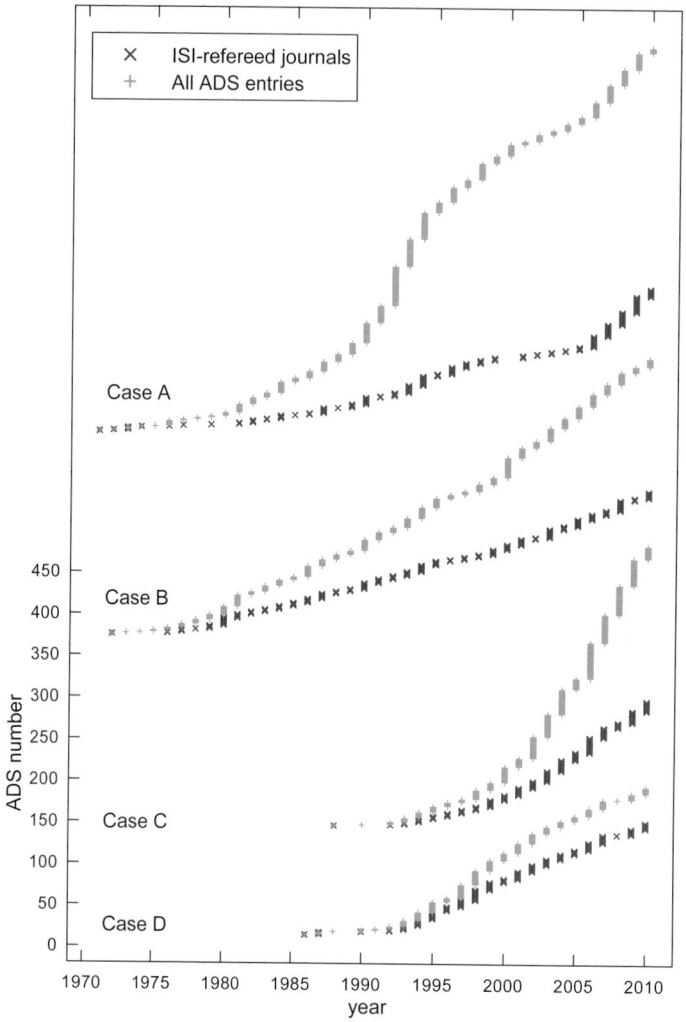

Fig. 5. Cumulative number of publications of four productive astronomers: ISI-counted journals (×, journal papers only), and ADS entries (+). The ordinate gives the ADS count at the moment when this graph was constructed. The curves for the various authors have been shifted vertically by arbitrary numbers.

SIMBAD data catalogs, and obituaries, see also the discussion on Figure 24. These astronomers' early career phases are characterized by a close match between both trends, but later on the ADS list strongly departs from the ISI-listed refereed journals. The four authors display different trends: A and C have steep "ADS slopes", with A showing irregular fluctuations in both samples of scientific output. Case B displays a more stable trend characterized by an unbroken steady continuity, and D shows a clear tendency of avoiding non-ISI papers. All one can conclude from these graphs is that all four scientists are indeed productive, though it is evident that **paper counts alone are a totally unreliable index for gauging any of these scientists' productivity**, see also Figures 8, 23 and 24, which compare prolific authors with *apparently* less productive ones.

5.2.2 Citation rates and citation index

Citations are references: every citation in the text body of an article corresponds (in principle) to a reference at the end of the paper. Abt & Garfield (2002) come to the conclusion that, for most of the journals in a given field, there is a linear relationship between the average number of references and the normalized paper lengths.[24] These authors list several pertinent parameters that shape citation counts, *viz.*:

1. the length of the paper: long papers tend to have more content;
2. publication in journals with high impact factors, because those are the journals more widely accessible;
3. rate of growth of the field: rapidly-growing fields quickly produce more papers that will refer to previous papers;
4. the nature of the research field;
5. the number of authors, because they and their colleagues tend to spread knowledge of their results to wider audiences, and
6. the location in the journal: the first article in the journal seems to produce more citations than subsequent ones.

The primary factor is the first one, and the last point, for sure, has become less influential in electronic publishing.

A citation index is a database of citations that keeps track of which publications in scientific journals cite which other papers. A very well known citation index is the ISI *Science Citation Index* (SCI), which is often applied in combination with the impact factor. Note that, just like **the impact factor is not a true measure of a journal's impact, the SCI does not necessarily yield a true picture of an author's value as a scientist**: there certainly is a correlation between citation and value, but it is far from a tight one.

Citation counts can also be used to rate projects and large instruments. Habing (2009), commenting on a paper that deals with the ESA Infrared Space Observatory (ISO) in the A&A special anniversary issue,[25] rightly asks the question what citation numbers really tell us. He counted at least 1600 papers based on ISO observations,[26] and he compares ISO with other large (space) instruments, according to two numerical criteria: the average citation number, and the fraction of papers with a citation number higher than fifty. Note that this criterion is not the classic citation rate, but the number of papers that carry the name or the acronym of the space facility in the title – hence, papers that do not mention the mission in their title, were not considered. The result is summarized in Figure 6 for CHANDRA, COBE, HIPPARCOS, HST, IRAS, ISO, ROSAT, SPITZER, UHURU, WMAP, and XMM. The "winners" seem to be UHURU, COBE and WMAP, with about three times more papers than the average of the others, and equally three times more papers with over 50 mentionings. Note also that these data are not homogeneous nor coherent, for example, some missions have

[24] Pages and paper lengths in this work refer to normalized 1000-word pages.

[25] A&A Volume 500, Issue 1, 2009.

[26] About equal to the total number of articles appearing annually in A&A.

short duration, and others go on for decades. Hence, this citation rate tells us something about the usefulness of these missions, but does not address the scientific value of their products, the papers.[27]

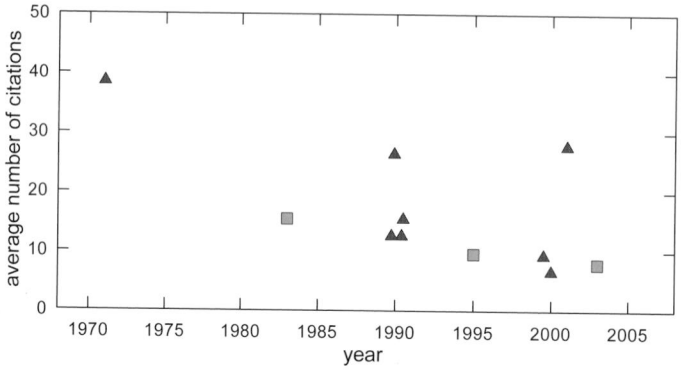

Fig. 6. Average number of papers with the name of a space mission in their title, as a function of date of launch. ■ are infrared satellites, ▲ represent other missions. Based on data from Habing (2009).

Condon (2009), under the title *No new observations, no new theories, enduring scientific value* (in the A&A special anniversary issue), draws attention to one paper by Kühr *et al.* (1981) that, unlike most highly-cited papers, presents no new observations or astrophysical theories, and contains "only" published and unpublished data. But that paper became "the" definitive catalog of strong radio sources at centimeter wavelengths, simply because these authors compiled all of the relevant radio and optical data scattered across the literature, brought the flux densities to a common scale, and selected the 518 strongest extragalactic sources, thus providing a **tool** to be used by all experts in the field – in a way contradicting the relatively low rank of such papers as illustrated in Figure 21.

Thus, **the total number of citations of an author says very little about the quality and usefulness of that author's paper**, especially if the collection includes some highly-cited papers. The mean number of citations per paper looks to be a better solution, but that average is biased if the sample contains several papers with very few citations. Another option is the h index, explained in the next Section.

5.2.3 The Hirsch h index

Hirsch (2005) introduced the so-called h-index to rank the scientific productivity and impact of a scientist. The index is based on the scientist's most cited papers, and takes into account one specific aspect of the *distribution* of citations that a scientist accumulated. A scientist with index h has published h papers, each of which has been cited at least h times. The Hirsch index is, by definition, an integer number. Cases A and B in Figure 5 have similar h-indices, cases C and D also have very comparable Hirsch indices (h is known to be gender-biased – Kelly & Jennions (2007) – hence A, B, C and D were chosen not to be of same gender).

[27]The first design version of Figure 6 had every mission labeled by its acronym, but then I removed the labels in order to avoid unintentional misleading, in case this graph is reproduced out of context.

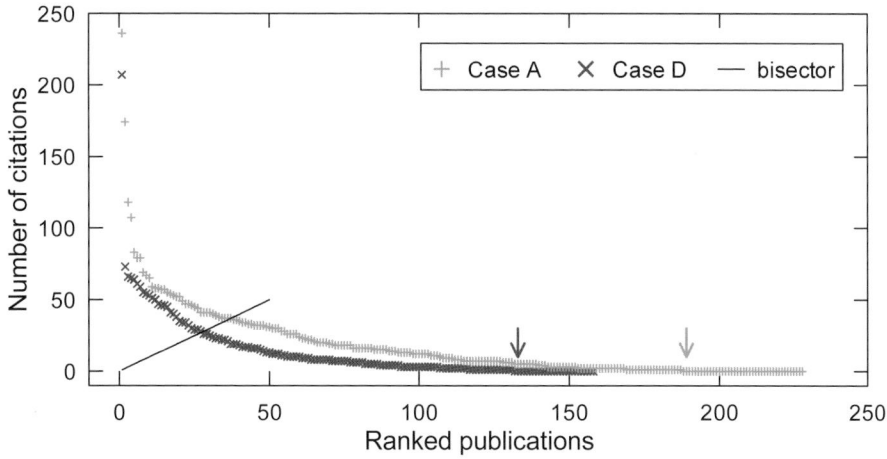

Fig. 7. *h*-index calculation for cases A and D, based on data provided by the ISI Web of Science. The vertical arrows indicate the onset of publications with 0 citations (13% for D and 18% for A).

Figure 7 illustrates the paper-ranking statistics for cases A and D, and allows an estimate of *h* by counting the number of data entries that lie above the bisector. The onset of publications with 0 citations is also indicated. The top scores of A and D are papers that are at least a decade old, thus well over the cited half-life – even exceeding the cited full-life – of the involved journals. A and D differ in *h* by about 10 units, and a quick inspection of Figure 5 might lead to the conclusion that the former is an overachiever and the latter is an underachiever. Note that the length of the research careers of these scientists differs by a decade, but that these two scientists also differ in "type", as is evident from the citation distributions. This difference in type can be the consequence of the working environment, but very likely it is to be ascribed to personal choice.

A useful section of both curves is what I would call the *threshold level*: the region just below the bisector that, if densely populated, will likely lead to an increase in *h* when only a few papers from that area receive a couple of additional citations. This region is mostly well-populated in the case of senior scientists, but notable exceptions exist, see case F in Figure 8, and in Table 1.

The *h*-index is practical because it is less influenced by citation peaks, and even by bibliometric errors. But it should be used with proper care and understanding, especially when applied in comparisons of veteran scientists with comparable *h*-indexed younger postdoctoral fellows. But the index is just a number, on a strongly-nonlinear scale (especially as it grows) and may not be used to conclude that someone is doing well because he or she is only a couple of *h* points short from a Nobel Prize winner, even in the same field, see also the footnote on page 263.

Figure 8 is an extension, on a logarithmic scale, of Figure 7, to which are added two top researchers (cases E and F) in the biological sciences: James D. Watson and William D. Hamilton (1936–2000), respectively. Jim Watson was awarded the Nobel Prize in

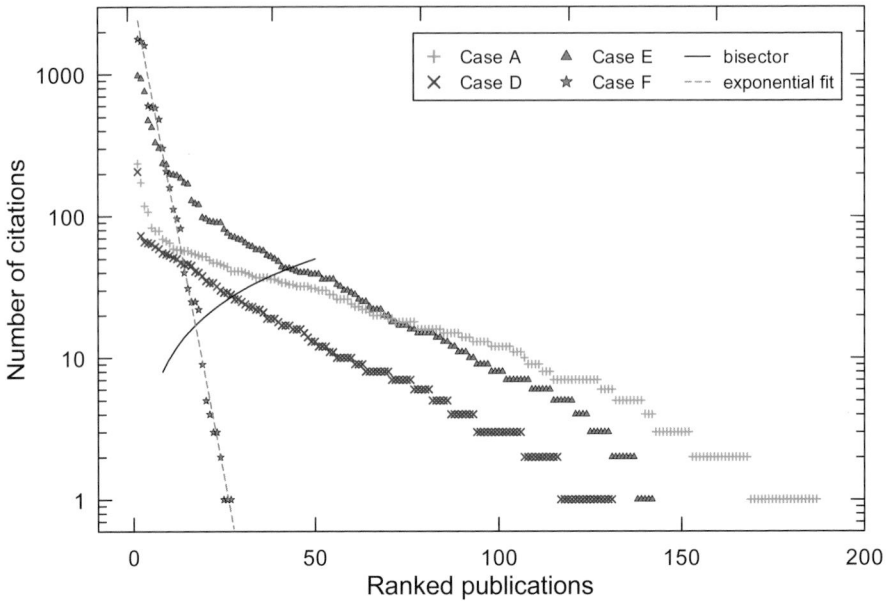

Fig. 8. Citation statistics for cases A and D, with two additional cases from biological sciences, on a logarithmic ordinate scale (based on data provided by the ISI Web of Science). Zero-citations were excluded. The straight line fitted to the citation counts of case F is the equation log(citation count) = $a + b$ log(rank number). The bisector yields the h-index.

Physiology or Medicine in 1962, jointly with Francis Crick and Maurice Wilkins. Hamilton, an evolutionary biologist who passed away 10 years ago, was hailed as *"one of the towering figures of modern biology"* (Angier 2000) and was regarded as *"unquestionably, the most influential evolutionary biologist of the second half of the 20th century"* (May 2001). He received the 1993 Crafoord Prize of the Royal Swedish Academy of Science, equivalent to the Nobel Prize. Hamilton is most interesting because he supported a theory considered by the scientific establishment to be unfashionable if not cranky[28] (Bozzi 2003). He may, in this respect, be compared to Sir Fred Hoyle (1915–2001) whose unorthodox theories (for example, the steady state theory) were also rejected by the astronomical establishment of his times.

Table 1 lists bibliometric parameters (rounded numbers) of the six case studies: total citation count, top score of the most-cited paper, the derived h index, the number of citations needed to promote the most-cited paper below the h threshold level to increase the author's h level by just one unit, and the scientific discipline for each case.

[28]One instance refers to the refusal of the Editors of *Science*, *Nature* and *Lancet*, to publish his comments about the oral polio vaccine (OPV) theory: that HIV-1 Group M arrived in our species via a live oral polio vaccine that had been accidentally contaminated by chimpanzee SIV (Simian Immunodeficiency Viruses).

Table 1. Bibliometric parameters of case studies.

Case	Total citation count	Top score	h index	citations to $h+1$	discipline
A	4000	230	37	1	astronomy
B	3200	280	29	1	astronomy
C	2300	120	25	1	astronomy
D	2200	210	27	1	astronomy
E	9000	980	43	1	biological sciences
F	8500	1780	19	11	biological sciences

Let us first compare Watson versus Hamilton (in my non-expert approach): Watson has more than 9000 citations, Hamilton has almost as many, but these two scientists with nearly the same total score differ by a factor of 2 in the top score, at the same time Hamilton's h index is way below Watson's, and he needs an unrealistic number of additional citations to increment h by just one unit. But Hamilton's accretion curve shows a most fascinating additional aspect that is absent in all five other cases: it is strictly linear (on a logarithmic scale), which means that there is an *exponential growth* of his body of citations. The other cases all show hysteresis (initial dormancy) and explosive rise (at the highest publication ranks). Watson's case is also interesting: at the time of the Nobel Prize award, he had only a dozen papers, with a total citation count of just below 3000 and a top score of 758, with $h = 11$. This means that it took almost half a century after the Prize to triple his number of citations, whereas the top paper's citations increased with a mere 30%. Hamilton is a vivid confirmation of the statement by Merton (1968):

> THE WELL-KNOWN FACT THAT A GOOD NUMBER OF SCIENTISTS WHO HAVE NOT RECEIVED THE PRIZE[29] AND WILL NOT RECEIVE IT HAVE CONTRIBUTED AS MUCH TO THE ADVANCEMENT OF SCIENCE AS SOME OF THE RECIPIENTS, OR MORE.

These citation accretion curves are very useful for monitoring the acceptability of a scientific idea, as in Hamilton's case, but their divergence illustrates that the use of one single number does not allow to fully quantify the value and production of a scientist (see also Sect. 18.6). Table 1 also illustrates the danger of comparing these numbers between fields as different from each other as astronomy and biology: A and B are very productive scientists, C and D – when corrected for the age difference – are well on their way to become of similar level, if not higher. But any direct comparison between these astronomers – or those figuring in Figure 20 – and the biologists, is out of place.

Figure 9 illustrates the correlation between h index and total number of citations for more than 200 authors and coauthors of a selection of highly-cited papers of one specific journal. The lower diagram reveals a more or less steady trend up to citation counts of about 2500, followed by a course with lower slope. A linear regression over the interval [500, 3500] yields a correlation coefficient of 0.56, which is relatively weak: the correlation is just a trend, of width $\Delta h \approx 15$, and with quite a number of outliers. The analysis

[29]The Nobel prize.

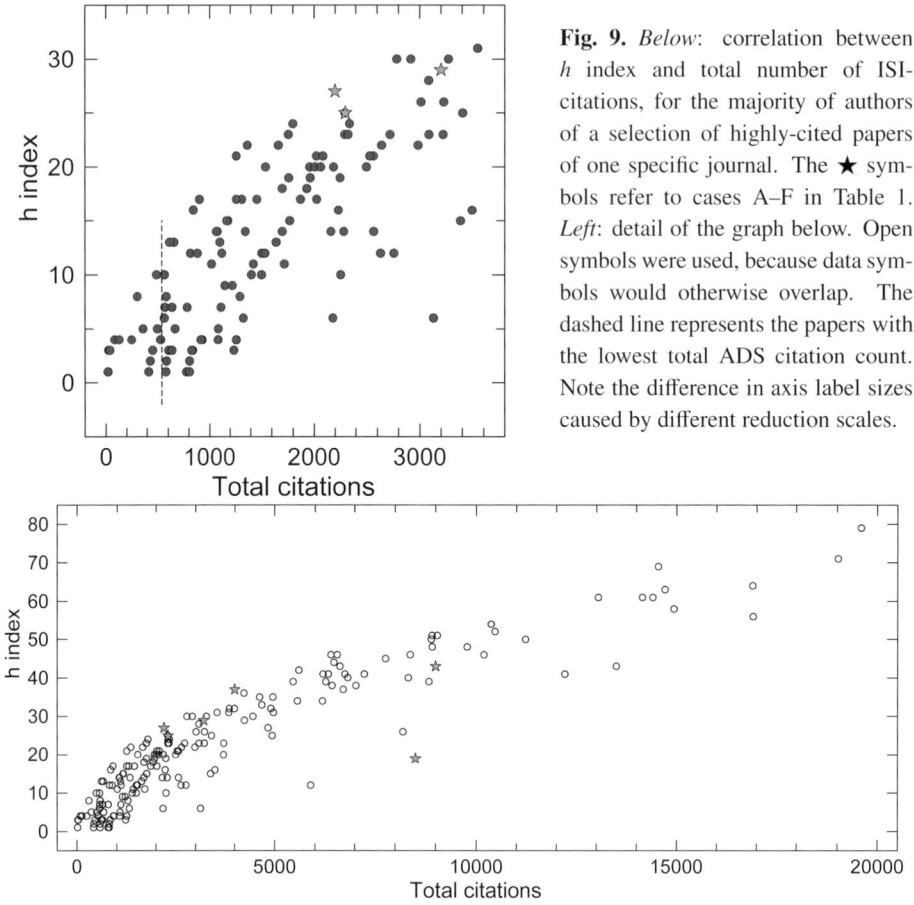

Fig. 9. *Below*: correlation between h index and total number of ISI-citations, for the majority of authors of a selection of highly-cited papers of one specific journal. The ★ symbols refer to cases A–F in Table 1. *Left*: detail of the graph below. Open symbols were used, because data symbols would otherwise overlap. The dashed line represents the papers with the lowest total ADS citation count. Note the difference in axis label sizes caused by different reduction scales.

also reveals that these papers embrace authors with very diverse h index and numbers of citations. The ISI database nevertheless returns several cases with even less citations, see also the same effect in Figures 19 and 20. Note, again, that the construction of these graphs was done with the mindset of a scientific administrator, *i.e.*, by searching on surname and initials, and by excluding subject areas outside astronomy and related fields, see also the discussion of this aspect in Section 18.1. In this case, the aspect of confusion is very strongly present in the several multi-authored papers describing space instruments, where many coauthors have a very limited number of papers dispersed over very technical areas like aerospace engineering and imaging techniques. When such an author has a very common surname, it is absolutely impossible to separate the true subject areas from the false, and no one who collects bibliometric data on more than 200 individuals will ever arrive at *exactly* the same result: **these measurements are, thus, not exactly reproducible**.

5.3 Should bibliometrics guide you in the decision where to publish?

There is no "where-to-publish" recipe that fits all papers and all authors of all scientific disciplines. A first aspect to verify is whether page charges are an issue. Once this obstruction is cleared, you may look for the journal that you think offers the best chance of your work being read (and cited): either a specialized journal in your field of research, or a journal that offers full *Open Access*. Check the journal website about what kind of research depth the journal requires: for example, a model of an uneventful binary, or just one new exoplanet, will not be a suitable paper for a high-level journal. Avoid starting out at the highest impact-factor journal and going down the impact-factor ladder, simply because this approach steals time from referees, Editors, and also from yourself in the first place (see Sect. 7.2).

Tenopir *et al.* (2010) conducted a survey of research publication characteristics and their relative values,[30] and concluded that the top in importance are the factors **relevance**, and **online availability at no personal cost**.[31] Then the journal itself and the fact that it should be peer-reviewed are the most important characteristics. The key findings of this study are listed below.

1. Topic of the article was ranked by all demographic groups as the most important characteristic that helps in choosing an article to read.
2. After topic, the next most important characteristics selected were online accessibility and source of article.
3. Author(s), type of publisher, and author institution were consistently ranked last.
4. Overall, online accessibility was considered more important than author reputation in the conjoint analysis, and much more important than type of journal.
5. Online access with a direct personal cost to the reader was rated as strongly negative; such a pricing model significantly reduces value to the user, and thus likelihood of use.
6. Articles from known top authors or unknown authors, are more likely to be read than those by known, but weak authors. Articles from top-tier peer reviewed journals or lower-tier peer reviewed journals are more likely to be read. Readers are less likely to read an article from a non-peer reviewed journal than from a non-journal source.
7. The highest rated conjoint profile was *"Written by an author I recognize as a top scholar, in a top-tier peer-reviewed journal, and available online at no [personal] cost"*; while the lowest profile was *"Written by an author I recognize as a weaker scholar, in a journal that is not peer-reviewed, and available online at some cost"*.
8. Readers from all disciplines ranked topic as the top characteristic. After that, there were some differences in the rankings of the other characteristics based on subject discipline. For example, respondents in engineering, science, and professional

[30] Involving 445 respondents from 12 countries, with a nearly equal number of male and female participants.

[31] The importance of accessibility was already demonstrated by Öpik (1977): good science is always read, whatever the prestige of the journal – **provided the paper is accessible to the community**.

disciplines placed more emphasis on journal title than other disciplines. The professional fields ranked online accessibility lower than any other discipline.

9. There was no difference in ranking of characteristics by gender or by age. However, for geographic location, there were two characteristics where the differences were statistically significant: source of article was more important to North Americans, while online accessibility was more important to researchers elsewhere.
10. There was no difference in article characteristic rankings between respondents with high or with low authorship. However, when removing topic and online accessibility, high authorship respondents ranked journal title as significantly more important.
11. Based on open-ended comments, there were indications that there are other characteristics of interest. These include readability, editing quality, and graphic design that clearly and efficiently supplements the text.
12. There are considerable differences by discipline in the number of articles read per month, although disciplinary differences were not as evident in the number of journal titles from which articles are read regularly. Younger scholars (those under age 40) reported that they read slightly fewer articles each month and that they read significantly fewer journals on a regular basis.

As Tenopir (2010) emphasizes, they were not testing any particular economic model: free to the user does not necessarily means *Open Access*, it could also mean that the users' library simply has a subscription.

6 Researcher misconduct

Before anything else, I wish to emphasize that research misconduct does not include honest error, and also that not all dishonestly derived findings are necessarily wrong. Misconduct may also be seen as an error, not in the sense of observational or computational, but simply as an *error of the system of science*. However, our scientific procedures (verification and refereeing, replication and validation) are designed to deal with "regular" error, not with fraud, and are thus not optimized for uncovering wrongful conduct.

Researcher misconduct mainly includes various aspects of data manipulation, but also features disruption of mutual trust. Researcher misconduct is separated from the next Section on author misconduct, because it does occasionally happen that an author is fooled by a team member in one way or another. *Vice versa*, a researcher (for example an observer, or a statistician) who contributes to a paper, may also be fooled by the leading author(s) in aspects of analysis and interpretation.

6.1 Amplification of work

Amplification of an experiment by reporting more work than you actually did, is a form of deceit that can have great consequences in the end. For example, if it is required to inspect raw data, then not doing so but saying that you did, is an unethical act. Or claiming that you analyzed a complete database, though only a small sample of the content was processed, is unacceptable.

6.2 On secrecy and secretiveness

Scientific research involves a high degree of competition that brings along the need for secrecy of ideas, and for concealment of unpublished results. This entails secrecy and confidentiality at several levels: the scientist, the research supervisor, the coauthors, the referees, the Editor – even the Publisher.

In a truly international project every team has its own motivation, its own background, and its own philosophy, and such diversity brings along its own problems. I witnessed two cases where different actions caused the same crushing results. The first problem concerned someone who revealed to a colleague outside the project which data in the literature he had used to formulate an assumption that was being tested by further multi-wavelength observations. The interloper repeated the calculations, and published the main conclusion as his own discovery. In another case, when collected data were analyzed by separate teams, one team presented final results as if they were coming from their data alone. The result of such actions are devastating: not only did they lead to pollution of the literature by adding trivial publications containing half-baked conclusions, they also took away any possible awards from those who eventually publish the original work. Do not forget that papers with anticipated results are seen as more valuable than later ones if the latter – even if they are of higher quality – only *seem* to repeat accepted results.

Secrecy in an industrial laboratory is a much stronger obligation than it is in our world of research in astronomy, but some aspects of *trade secret law*[32] are equally applicable to us. A trade secret is described as a formula, practice, process, design, instrument, pattern, or compilation of information that is not generally known, and by which a research team can obtain an advantage over competitors.[33] The business world protects trade secrets by demanding their employees to sign a *non-compete clause* or *non-disclosure agreement* that should prevent the employee from collaborating with a competitor, or from disclosing confidential information to outsiders. Although such clauses are very rare in our research discipline (where economic stakes, in general, are astronomically low), non-competition and non-disclosure are usually implicitly understood when a scientist accepts a research contract. This does not mean that you are not allowed to speak with colleagues from competing groups who are personal friends, but **it means that you must not reveal the exact details of the teamwork to outsiders**. In particular, you are not allowed to share data with outsiders, even less so to include outsiders in your project without consulting your colleagues and supervisors in the first place.

A very specific situation is the case of a researcher leaving the lab to join another team, which entails restrictions on data rights, see also the discussion on copyrights on page 206.

6.3 Forging

Babbage (1830, see Sect. 1) defines forging as publishing observations that were never made – in other words: fabrication of data. Forging is mostly done by the person who

[32] Not to be confused with patent law.

[33] Description adapted from *Wikipedia*, by replacement of the terms *business* and *economic*.

carries out the research work. Babbage describes one of the earliest such cases found in astronomy: the *D'Angos affair*. Jean Auguste D'Angos (1744–1833) claimed to have discovered a comet on April 11th 1784, and sent the elements of its orbit to Charles Messier (1730–1817) in Paris (Jones 1991). In 1786 D'Angos submitted 14 "observed" positions of "his" comet for publication in the *Leipziger Magazin für reine und angewandte Mathematik* (Hindenburg 1786). The short-lived journal *Correspondance Astronomique, Géographique, Hydrographique et Statistique*, edited by Franz Xaver von Zach (1754–1832) published in 1820 a paper by J. F. Encke, entitled *"Imposture astronomique grossière du chevalier D'Angos devoilée"*.[34] Encke (1820) explains, in a very systematic way, all the problems and shortcomings of D'Angos' data, *viz.*:

- Charles Messier was never able to observe this comet;
- the comet was very faint, nevertheless he had observed it at magnification 300;
- the reported positions were ecliptic longitude and latitude, instead of the usually provided observed equatorial coordinates;
- none of these positional measurements fitted the orbit given by D'Angos;
- the residual errors (observed minus calculated) were inconsistent;
- D'Angos, when asked for, refused to provide the raw data, and
- his original orbital elements fitted the observed positions only if the comet's distance to the Sun was multiplied by a factor of 10, what showed that D'Angos had calculated – rather than observed – these cometary positions.

This is a case where replication is not possible, and where the only direct evidence is the fact that Messier could not confirm the existence of the comet. Encke also mentioned that this was not the first time that D'Angos had reported forged observations. Evidently, D'Angos knew the methods, though claimed extreme accuracy beyond the capabilities of the instrument, committed errors, refused to render the raw data, and destroyed his records. In short, his "data" were merely the result of backward-engineering from a model.

An interesting outcome of this matter was that one year after Encke brought to light the D'Angos forgery, the Hungarian astronomer Johann Pasquich observed a new comet, and sent his positional data and orbital elements to the newly-founded *Astronomische Nachrichten*. His assistant, however, accused Pasquich of fraud (Patkós 2004), and the accusal was eventually investigated by five famous astronomers (Gauss, Olbers, Bessel, Encke and Schumacher), who came to the conclusion that the accusation was unfounded, which led to the rehabilitation of Pasquich – see Schumacher (1824), and Gauss (1824).

A quite similar case, with the same outcome, concerns the alleged revision of observational data by the Jesuit astronomer Maximilian Hell (1720–1792) on the 1769 Venus transit that he observed from Vardœ (Norway). Hell delayed publishing his observations, and was later accused of withholding his data until all other data were published, with the intention of altering his own results. In 1835 Carl Ludwig von Littrow (1811–1877)

[34] *Flagrant astronomical fraud by knight D'Angos uncovered.*

published Hell's journal, and attacked Hell saying that he had adjusted his data to agree with those published by other observers (see Botez 2004). Half a century later, Hell was vindicated by Simon Newcomb, in an elaborate paper that demonstrated, step by step, that Littrow got it wrong (Newcomb 1883). In particular, Newcomb's investigation showed that Littrow was ill-placed to scrutinize the colored inks of the original manuscript, simply because he was color blind.

6.4 Trimming

Section 11 of Paper II (*Lying with graphs*) dealt with outliers. Babbage (1830) cites the opinion of Jean-Baptiste Joseph Delambre (1749–1822) that it is essential to publish all observations that have been made in an experiment, for two reasons: because it is important to know the aberrant data, and because *"The uttermost care cannot protect the most experienced observers* [against deviant data], *and he who would produce only perfectly-fitting angles, would have been remarkably well served by the circumstances, or would not have been sincere."*[35]

Trimming, according to Babbage's definition, *"consists in clipping off little bits here and there from those observations which differ most in excess from the mean, and in sticking them on to those which are too small"*, a manipulation that conserves the mean values, and is difficult to detect. Trimming, nowadays, is rather confined to (automatic or manual) clipping of data that differ most in excess from the mean, see also Section 11.4 in Paper II.

6.5 Cooking

Babbage (1830) describes cooking as *"an art of various forms, the object of which is to give to ordinary observations the appearance and character of those of the highest degree of accuracy. One of its numerous processes is to make multitudes of observations, and out of these to select those only which agree, or very nearly agree"*. Cooking is the selection of only those data that achieve agreement, and is often the consequence of pressure by a preconceived hypothesis or theory, as Boring (1962) explains:

> To the observing scientist, hypothesis is both friend and enemy.

Cooking is much worse than trimming, because the probability that the outcome is plain wrong is higher than when clipping data. Cooking, just like trimming, is difficult to detect.

One likely case of cooking is the set of geographical latitude measurements of Edward Sabine (1788–1883), who conducted a series of experiments to determine the true figure of the Earth, about whose results Babbage said: *"On whatever subject Captain Sabine*

[35] Free translation of *"Mais, quel que soit le parti que l'on préfère, il me semble qu'on doit tout publier. Ces irrégularités mêmes sont des faits qu'il importe de connoître. Les soins les plus attentifs n'en sauroient préserver les observateurs les plus exercés, et celui qui ne produiroit que des angles toujours parfaitement d'accord, auroit été singulièrement bien servi par les circonstances ou ne seroit pas bien sincère"* (Delambre & Méchain 1806).

touched, the observations he published seemed by their accuracy to leave former observers at a distance".

Whether Babbage is right, or whether Sabine was honest, is not the issue here, the point is that "statistically too smooth" data always raise suspicion: **baffling accuracies are often a signature of fraud**. Figure 10 shows box-whisker plots for Sabine's 64 geographical-latitude measurements with a repeating circle of 6 inch diameter, and for 1318 observations with a 16-inch instrument from three other observers: although Sabine's body of data was only about 5% of his colleagues' database, and was acquired with an apparatus of barely half the size of the competitors' instrument, his internal scatter is much lower than the scatter of the others.

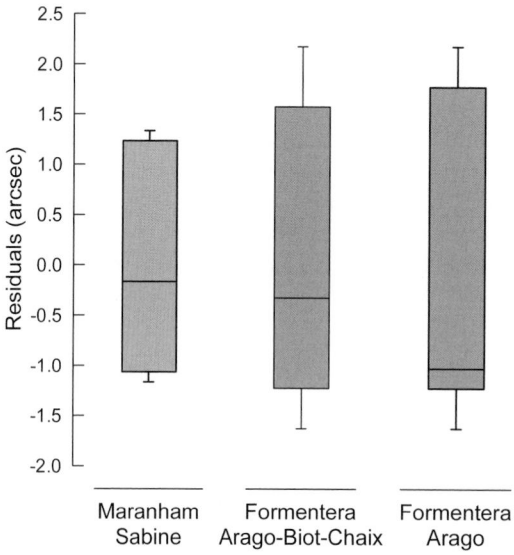

Fig. 10. Box-whisker plots for geographical latitude measurements with a 6-inch instrument at Maranham (Maranhao, Brazil), and with a 16-inch instrument at Formentera (Spain). The 66 Maranham data were collected by Edward Sabine, the leftmost Formentera box is based on 1318 observations by 3 observers, and the rightmost box on 860 measurements from one single observer (François Arago). Note that the median differs from the mean (which is 0 for all plots, since the data are differential with respect to the average). Based on data listed by Babbage (1830).

Captain Sabine was by no means the first to nudge his data, as shows the story of Robert Holmes, a 17th century English sea captain, who carried two pendulum clocks on a voyage to Africa. On their return trip, the four ships were running out of potable water. Three of the four captains had determined their longitude with the classical calculations and differed by more than 100 leagues[36] from the position based on the pendulum clocks, what led Captain Holmes to trust that they were only some 30 leagues distant from the Isle of Fuego (Holmes 1666). Indeed, the ship's position turned out to be exactly right, although the inventor of the clocks, Christiaan Huygens, had his doubts and asked the Royal Society: *"I have to confess that I had not expected such a spectacular result from these clocks. To give me ultimate satisfaction, I beg you to tell me what you and your colleagues at the Royal Society think of this Relation, and if the said Captain seems a sincere man whom one can absolutely trust. For it must be said that I am amazed that the*

[36]A nautical league was about 5.6 km.

clocks were sufficiently accurate to allow him by their means to find such a tiny island".[37] Jardine (2010) checked the ship's journal, and found that Holmes had fudged his observation log, so it was, again, the too-precise nature of the match between his calculated position and the geographical location, that led his peers – including the inventor of the instrument – to raise suspicion. In both cases **the cheating was revealed not by the data themselves, but by the statistical signatures of the observational errors.** We can only guess about the motives for these actions, though it is not proven that the motivation – at least in Sabine's case – was deliberate cheating.

The opposite also occurs: increasing error estimates, in order to be safe – a safety that would be illusory, since a crucial discrepancy, which might otherwise reveal some basically important new fundamental fact, may consequently get buried and lost (Franklin 1989).

The only way to eliminate forging, trimming and cooking altogether, is to make all crucial data available along with the publication that brings out the analysis and the results.

6.6 Data manipulation

Feynman (1974) gives one example of how all of us can fool ourselves: Robert Millikan and Harvey Fletcher (see Sect. 12) measured the electric charge carried by a single electron with the famous oil-drop experiment in 1909–1910, and obtained $e = 4.774 \pm 0.009 \times 10^{-10}$ esu[38] (Millikan 1913). This elementary charge is a fundamental physical constant that equals $1.602176487 \times 10^{-19}$ coulomb, or $4.80320427 \pm 0.00000012 \times 10^{-10}$ esu. Millikan's result, though very *precise*, was not *accurate*, and was too small by about 0.6% – partly because he used too small a value for the viscosity of air.[39]

Feynman points out that it is interesting to look at the history of measurements of the charge of the electron after Millikan: scientists systematically obtained values for e above Millikan's, and

> THEY THOUGHT SOMETHING MUST BE WRONG – AND THEY WOULD LOOK FOR AND FIND A REASON WHY SOMETHING MIGHT BE WRONG. WHEN THEY GOT A NUMBER CLOSER TO MILLIKAN'S VALUE THEY DIDN'T LOOK SO HARD. AND SO THEY ELIMINATED THE NUMBERS THAT WERE TOO FAR OFF.

In other words, it is the scientific society at large that seems very susceptible to being directed by this unscientific approach to scientific experiment, and its outcome. There is certainly no suggestion of fraud here – the involved phenomenon is called a *bandwagon effect* by Franklin (1989): **although each measurement was honestly made, they were, except for the first, conducted in the light of previous results.** A most recent astronomical example of the bandwagon effect is documented by Schaefer (2008), *viz.*, concerning the dichotomy between independent measurements of the distance of the Large Magellanic Cloud before and after the *Hubble Space Telescope Key Project* (see Fig. 44 in

[37] Letter 1325, Christiaan Huygens *Œuvres Complètes* Vol. 5. Translation from Jardine (2010).

[38] esu: electrostatic units, or statcoulomb in the cgs system of units

[39] That, as he points out, enters his calculation with a power of $\frac{3}{2}$.

Paper II) adopted a distance modulus of 18.50 ± 0.1 in 2001.[40] Schaefer alerts workers in the field that 31 independent papers after 2002 yield distances that cluster *too* tightly around the HST-based value, and sees strong evidence for the presence of a bandwagon effect.

Rosenthal (1976) calls attention to a milder form of deceit, *viz.*, the phenomenon that students conducting laboratory experiments were found eager to deliver a "better" report, where better is defined in terms of conforming to the outcome expected by the supervisor. There is only a small step from this situation to service observing in astronomy, where the observer delivers the data, the receiver sometimes not knowing how the data were obtained, especially if no control or calibration data are provided. I have witnessed a situation where an observer had not opened the dome during two consecutive nights because of poor (non-photometric) weather, and after having been spurred by the remote supervisor, stated *"if they want data, they will get data"* and consequently opened the dome and carried out the measurement, despite the fact that the sky was even worse than on the preceding nights. Though such behavior cannot be classified as data cooking, because its effect is opposite (errors will augment), it is nevertheless unacceptable. This example shows that also modern instruments not always eliminate observer effects.

Babbage (1830) explains the steps required for proper handling of experimental data:

1. the first step in the use of every instrument, is to find the limits within which its user can measure the *same object under the same circumstances*;
2. it is only from a knowledge of these limits, that one can have confidence in the measures of the *same object under different circumstances*;
3. and after that, of *different objects under different circumstances*.

This boils down to knowing the instrument, to analyzing the control data in exactly the same way as the science data, and to describing the control data with the same rigor as the scientific data are reported – see also Feynman's opinion formulated on page 181.

6.7 Image manipulation

Charles Percy Snow (1905–1980) was a research physicist who later turned to administration and writing. He is the author of *The Two Cultures and the Scientific Revolution* (Snow 1959), in which he describes the culture of the scientific world and the culture of the artistic world. In his (less known) novel *The Affair* (Snow 1960), he describes what happened to a young Cambridge science Fellow who is charged with scientific fraud – in fact, image manipulation – and who was eventually dismissed from his college. I refer back to this case on page 233.

Paper II discussed deception – mostly inadvertent – by improper image design and construction. Image manipulation is defined as the *altering of a taken image*.[41] Image alterations, in astronomical research, include misrepresentation, brightness and contrast adjustments, accentuating or diminishing parts of the image, and cleaning up backgrounds.

[40] I am indebted to Dr. Laszlo Szabados for drawing my attention to this case.

[41] Another misconduct, copying images without revealing the source, belongs to the discussion on plagiarism.

Astronomy, moreover, differs from other sciences (like biology) in the sense that it is generally not possible to duplicate the experiment only for the sake of getting a better image.

Beware, though, that many image alterations are quite legitimate, and are considered fraudulent only when the manipulation is not documented. Rossner & Yamada (2004) underline that it is important to understand where the ethical lines are drawn between acceptable and unacceptable image adjustment, and they provide some general guidelines for the proper handling of digital images. The severity of a manipulation depends on **"whether the alteration deliberately changes the interpretation of the data"**. CCD images, for example, undergo substantial basic transformations, including cleaning backgrounds of unwanted signals, such as cosmic-ray hits. Automatic adjustments to image data (applied to the complete image) are difficult to trace, but manual changes on the pixel level can be revealed a posteriori by applying relatively simple algorithms. The same is the case with rescaling, *i.e.*, bringing sub-images to the same resolution through interpolation. Such transformations change the data content, and should be documented in the publications. Rossner & Yamada (2004) conclude

> For every adjustment that you make to a digital image, it is important to ask yourself, 'Is the image that results from this adjustment still an accurate representation of the original data?' If the answer to this question is 'no,' your actions may be construed as misconduct.

This is very true for many astronomical images where the touch up was driven by the desire to present aesthetically pleasing pictures.

Parrish & Noonan (2009) present several cases of misconduct associated with image handling, as well as some methods for detecting such manipulations. Image manipulation is mostly revealed by coauthors, grant application reviewers, alert referees, meeting attendees, and journal Editors. Applying image forensic software may one day become as regular as applying plagiarism detection software is today.

7 Author misconduct

7.1 Lying about publication status

Lying about publication status – for example, referring to a submitted paper as an accepted one – is mainly a problem of project reports and of grant applications, but it also happens in research papers. This is not a good practice, because you never know that the person who reads such false statements may be involved in the refereeing of the reference that is alluded too. A very common mistake that occurs in grant reports is "selling" a non-refereed paper as a refereed publication: evaluation panel members easily spot such "errors", with often fatal consequences. More generally, lying to your peers (supervisor, examiner or referee) is never a good idea, because it is very difficult to invent "clever" lies, not to speak of the risk that lies will cast doubt on about everything else that is presented in your work.

7.2 Cascade submissions

Sending a manuscript from one journal to the next, descending the impact-factor ladder until the paper – after many iterations and input from unknown referees – gets accepted after all, is an unacceptable practice. The system leads to a tremendous overhead, because researchers spend their time trying to publish their product, rather than moving on to the next research topic (see the advice in Sect. 5.3). Not to speak of the loss of time by referees and Editors: in one such case I became referee for the second time – after seven months during which the paper had descended from the top level shown in Figure 1 to below the bottom line.

7.3 Omitting author names

Denial of authorship, unfortunately, occurs on a regular basis. Arguments for omission are either not given by the principal authors, or some kind of flawed market rhetorics is used to "prove" why this or that author should be off the list. The best approach in authorship negotiations is to demonstrate to which part of the work you have actively contributed. Vice versa, when an authorship claimant asks you to be included, then gently ask that person to indicate what was his or her contribution to the paper. This system, of course, does not work in teams applying reciprocity in some kind of gift economy, see Paper I for more details.

Names of support staff (data support, technicians, night assistants and telescope operators, etc.) are usually listed in the acknowledgments, but when their impact is significant, coauthorship must be considered (see Sect. 9.10 in Paper I).

7.4 Incorrect authorship order

Rules and guidelines on authorship order are discussed in Paper I. Authorship order is a political (and also a cultural) aspect, and reflects the power game inborn in the conduct of science (Scott 1997). In a survey to assess readers' perceptions of authors' roles based on authorship order,[42] Bhandari *et al.* (2003) noted a 90% level of **endorsement of the proposal to rank the order of authorship according to the amount of work done** – that includes the writing of the manuscript – whereas 80% disagreed that authorship should be based on seniority. Furthermore, there is large variability in the interpretation of the roles of the second, third, and fourth authors. One respondent provided the comment

> WHATEVER THE ORDER, THE SIGNIFICANCE OF THE ORDER SHOULD BE UNDERSTOOD BY ALL. IT IS NOT THE ORDER BUT THE MEANING OF THAT ORDER THAT IS CRITICAL.

In fact, what that person stresses is the need to **understand** the order of the names in the authorline in the same vein as the reader wants to understand the scientific content of the paper. In other words: **the list of authors, and the ranking of the names, is part of the information content of the paper.**

[42] A fictitious study title with fictitious authors was created for this investigation.

7.5 Adding noncontributing authors

Another form of dishonesty is to submit a paper coauthored by one or more senior scholars without their knowledge: this is what happened in a case I refereed, and where the manuscript and the science were of such low level that letting it pass would have been an insult to these innocent "coauthors". Why do people do such things? I can only guess, but I think the submitting author expects that the referee might be a colleague and personal friend of the ignorant senior coauthors, and thus hopes for a lenient treatment of the manuscript. Leading journals have strong policies on this issue, and even require the corresponding author to declare that all coauthors agree with the submitted and all subsequent revised versions of the manuscript. The journal *Science* states that

> ALL AUTHORS MUST AGREE TO BE SO LISTED AND MUST HAVE SEEN AND APPROVED THE MANUSCRIPT, ITS CONTENT, AND ITS SUBMISSION TO SCIENCE. SCIENCE WILL SEND AN EMAIL TO ALL AUTHORS TO CONFIRM RECEIPT OF EACH PAPER. SUBMISSION OF A PAPER THAT HAS NOT BEEN APPROVED BY ALL AUTHORS MAY RESULT IN IMMEDIATE REJECTION WITHOUT APPEAL.[43]

Submission to a *Nature* journal is taken by the journal to mean that **all the listed authors have agreed on all of the contents**, and that the corresponding author is responsible for having ensured that this agreement has been reached.[44] Some journals make such a declaration obligatory through the online submission system: submission is simply not possible as long as that statement is not flagged. I refer to Section 7.5 for additional thoughts on the problems associated with honorary authorship.

Including coauthors without their consent could even lead to legal prosecution, especially when you include someone else's name in an application for a grant or an observing proposal, and it is very unpleasant – if not humiliating – for the duped to find out to have been unwittingly dragged before a peer-review panel with a proposal of low standards.

7.6 Refusal to accept responsibility

Rennie *et al.* (2000) mention authors' refusal to accept responsibility for their articles, despite their ready acceptance of credit. The problem increasingly occurs in papers signed by dozens of coauthors, and is not always the consequence of ill intent or dishonesty. Part of the phenomenon may be attributed to the mere psychology of team work, *viz.*, the *groupthink syndrome* as described by Janis (1972): the dynamics of large groups, and the power it has in swaying opinions of individual members of a team, leading to a decrease in mental alertness of the members of the group. This author identifies several symptoms that I also see operational in the writing process in those cases where more than a handful of authors are involved, *viz.*:

- an illusion of excessive optimism, shared by most or all the members;
- an unquestioned belief in the group's inherent morality;

[43] `http://www.sciencemag.org/about/authors/prep/gen_info.dtl#authorship`
[44] `http://www.nature.com/authors/editorial_policies/authorship.html`

- reduction of self-censorship with respect to the apparent group consensus;
- direct pressure on any member who expresses strong arguments against any of the group's stereotypes, and
- a shared illusion of unanimity concerning judgments conforming to the majority view, augmented by the false assumption that silence means consent.

When these symptoms are present, the team members perform their collective tasks (*i.e.*, verifying each others contributions, proofreading, etc.) ineffectively, and this will most likely result in papers of lower standard.

7.7 Abusing the referee

Section 10.1 of Paper I referred to the attitude *"we have done enough now and any remaining errors will be flagged by the referee"*. This clearly is an abuse of the referee and of the refereeing system. Another form of ill-treatment of the referee is to accept valuable advice without even caring for a proper acknowledgment. The worst attitude, however, is systematic use of the referee only as a means to an end, *i.e.*, to get the paper published. It regularly occurs that upon reviewing a second or third paper of a same author or group of authors, the referee notices that the same type of fundamental error is repeated again and again. This looks much like the referee is seen as a throwaway advisor of one single moment, from whom nothing needs to be learned.

7.8 Misquoting, improper attribution and wrongful citation practice

Misquoting (see Paper I) is repeating something someone has said in a way that is not accurate. Deliberate misquoting is a frequently used tool in politics, even by the media, and contributes to inaccurate reporting and to the rewriting of history. Misquoting also occurs in scientific papers, and it is either intentional or inadvertent. Misquoting of the first kind is an unethical act. There are basically two situations where inadvertent misquoting may occur: citing out of context (happens when you only read the abstracts, and quote from this limited information), and when quoting in translation (it is, therefore, good practice to include the original quote in a footnote, together with your translation).

Improper attribution refers to assigning a discovery to someone else – including misappropriation to yourself. The latter form definitely is an unethical act, the former is sometimes linked to the so-called *Matthew effect* in science, see Section 18.5.

Wrongful citation practice is any unjust way of citing, and Paper I dealt with some of the consequences. The following list gives some of the most frequently applied practices:

- citing only those references that support your conclusions;
- claim that cited references support your conclusions, whereas they do not;[45]

[45] I have recently resigned as a referee because the authors refused to remove a reference that did not even touch on the data they presented.

- citing only papers of your buddies and close colleagues;
- citing all your own papers, but only the last one of your adversary;
- citing papers while having read only the abstracts;
- deliberate misreferencing of other people's work in a manner that magnifies one's own work (see Maddox 1993), and
- not mentioning that your previously published result has been disputed in the literature.

7.9 Hoaxes

Babbage (1830) describes an example of the publication of the discovery story of a new family of shellfish – of which the generic name was borrowed from the discoverer's surname – however, of a completely fake makeup. This hardly seems possible nowadays, although Philip Davis, a PhD student in scientific communications at Cornell University, and Kent Anderson, an Executive Director at the *New England Journal of Medicine*, submitted a fake manuscript to *The Open Information Science Journal* at the end of January 2009. The nonsensical article had been composed by a computer program, and the journal claimed that the manuscript had been peer reviewed, and even requested payment of USD 800 in fees (Grant 2009a). Another, though even more brazen, was Sokal's (1996) hoax article referred to in Paper I. As stated in Paper I, though hoax papers are written to expose misconduct, they are not appreciated, because the method of the protest violates the principle of mutual trust.

7.10 Copyright infringement and plagiarism

McSherry (2001) explains the difference between *copyright infringement* and *plagiarism*: these acts differ with respect to the nature of the violation. Copyright infringement, on the one hand, occurs whether or not there was any deliberate effort to deceive, simply because the copyright holder's (economic) interest has been damaged. Plagiarism, on the other hand, is directly concerned with intent, and is not a legal term (Stearns 1992). In short: **copyright is about money, plagiarism is about truth**.

Copyright infringement. There are two general confusions on copyright issues:

1. the concept of *public domain*, and
2. the question: which copyright law is applicable?

Google Books is very clear: *A public domain book is one that was never subject to Copyright or whose legal Copyright term has expired.* Whether a book is in the public domain may vary from country to country. The first US copyright law (1790), protected books, maps, and charts if they were created by residents or citizens of the United States. The term of their exclusive right was a mere 14 years, with the right of renewal for 14 more years (Berry 2000). The latest revision of the law gives authors who published after 1978 a copyright term of life plus 70 years, and authors who published earlier get a term of life plus 90 years.

A special form of copyright infringement is *dataright infringement*, in particular when a researcher leaves the lab or the research institute, and where the question is: which data may you consider to be truly yours? Unless the data are public domain, datarights reside in practice with the principal investigator or with the research-grant holder. Some thesis supervisors allow their postdoctoral fellows to take data along when they move to another campus, others do not.

What about giving data to someone, in good faith, and then finding out that this person does not fulfill the promises? Asking the data back makes no sense in our digital age, although copyright or user rights can be withdrawn at any time.

Plagiarism. This is the use of ideas or words of another as your own, without crediting the source.[46] Plagiarism does not necessarily affect the quality of scientific information – unless the plagiated information was wrong in the first place.

Although the term plagiarism is thus pretty well defined, it is often wrongly understood, and the perception of the act frequently depends on the context in which it occurs. The drama of this situation is very well portrayed in Skármeta's[47] (1985) novel, through the enamored but illiterate Mario, who tries to win the heart of his beloved Beatriz by offering her a poem that he plagiarized from the famous poet Pablo Neruda. When the latter reproaches him with plagiarism, Mario exclaims[48]

> POESY IS NOT THE PROPERTY OF WHO WROTE IT, BUT OF WHO USES IT!

Plagiarism – the act of wrongful copying – refers to writers and scholars, and involves three elements to be considered:

1. the amount of copying;
2. the attribution of authorship,[49] and
3. the intent.

Plagiarism is one of the most severe ethical problems encountered in scientific publishing, and it is also seen as an assault on an individual's academic rights: McSherry (2001) mentions a senior researcher who describes his experience with plagiarism as "intellectual rape".

That hard-core plagiarism also happens in astronomy, is illustrated by the following story. Several years ago I was asked to referee a paper that criticized a work by a deceased colleague. To my surprise, I found that 80% of the Abstract and of the Conclusions were copied *verbatim*[50] from the paper that was criticized, and several other paragraphs were literal copies as well. Even a Table was copied to an unacceptable degree, including the typos that the typesetter of the vintage paper had made. My advice, as referee, was *"The scientific content of this paper does not correspond to the level of quality expected by this*

[46] Note that unlawful use of copyrighted material is also called *piracy*.
[47] Antonio Skármeta (1940–) is a Chilean writer.
[48] Translation of *¡La poesia no es de quien la escribe, sino de quien la usa!*
[49] That is, hiding the relationship between original author and her/his creation
[50] *verbatim* means *word for word*.

journal. Publication of this paper will lead to legal problems with the copyright holder". Needless to say, the paper was rejected.

Scientific writing is a commitment to intellectual honesty. That means that – besides being honest – you are supposed to be honest with the data, report honestly, and cultivate an honest attitude of citing the sources that you use. Thus, if a copied passage is properly credited, it is not considered plagiarism, although it can be unlawful doing so without permission. Paper I discussed the so-called *Principle of Fair Use*, a privilege that allows users to copy without permission for non-commercial purposes of criticism, research and teaching. Thus, when citing a few sentences, provided you give due credit, no permission is needed. But the principle is vague and when in doubt, it is better to ask for permission.

Some allegations of plagiarism are difficult to trace, for example Luyten (1981) presents a tabular description of striking similarities between his (1974) paper in *Science* on the technical description of a measuring machine, and a paper published by his colleagues in 1980. It is also generally acknowledged that the refereeing system is not designed to systematically detect plagiarism, though some journals provide the referee with the submitted TEX files, and the structure of these files often provides a very clear insight in the way the file was constructed: when many paragraphs (and references) are flagged out, there is a great probability that the paper has already been submitted or published elsewhere. Especially the format of citations and references, and their ordering, can tell the reviewer where the paper was submitted or was published before.

Mapmakers, and other graphical artists, have a very original way to discover piracy, *viz.*, by adding deliberate falsification – for example, subtly placed trap streets[51] or other fictitious marks (Mommonier 1996, see also Sect. 11 in Paper II). Fortunately, we may expect dark times ahead for poachers of texts and ideas, since academic institutions now develop powerful software for cross-correlation of published material through a search for identical passages. A practical example of plagiarism hunting, based on the papers collected in these Proceedings, is given in Section 16.5.

A milder form of deceit is presenting a compilation of data or references as if it were your own work, whereas the list was simply copied or downloaded from another source. As a rule, **you should only list those references that you have actually consulted**, not copy lists from other places. Not only because it is not fair, but because this form of copying perpetuates and propagates bibliographic transcription errors that are always present in such lists (see Paper I). Such errors contribute to confusion, and to bibliometric bias, as explained in Section 18.

So-called *innocent plagiarism* occurs when an author reports "new" facts or discoveries that were already published by others (this also includes the phenomenon of *rediscovery*). Finding out, at the end of the road, or even during the publication process, that the fruits of your labor are not really new, but have been obtained already by someone else, is a very sad experience. If you find yourself in such a painful position, then the best solution is to explain the situation to the reader. I was recently involved in such a case, where

[51] Fake streets on a map, with the aim to trap copyright violators.

the first author – in response to doubts formulated by a referee – elaborated a fresh statistical approach which, upon reading, I found that a mathematically much more detailed method had already been published in 2009. A similar situation occurred to Lombard & Koen (1993), in a sequel paper to Koen & Lombard (1993), who solved the problem very correctly by stating *"After completion of this paper, the authors became aware of much earlier work (Sterne 1934) dealing with the same topic. Sterne's paper appears to have been largely overlooked ... "*.

Another type of involuntarily plagiarism is the so-called phenomenon of *cryptomnesy*, that refers to remembering a message (idea, concept, title, etc.) while forgetting its source: a mixture of remembering something very well, and of totally forgetting something else. Draaisma (2005), in his college inaugural lecture, underlines that no one is immune to cryptomnesy. The phenomenon is the opposite of *tacit knowledge* and *explicit knowledge*, *i.e.*, knowledge that is very widespread, such as, for example, the statement that the orbit of a binary is elliptical, which can be made without referring to Kepler's original paper.

7.11 Self-plagiarism, text recycling and paraphrasing

A very special form of plagiarism is *self-plagiarism*, also called *autoplagiarism* – that is, re-using sentences and complete paragraphs from your own published works, even redundant duplicate publications. Text recycling is sometimes unavoidable, for example when describing telescope and detector characteristics, algorithms, and software in series of papers.

The question *"how can you steal from yourself?"* is often asked. True, you cannot steal from yourself, therefore, self-plagiarism may seem a smaller offense than stealing from someone else. But when authors cite text from their own previously published works without mentioning or citing it, it misleads the reader, and in some sense, takes from the scientific community, see the list of consequences in the next Section. And what if a honorary coauthor reproduces text from a source that he or she gratuitously coauthored? In principle, it is plagiarism, in practice it is autoplagiarism, but formally there is nothing wrong – that is, **as long as the original source is quoted**.

Paraphrasing is expressing the same idea with other words – an action that is not evident, because some statements are already so well expressed that it is just very hard to modify a few words without losing some of the original meaning. Babbage (1830), for example, explains why he did not paraphrase, but instead kept to the original wording: *"it is impossible, even for the most candid, in all cases, to convey precisely the same sentiment in different language ... and I thought it therefore more fair towards those from whom I differed, as well as to the public, to give the precise words."* So, paraphrasing is ineffectual, and it is unfair both in taking the content of a message and in expressing it differently (which may even lead to misquoting). On page 243, I list Gunsalus' (1998) "six rules and eight steps" in a summary format that does not match the source word for word, though it contains verbatim copies of some headlines: such paraphrasing is entirely supported by a reference to the source, and does not fall under the denominator plagiarism.

7.12 Dual and redundant publications

Dual publication is the publication of the same article in more than one journal. There is some controversy over the question if such duplication is acceptable, even when the source – yourself – is cited. Plain duplicate papers create confusion, see the discussion on the duplicate publication of Belton *et al.* (2007) in Paper II. Not to speak of the bias introduced in the citation count and in its secondary bibliometric indices when papers are counted twice, or when citations are not counted because the "wrong" version (*e.g.*, in the case of Belton *et al.* 2007) is cited. Millikan (1910, 1911) is a typical case of duplication that is quite common nowadays: his first paper was a 12-page version in *Science*, the second paper was a 49-page elaborate account of the complete oil-drop experiment.

Journal editorial instructions usually mention that manuscripts submitted for publication should not be submitted to any other journal; the Editors also expect to be informed when a submitted manuscript has previously been rejected by another journal. Simultaneous submission of the same paper to multiple journals is not acceptable. Kennedy's (1985) safe rule is:

> IF IT'S BEEN IN PRINT, AND ESPECIALLY IF IT'S BEEN PEER REVIEWED, IT SHOULDN'T BE PUT IN PRINT AGAIN.[52]

I would say that plain unnecessary duplication should be avoided, but sometimes mild repetition may be useful from the educational point of view: this book, for example, contains several passages and graphs that were previously published in a scientific context, and are reproduced here (with permission) for didactical emphasis.

In 1969, the *New England Journal of Medicine* instituted the *Ingelfinger rule*, an embargo designed to keep scientific findings out of the media until peer-reviewed and published in the journal. The rule's founder and namesake, the journal's Editor Franz Joseph Ingelfinger (1910–1980), stated that no article he printed should have been published elsewhere beforehand (Toy 2002).[53]

If duplication of content is not matched by complete duplication of the author list, then the problem of assignment of responsibility arises. Broad (1981) touches on that question, and mentions the case of a paper authored by four researchers that appeared, during the same month, in two separate journals – see Potkin *et al.* (1978) and Berger *et al.* (1978). However, despite the fact that both articles presented opposite conclusions, the papers had two authors in common!

What about duplicate or dual publication of the same content in different languages? Wen & Gao (2007) published the viewpoint that *"Instead of being forbidden, it*[54] *should be encouraged so as to maximize the effectiveness of academic communication and equalize the rights of creating, distributing and accessing knowledge"*. These authors do not consider such action as a violation of academic ethics, because they maintain that papers

[52] *In print*, obviously, also includes electronic publication.

[53] And, furthermore, that a scientific claim should always be verified before dissemination to the public. Thus, the rule prohibited scientific authors from releasing data to the mass media before the journal's date of publication, a rule that is still honored by some high-impact journals.

[54] *i.e.*, to submit a paper in English for international journal publication, based on a published paper in Chinese.

in different languages are not duplicate publications, because different languages involve different norms of writing. In other words: not translating papers is viewed as leading to academic inequality and to linguistic unfairness.

The most pertinent question in this discussion is: *who translates?* A translator uneducated in the narrow scientific domain of the scientific text, unavoidably, will change content. Unless the translation is done by a scientist, but that can only be done by someone well-trained in the specific research field of the source paper. And how true can such translation ever be?[55]

I have another objection: producing duplicates takes time and effort, and this activity may appear, psychologically, as a research substitute, a form of sublimation of non-creative energy leading to apparent bonus and profit in bibliometric counts, at the cost of innovative research.

As already said above, duplication is acceptable, to some extent, if the duplicated material appears in a different context, for example research graphs and tabular material serving a second life in educational texts. Duplication stands on the precondition of truthful citing and quoting: always mention the original source of Figures, never reproduce a copy published by someone else, and never ever refer to one of your former published copies as the original source of any graphical information that was copied or modified, see Paper II. In summary, paper duplication should be avoided because:

1. it leads to inflation of the literature;
2. it biases the publication record;
3. it steals time from the composing author(s);
4. it steals time from the referees;
5. it forces researchers to read more;
6. money is lost in page charges;
7. more librarian's money is spent on subscriptions;
8. journal space is limited;[56]
9. it entails copyright issues;
10. it leads to author disputes if not all authors are included in the duplicate, and
11. it biases the impact factor of journals.

8 Referee misconduct

Let me, again, give a warning at the beginning of this Section: referee misconduct, no doubt, exists in many forms. Nevertheless, the helping role of referees cannot be overestimated, as Maddox (1984) remarks:

> ALL JOURNALS KNOW OF PAPERS THAT HAVE BEEN ENORMOUSLY IMPROVED BY THE CONTRIBUTIONS MADE BY REFEREES.

[55] Paper I discusses such a case: the problem of Eddington's incomplete translation of Lemaître (1927).
[56] Some journals' annual number of pages is contractually fixed.

Referees are no more no less than *gatekeepers*, assisting the Editors keeping the system going by helping them to determine what gets published. Referees are, of course, authors wearing a different hat (Lawrence 2003).

From my experience, I have the impression that students are not really familiar with the role of the referee, and that the referee is quite often confused with the Editor. A scientific referee is a person who reviews a paper and recommends that it should or should not be published, and so helps separate the wheat from the chaff. One extreme view is to see the referee as a censor, one who deletes harmful material; the opposite view is to see the referee as a kind of patient corrector, an invisible person who corrects and remedies poorly-written manuscripts.

The exact role of the referee is expressed in Feynman's *"you are the easiest person to fool"*, which means that when you have an idea, you yourself do not have all the means to examine it, but when someone else formulates an idea or hypothesis, then you – as outsider – have supplementary prospects to probe these views. Moreover, guesswork and checkwork, or the distinction between the creative and the critical sides of thinking, is not obvious – and these modes of thinking are often quite unequally developed in one single person (Medawar 1990). Moreover, any form of critical self-evaluation is, by definition, part of an internal circuitry, whereas a screening by an (independent) outsider is always part of an external process: here enters the referee.

Paper I described several aspects of refereeing. The peer-review system is entirely based on the assumption that all parties – *i.e.*, referee, editor and author – are truthful. In particular, confidentiality is key to the peer-review process (Heywood 2008). Smith (1997), on peer review, says

> WE KNOW THAT IT IS EXPENSIVE, SLOW, PRONE TO BIAS, OPEN TO ABUSE, POSSIBLY ANTI-INNOVATORY, AND UNABLE TO DETECT FRAUD.

Some of his points are sometimes, but not always, true. Wilson (2002) assigns the most serious problem with the review system to **neglect of duty** by

1. Editors who refuse to take responsibility, preferring to operate by majority vote of the referees – thus, Editors that do not judge the referees, and

2. referees who do not carefully read nor evaluate the manuscripts, leaving referee work to the Editors.

It is very important to learn to understand how the refereeing system works, and how to handle it for your own profit. And it is crucial to acquire the skill of coping with the good and the bad sides of the refereeing system, and to refrain from abusing the system. The following Sections deal with some problems related to wrongful conduct of referees.

8.1 Breaching confidentiality

Passing a manuscript under review to a colleague – or to a student, see the case described on page 215 – is a flagrant violation of the principle of confidentiality, and is totally unacceptable.

8.2 Hasty or superficial revision

Scientists work under the pressure of time, and this time pressure may be the reason why some referee reports show signs of hasty composition. Smith (1997) cites an interesting experiment, where 8 deliberate errors were inserted in a paper about to be published in the *British Medical Journal*. The manuscript was sent to 420 potential reviewers, of whom 221 responded. 16% of them did not spot any of these deliberate errors, and nobody spotted more than 5 errors, with a median value of 2. Hasty and superficial revision is the more appalling in the case of manuscript rejection without thoughtful reasons.

8.3 Long revision times

This refers to misappropriation of priority by delaying the publication in order to achieve personal profit. An example is Hamilton (case F in the discussion on page 190), who submitted an important paper to a biology journal, where the referee demanded major revisions, even splitting the paper in two companion papers, which took Hamilton nine months to fix. Meanwhile, the referee submitted a quick-fix three-pages paper to *Nature*, in which he created a new catch-phrase for the very process that Hamilton had previously outlined, including the main results of the paper in revision (Dugatkin 2006). So, the referee had slowed down Hamilton's paper, meanwhile accelerating his own opus – a clear case of conflict of interest (see Sect. 8.7). I have seen the same happen in astronomy, though only once.

8.4 Professional arm wrestling and taking revenge

It happens that referees engage in some kind of professional arm wrestling only for the sake of wrestling – just like some attendees at meetings interrupt speakers seemingly just for the pleasure of interrupting. Such behavior by a referee is often accompanied by unreasonable demands. Obviously, wrestlers and avengers should be bypassed – by the Editors as well as by the authors. Lawrence (2003) advances another aspect:

> REVIEWERS ARE NOW BEING ASKED TO JUDGE WHETHER A MANUSCRIPT CONSTITUTES A *Science* PAPER – IS IT SUFFICIENTLY EXCITING TO INTEREST THE 'GENERAL READER'? THIS PARTICIPATION IN EDITORIAL DECISIONS GIVES REVIEWERS OPPORTUNITIES TO PUNISH AUTHORS THEY DO NOT LIKE, SETTLE OLD SCORES AND HOLD UP COMPETITORS.

The same author illustrates the situation by mentioning a (fictitious?) case where one referee is furious that his student's article was rejected by one journal, and takes revenge on similar work that he is reviewing for a competing journal.

8.5 Silencing dissent

Peer pressure seems quite effective in silencing dissent (Hetherington 1983) and in supporting dogmatism: as defined by Öpik (1977), dogmatism is the refusal (by adherents) to consider alternatives to their own doctrines, and the rejection of criticism beforehand. Barber (1974) illustrates the issue of biases within an accepted paradigm – but also the biases that are inherent to the paradigm itself – by referring to Mitroff (1974), who interviewed 42 lunar scientists, asking questions such as, *"Do scientists have to be committed to their ideas?"* and, *"Is commitment a threat to objectivity?"*. Mitroff reported[57]

> OF THE 42 SCIENTISTS INTERVIEWED, EVERY ONE INDICATED THAT HE THOUGHT THE NOTION OF THE PURELY OBJECTIVE, UNCOMMITTED SCIENTIST WAS NAIVE ... TO THE CREDIT OF THESE SCIENTISTS, THEY NOT ONLY FREELY ACKNOWLEDGED THEIR BIASES BUT ALSO ARGUED THAT IN ORDER TO BE A GOOD SCIENTIST, ONE HAD TO HAVE BIASES. THE BEST SCIENTIST, THEY SAID, NOT ONLY HAS POINTS OF VIEW BUT ALSO DEFENDS THEM WITH GUSTO. THEIR CONCEPT OF A SCIENTIST DID NOT IMPLY THAT HE WOULD CHEAT BY MAKING UP EXPERIMENTAL DATA OR FALSIFYING IT; RATHER HE DOES EVERYTHING IN HIS POWER TO DEFEND HIS PET HYPOTHESES AGAINST EARLY AND PERHAPS UNWARRANTED DEATH CAUSED BY THE INTRODUCTION OF FLUKE DATA. THE OBJECTIVITY OF SCIENCE IS A RESULT, THE SCIENTISTS SAID, NOT OF EACH INDIVIDUAL SCIENTIST'S UNBIASED OUTLOOK, BUT OF THE SCIENTIFIC COMMUNITY'S EXAMINATION AND DEBATE OVER THE MERITS OF RESPECTIVE BIASES.

8.6 Anonymity and its effects

Referees either operate anonymously, or waive anonymity by signing their reports, see Table 2 for the different types of open and closed reviews. A&A uses the two modes listed in the first column (depending on the wish of the referee). Editors should pay close and continuous attention that the referee's wish for anonymity is respected, and make sure that the identity of the reviewer is not revealed by the metadata included in the file properties of the referee report document.[58]

Table 2. Open and closed referee or peer review systems.

	Referee knows author names	Author names hidden from referee
Authors know referee's name	OPEN REVIEW	—
Referee's name hidden	ANONYMOUS REVIEW	BLIND REVIEW

Babbage (1830), referring to some anonymous attacks on the conduct of various scientific bodies and of their officers, concludes:

> IF A FACT IS TO BE ESTABLISHED BY TESTIMONY, ANONYMOUS ASSERTION IS OF NO VALUE; IF IT CAN BE PROVED, BY EVIDENCE TO WHICH THE PUBLIC HAVE ACCESS, IT IS OF NO CONSEQUENCE (FOR THE CAUSE OF TRUTH) WHO PRODUCES IT.

[57] Mitroff's paper is not available online, so I quote from Barber (1974).

[58] This problem also existed in the past: some referees were not aware that the paper on which they typed their unsigned report carried the watermark of their university (Duerbeck 2011).

Öpik (1977) is even stronger in his opinion: *"Anonymity in refereeing is like kicking somebody in the dark, without a chance of response; it 'protects' the reviewer but not the author. The sooner this scourge of anonymity is abandoned, the better for the honest pursuit of research. If fewer referees can be found when there is no anonymity, it will be only to an advantage: those who consent will be a more qualified selection for the job of critics."*

It is a fact that anonymity leaves more room for cheating and dishonesty – including impoliteness – from the side of the referee: sheer anonymity simply gives a faceless person a different kind of power.[59] On the other hand, namelessness does not mean unconditional protection: it has occurred that Editors first offer anonymity, and then pull it away – not to speak of editorial archives that, one day or another, become publicly accessible. Anonymity certainly leads to most devastating consequences in conflict-of-interest situations, see Section 8.7. From my own experience, I always found it essential to sign my referee reports, although a couple of times this has led to tensions with colleagues who do not understand the philosophy of the process. I only request anonymity when I find signs of misconduct, or when the paper is an absolute and irreparable disaster.

Lawrence (2003) comments on an aspect of anonymity that is not often mentioned:

> UNLIKE THE AUTHORS WHOSE NAMES ARE OUT THERE, ANONYMOUS REVIEWERS WILL NOT BE HELD TO ACCOUNT IF THEY MAKE A MISTAKE.

It is certainly easier for author-referee communication if both sides know who is who.

8.7 Conflicts of interest

A *conflict of interest* is a conflict between the private interests and the official responsibilities of a person in a position of trust.[60] As said before: trust is the keyword in the peer review process and in the subsequent editorial decisions. Conflict of interest occurs when the reviewer possibly can gain or lose from the decision whether or not – and when – to accept a paper for publication.

Lawrence (2003) states that roughly half of the submission letters he receives request him not to use certain reviewers, often because of conflicts of interest, and because of fear of misuse of the information in the paper – although he thinks that this request is sometimes a ploy to avoid sharp-eyed and critical referees.

The *British Medical Journal* guidance for reviewers explicitly states that their Editors should know about any competing interests that reviewers may have, so they ask reviewers if they have accepted any reimbursement for attending a symposium, a fee for speaking, or research funds and fees for consulting, from an organization that may in any way gain or lose financially from the publication of the paper under review. They also ask if the reviewer has been employed by an organization that may in any way gain or lose financially from the publication of this paper, or any other competing financial interests. In addition, they ask authors to reveal close relationships with, or a strong antipathy to,

[59] As can be seen on many internet forums that have become havens for nastiness.

[60] *Merriam–Webster Collegiate Dictionary.*

a person whose interests may be affected by publication of the paper under review.[61] In astronomy there are few such occasions where a referee may have a financial stake or a relationship with commercial enterprises, hence conflicts of interest are mostly entirely of academic character, *i.e.*, a matter of scientific credit and priority.

As a matter of illustration, I copy a mail that I recently sent to the Editor of a high-impact astronomy journal: *"I would certainly be willing to review this manuscript, but I hesitate because of a possible conflict of interest: I have observed that star at half a dozen occasions ... but have not analyzed the data yet"*, and the answer *"No, I don't really think this is a problem, as long as you feel that you can be impartial"*. That is exactly the issue: **to remain impartial**. But since a referee is supposed to be an expert in the matter under review, the line between interest and conflict of interest may be quite thin at times. For example, some journals allow the referee to oblige authors to release their data, so when the journal's data policy is driven by the referees, the Editor must exercise great care to preserve the integrity of the complete editorial process.

I finish this Section with an account of the dreadful consequences of a conflict of interest that I personally lived through in the early nineties, when adequate electronic mail facilities for sending documents were still nonexistent. A PhD student from a foreign university received a grant from my government, and came to my institute for a three-months working visit, with the intention to compose a paper based on data collected by three observers: myself, the student, and a senior colleague from still another country (hereafter A, B and C, respectively). The first (and also a revised second) version of the manuscript, authored ABC, was composed by the time the student returned to his home institute. Three months later, B offered to include some calculations from his PhD thesis stating *"the result is not new, and I am not going to publish it anywhere, except of a short note in the conference paper"*. To my great surprise, another three months later, B's PhD supervisor (hereafter called X) handed me two manuscripts: a data paper (authors ABC) describing the project but without any conclusions, and a science paper, authored BA. The manuscript had been retyped from the printout of the last version of the manuscript on which we had worked together for several months.

I was kindly told that the main reason why I deserved coauthorship for the BA-authored paper was for all the time and energy I had invested in financial support and hospitality of the student. Of course I disagreed, not in the least because of my deep regret for all the time and energy I had put in the composition and in the typesetting of texts and Tables in the original version, because of the total absence of any respect for the division of labor, and for reducing my text to a pure datapaper, while my analysis and discussions had been dumped – yes, the terrible feeling so well expressed by Melanie Safka as referred to in Section 13.20 of Paper I. So, as requested, I submitted the ABC data paper, which was accepted in less than three weeks, and submitted my own analysis in a short paper authored by A only. Author B submitted within weeks to the same journal.

The first referee report of my paper came after about one month, and was very negative. I complied with all comments, even thanked the anonymous referee. After several

[61] `http://resources.bmj.com/bmj/authors/checklists-forms/competing-interests`

weeks of silence, I phoned the editorial office, and was told that it indeed rarely happens that a referee's second reply to such a short paper takes such a long time. Finally, two months after the first report, I received a second referee report, which was again quite negative, but that revealed two strange aspects. First of all, the English language was much less polished than was the case with the first report. Second, it contained the passage *"it is difficult to discuss about the thesis which is not widely available"*, referring to a reference that I had insisted to include. This very specific statement, and the syntax of that sentence, plainly confirmed that it was student B who was my anonymous referee, because we had discussed that issue before, and because he had been in a position to consult that very thesis work, of which I had been a jury member.

I only was puzzled why I had not recognized the situation at the moment I read the first referee report: the reason, simply, was because that report had been edited by the supervisor of B. On approaching X, he replied *"I never told you I didn't correct B's English in his first referee report"*. Thus, B had refereed a manuscript that he helped writing half a year before, attacking A with objections that were never raised during the six months preceding the debacle. And during this three-month refereeing process, B – as well as X – sent me several friendly emails on other matters we had been working on.

Needless to say, the collaboration with B and X was ruined forever. We had differences of opinion when writing the first version, but these were not profound: they were on the level of citing and (mis)quoting, on some verifications of numerical calculations, and on the question whether the model accounts for all the data to 100%, or whether some unaccounted residual noise remained in the data – differences of opinion that are quite natural in any scientific collaboration.

I waited till B received his PhD degree, and took the time to write him a letter that underlined the absurdity of such behavior inside a small active team as the one we had:

> BECAUSE IT DOES NOT WORK: A TWO- OR THREE PERSON TEAM CANNOT THRIVE ON SECRECY AND LACK OF OPENNESS, AND – ABOVE ALL – BECAUSE SMALL TEAMS ARE SUCCESSFUL MAINLY DUE TO THE COMPLEMENTARITY OF THE MEMBERS.

9 Grant-reviewer misconduct

There is some confusion in nomenclature: though reviewer and referee are often used as synonyms, I here use reviewer to refer to grant proposal or project reviewer, a very specific type of refereeing that is also called *peer review*. Peer review is even more of a black box than article refereeing.

Plagiarism during the review process of grants, is the absolutely worst form of research misconduct: it is the fourth type of violation of the standard of honesty, and is now formally part of the definition of FFP (*Fabrication, Falsification and Plagiarism*, see Kaiser 2004), and certainly is a fraud that is not as easily detectable by an alert referee as catching a dual submission. McSherry (2001), for example, mentions the case of one applicant who received a very negative grant review, and who had to review a proposal by a competitor later on: that proposal contained verbatim sections from his own original proposal. Thus, the reviewer first gave a negative evaluation, but then stole the ideas from the proposal, and used them for an application of his own. The victim, in that case, decided not to fight the case because

> YOU ONLY END UP SMELLING WHEN YOU GET IN A FIGHT WITH A SKUNK.

That desperate attitude of not speaking out is typical of victims of such actions, see my own attitude described on page 246, and also Section 17.2. Note that similar processes may also operate in telescope time allocation committees.

10 Editorial misconduct and editorial mismanagement

Editorial misconduct is much more difficult to detect than author misconduct: editorial accountability is a matter of considerable complexity, because Editors have lots of freedom – though that liberty is not unbound. As Marcovitch & Williamson (2003) put it:

> EDITORS ARE LIKE EMPERORS: THEY MAY SEEM TO HAVE ABSOLUTE POWER, BUT ARE ACTUALLY CONSTRAINED BY THE NEED TO ACCOMMODATE A HOST OF COMPETING INTERESTS, INCLUDING AUTHORS, READERS, OWNERS, LEARNED SOCIETIES, AND FINANCIAL CONTROLLERS. THE EDITOR MUST HAVE THE FINAL DECISION, BUT IDEALLY, HE OR SHE WILL SEEK ADVICE FROM WISE AND TRUSTED COLLEAGUES.

The principal aspects of editorial misconduct are related to financial conflicts of interest, *i.e.*, when Editors have a stake in their journal through the Publisher or owner of the journal (for example, when the publication is owned by a for-profit enterprise). Such conflicts of interest are very rare in astronomy, though academic conflict of interest or personal rivalry – as is the case with peer reviewers and referees – cannot be excluded. Tennant (2006) underlines that, whatever the business model of the journal (see also Sect. 16.3), **editorial integrity should be characterized by courage, openness and transparency**.

By no means all editorial mishaps can be classified as outright misconduct: there exists a wide spectrum of misdemeanor, going from high- to low-profile misconduct and misbehavior, encompassing non-performance, incompetence, even systematic suppression of undesired research. The subsequent Sections are not explicitly labeled in terms of misconduct, but are rather a list of problems related to editorial management in the broadest sense.

10.1 Non-responsive or inactive Editors

Inaction is understood as absence of adequate response to correspondence, especially to complaints – even failing to provide appropriate procedures for complaining or appealing against editorial decisions. Inaction may appear as low-level damage, although its power is substantial, considering that the harm is inflicted upon a large number of authors. Wrongdoing is thus not restricted to taking an action, but includes all aspects of "not acting at all".

Kozak (2010) exposed the situation in which an author submitted a manuscript to a journal, but no acknowledgment of submission was received. After two weeks, the author asks if the manuscript was received – no reply. After a month another inquiry – no reply. The same after two and four months. The author writes the Editor *"I don't know what to do: I cannot submit the manuscript elsewhere ... "*. This is not a fictitious case, but an example taken from real life. Of course, the author has only one way out: withdraw the manuscript.

Editors work under heavy time pressure: Henry Gee (Editor at *Nature*), for example, receives about 600 manuscripts per year to read *"of which we'll have space to publish only the tiniest fraction"* (Gee 2006). Nevertheless, authors and referees are crushed too, so that argument does not justify the systematic absence of procedural measures for tackling editorial offices. Horton (1998), based on three instances in which Editors had clearly breached accepted standards of behavior – one of which was a conflict of interest – proposes that the appointment of an *ombudsperson*[62] should be considered by Editors of all scientific journals.

Another instance that calls for attention is the case of post-acceptance rejection of manuscripts. Post-acceptance discovery of a major flaw in a paper, or cheating, or duplicate submission, is an ethically justified reason for retracting a paper that was previously accepted. In a summary of a discussion on a case of post-acceptance rejection of a manuscript,[63] several authors and Editors expressed their strongly negative opinion about the rejection of a specific manuscript months after its acceptance. The rejection was for technical reasons (*i.e.*, editorial backlog), nevertheless the situation was judged unethical. Along the line of Horton (1998), the forum suggests that

> JOURNALS SHOULD CONSIDER HAVING AN OMBUDSMAN WHO CAN EVALUATE AND INTERVENE IN DISAGREEMENTS BETWEEN THE EDITOR AND AUTHORS, AS WELL AS HAVING A WRITTEN POLICY ON THE PROCESS FOR APPEAL OF EDITORIAL DECISIONS.

The *Lancet*, for example, appointed an ombudsperson, whose concerns are maladministration, *i.e.*, avoidable delay, bias or unfairness, discourtesy, failure to follow proper procedures, and mistakes in the handling of claims.

10.2 Inappropriate modifications of manuscripts

It occurs that authors feel offended when confronted with the outcome of the language and copy editing at the editorial office, because some changes may appear to somehow curtail the freedom of expression of authors. As such, overcorrecting a manuscript for what might be perceived as a better result, is to be avoided. Langdon-Neuner (2010), for example, laments receiving a proof back from a journal: the authors of the article had used the word "done" on several occasions, but the copy editor had replaced all "done" with "performed" or "conducted", and she rightly asks if the Publisher *"has a style guide that classes 'done' as a dirty word?"* Authors object to such changes, simply because **authors have the right to object to changes that modify the original intention or meaning of their message**. Nor do authors appreciate direct editing of the source LaTeX file, except

[62] An impartial intermediary who mediates between authors and Editors.

[63] `www.wame.org/resources/ethics-resources/post-acceptance-rejection-of-a-manuscript`

for minor issues such as formatting and consistency of abbreviations that typically belong to the realm of copy editing.

Section 13.21 in Paper I explains the levels of editorial changes that authors may encounter, but also mentions a severe form of modification: **outright censoring**, *i.e.*, forcing an author to modify content, or simply modifying manuscript content without the author's consent, or without the author's knowledge. Such form of interference usually does not come from Editors of peer-reviewed journals, and if it happens, it will certainly be seen as highly unprofessional. But censoring may occur in conference proceedings where, quite often, the Editor is one of the organizers of the meeting. **Even if you express views that are directly opposed to the conference consensus, no Editor may claim the right to censor or to ban your contribution**. There are just enough genteel ways to address opposing opinions, just as there are also many ways to expose such editorial misbehavior.

10.3 Proximity between authors and journals

The degree of proximity between authors and journals (journal-related authors), can possibly be a problem. Campanario (1996) finds that some journals are better for journal-related authors than for other authors: articles by editor-authors or referee-authors were found to be significantly longer than articles by other authors. Publication practices of Editors in their own journals was assessed by Bosnjak *et al.* (2011), who come to the conclusion that the majority of Editors did not misuse their own journals for scientific publishing and academic promotion. These authors, however, stress the need for greater transparency in the management of editorial conflict of interest in academic and scholarly journals.

The best remedy against possible misuses is to archive the minutes of editorial decisions, appoint Editors by open advertisement, and make sure that Editors are not overwhelmed by other commitments such as teaching and management.

10.4 Forging, bypassing or shortcutting the refereeing process

Bypassing the refereeing process unfortunately happens, as is shown by the following example of an *Open Access*, peer-reviewed, online journal with 7 Editors, 36 Editorial Board members, and an impact factor of 2.4. This journal's Table of Contents identifies those articles that have been especially highly accessed, relative to their age. Hon *et al.* (2010) was such a highly-accessed 2-pages paper, written in excellent English, entitled *Influenza or not influenza: Analysis of a case of high fever that happened 2000 years ago in Biblical time*. The "analysis" consisted of an interpretation of half a dozen verses of the Bible concerning a woman with fever, who was cured when touched by Jesus Christ.[64] The authors diagnose the illness, and ponder whether the virus is influenza, avian or swine flu, parainfluenza, or other respiratory viruses – next to considering whether the illness was inflicted by a demon or by a devil. The paper was well-structured, and the

[64]The gospels Mark 1:29–33, Matthew 8:14–15, and Luke 4:38–39.

three authors specified their contributions as *"KLH conceived of the study, and was the principal author. TFL and PCN advised and reviewed the manuscript. All authors read and approved the final manuscript."* There were 9 references, including 7 references to the first author and one to The Holy Bible.[65] The paper appeared first on the top-twenty most accessed articles for the last 30 days (on August 30, 2010), with more than 13 000 downloads over the last year (top 11 of all times, for that journal). That case called for retraction, and the Editor-in-Chief of the journal issued a declaration that he and the authors *"acknowledge that the article does not provide the type of robust supporting data required for a case report and that the observations made are mostly speculative. As such, the article does not fulfill the criteria for a case report and does not meet the high standards expected of the journal"* (Garry 2010).

Some time ago I received a paper to review for an ISI-listed astronomy journal with a larger-than-two impact factor, and I reacted within 24 hours: *"I looked at the paper and have doubts that this is really a 'full length research article', it rather looks as something for a meeting on the history of astronomy"*. Exactly two weeks later, I received from the "Journal Manager" a message that my review assignment will be terminated, and that the paper was accepted because *"a longish discussion has taken place regarding the suitability of the paper for the journal, not regarding its content (where the editor is sole judge), but regarding its format"*. I promptly expressed my surprise about the acceptance of this manuscript in *not-regarding-its-content mode*, without regular peer review. Two weeks later (*i.e.*, one month after I received the invitation to review), I received an apology from the Editor on the internal confusion at the editorial office. The paper now features as a refereed paper in the ISI database, with 0 citations so far.

Both examples clearly illustrate that some peer-reviewed journals occasionally short-cut the refereeing process.

10.5 Manipulation of the journal impact factor

No journal Editor will deny being pleased when seeing the impact factor of the journal to be on the rise, and most Editors will take measures to prevent the impact factor from going downhill. Journal Editors often seek to reduce the denominator in the impact-factor calculation, or court authors (as mentioned in Sect. 5.1.1). But some other interventions to boost the factor are recognized as corrupt, for example, suggesting to authors that they include references to the same journal as to which they submit. The following list is only a sample of statements that document this situation.

- *I submitted an article to a leading journal last year (with IF > 5). The article was accepted, but the Editor asked for 2 references from his journal to be included to clarify a point we made in the Introduction. One of the suggested references was work from the Editor and his group!* (Sieber 1998);

- *In addition, the journal editors make the following request from you. Please review the past three years of our journal, and, where appropriate and relevant, cite all*

[65] *New King James Version*, Nashville: Broadman & Holman Publishers 1982.

articles that would be relevant to your paper's subject. We appreciate your efforts in citing relevant previous studies from our journal. Thank you. (letter from Editor, Pierson 2003);

- *I should like you to look at some recent issues of the Journal of* [name deleted] *and add citations to any relevant papers you might find. This helps our authors by drawing attention to their work, and also adds internal integrity to the Journal's themes* (letter from Editor, Monastersky 2005).

The *Committee on Publication Ethics* (COPE) states that it is corrupt to artificially inflate a journal's impact factor by suggesting to authors that they include references to secure publication.[66]

10.6 Breaching confidentiality

In the editorial process, confidentiality means that

1. the Editor should not reveal the content of submitted articles to outsiders, and
2. an Editor, when suspecting misconduct, should not disclose any information – unless the journal, in its instructions and guidelines to authors, explicitly reserves the right to pass information to third parties if misconduct is suspected.

Heywood (2008) emphasizes that Editors need to exercise particular caution when dealing with allegations of unethical behavior or suspected misconduct. All correspondence with an Editor is considered confidential,[67] but *libel* (false charges or attacks on the good name of someone) in a letter or email, a referee report or book review (especially when it deals with dubious conduct) can lead to a lawsuit – whether the defamation was intended or not.

Authors can bring a libel claim in any country where publication has taken place, or where the paper has been downloaded, and libel laws – like copyright laws – are different everywhere. In England, for example, libel laws are very claimant-friendly, the opposite of the US where the laws are on the side of the defendant. Heywood suggests that journals and their Editors manage the risks of claims for breach of confidence by providing

- clear guidance notes for Editors, authors and reviewers on the journal's policy on confidentiality;
- guidance for Editors on handling hostile reviews and allegations of misconduct;
- a libel policy;
- clear guidance for reviewers on avoiding making libelous or personal comments;
- an email policy, and
- an effective data-protection policy.

[66] COPE Report 2003: `www.publicationethics.org.ukscope` edited by C. White, BMJ Books 2004.
[67] Email messages, though, possess embedded dangers, notably the Reply to All button.

Referee forms often include a box for confidential comments to the Editor: referees can use this space without much worry about tactfulness, because the Editor is bound by discretion, simply because that confidentiality is explicitly offered, and thus – in principle – ensured. It is thus the Editor's task to transfer these opinions to the author in a way as diplomatic as possible.

A while back, I was asked to review a revised manuscript of a case where the referee was not available to inspect the revised version. The invitation to review mentioned that the report of the first referee was attached, and I accepted the invitation within 24 hours, but received no reply. One month later, the editorial assistant let me know that *"Your services on manuscript X are no longer required ... Because we did not receive a reply to our request for a review of ... we assume you must be very busy or on extended travel. We will select an alternate reviewer in order to keep reviewing delays at a minimum"* – that is, exactly one month after I had answered that very request to review! This event shows four fundamental shortcomings in the editorial process of that (ISI-listed) journal:

1. a revised manuscript should go back to the first reviewer, because only that person can judge whether the revision is adequate;

2. sending a revised document to a second person is prone to cause additional delay, because the second reviewer certainly will bring up other issues – hence my answer to the Editor: *"I think that the fastest – and fairest – way to get the revised version verified, is to ask the first referee again"*;

3. sending a confidential document together with the invitation to review is a breach of confidence, and is just not done, and

4. assuming that inefficiency and non-respondence is a trait of others.

The question in such cases always is: who is to blame, the editorial office, or the referee? In this case it was certainly not the referee.

11 Publisher misconduct and mismanagement

11.1 Manipulation of journal content

This issue is not of much concern in the astronomy journals I know of, but manipulation of journal content does occur in for-profit publications, especially in the biomedical and pharmaceutical sciences.

The creation of fake journals is the most extreme form of manipulation of content. One medical publishing company published during 2000–2005 six fake "scientific journals", which really were marketing journals paid for by pharmaceutical companies – the journals carried the name of the Publisher, but not the name of the sponsors. One publication had the look of a peer-reviewed journal, but contained only reprinted or summarized articles – most of which presented data favorable to the sponsor's products (Grant 2009b). So the sponsor had created a journal only to serve as a marketing tool. The fraud came to light through a civil suit filed by a patient who suffered a heart attack while using a drug that had been positively described in that "scientific" literature.

11.2 Manipulation of journal impact factors

Section 10.5 also touches on the Publisher, because some commercial Publishers pressure their Editors to inflate the impact factor by artificially increasing the number of citations. Some commercial Publishers push their Editors to increase journal content by lowering the level of quality of the journal. Note that lowering the level of quality, unfortunately, does not necessarily mean that the impact factor will decrease, simply because poor papers are chiefly cited for their incorrect content, not to forget the impact of self-citation (see Sect. 18.5). One former Editor of a chemistry journal recently told me that the Publisher of his journal left him the choice: lower the standards and accept more papers, or resign. He decided for the second option.

12 Research supervisor misbehavior

For the title of this Section I intentionally use the term *misbehavior* instead of *misconduct*, because the former covers a wider spectrum that includes lesser forms of misdemeanor that do not necessarily transgress moral or civil laws. Moreover, this is not the proper place to deal with supervisor misdemeanor in the broadest sense, so most of my comments bear only on aspects of scientific writing.

The nomenclature on supervisors is not very well defined: the term mostly refers to PhD thesis promotors, but also includes senior and junior supervising postdoctoral fellows, and *tutors* or *mentors*.[68] The functions of supervisor and mentor are, ideally, combined in one person, and mentoring should also incorporate the transmission of ethical standards from senior to junior scientist.

Weil (2001) contends that the mentor is, first, a living demonstration of how one pursues work in the field by embodying the standards of responsible conduct. However, merely presenting a model of appropriate conduct is not sufficient, and the inexperienced person needs additional instruction to appreciate what the norms of responsible conduct are, what in the mentor's conduct should be imitated, and why. This is accomplished through nurturing activities that require intense interactions in both directions. Weil (2001), moreover, explicitly stresses that advising and counseling inexperienced young scientists in unethical or illegal practices, is to be qualified as unethical or illegal, and she consequently coins the term *pseudo-mentors*. Examples abound: nurturing internal competition – especially when several students work on a same topic without clear guidelines, or force students to submit irrelevant posters at meetings, or foster paper slicing, or "helping" young researchers with referee reports (the case described in Sect. 8.7) may very well lead to the apprentice accepting this attitude as a normal and traditional part of the trade to win fame.

A historically most interesting story about the relationship between supervisor and student, about the way they shared authorship, and on how the times have changed, is told

[68]The term *mentor* comes from Greek mythology: when Odysseus left for the Trojan War, he asked his friend Mentor to provide parent-like supervision of his son, Telemachus.

by Fletcher (1982).[69] He had been working with Millikan on the oil-drop experiments,[70] the results of which were to be Fletcher's thesis. When the first paper was written up (Fletcher did more of the writing than his supervisor), Millikan came to Fletcher's apartment to discuss who was to be the author of that paper, and of four other papers that were taking shape. Using the argument that Fletcher should be the sole author if a published paper is to be used for his PhD, the five papers were authored M, M&F, F, M&F, and F. The last paper of this series became Fletcher's PhD, the first paper (Millikan 1910) led Millikan to receive the 1923 Nobel Prize in Physics.[71] Fletcher (1982) writes that he was obviously disappointed, since he had expected to be a joint author, though he clearly states that he had no bad feelings towards his supervisor, and that he had not been unfairly treated by Millikan.

Lee *et al.* (2007) contend that having a good mentor early in your career can mean the difference between success and failure in any field. But a private mentor is just not enough: the young researcher should be surrounded by an environment that fosters and protects ethical behavior, and that follows the norms that should govern the supervisor–student relationship. This rationale was elaborated by Goodstein (2002), in an open letter to the California Institute of Technology (Caltech) community, in which he presented a transmuted version of the biblical Golden Rule that is the root of all principles of ethical behavior

DO UNTO OTHERS AS YOU WOULD HAVE THEM DO UNTO YOU[72]

as *"No member of the Caltech community shall take unfair advantage of any other member of the Caltech community"*. Goodstein states that this rule is not adequate, because it does not explain how to deal with the world outside the own institute. Caltech, consequently, developed a policy on research misconduct, see Section 16.4 for a description of its principal elements (*i.e.*, definition of research misconduct, and the procedures of investigation).

The main impression I got from speaking to students,[73] is that many supervisors supply good working and funding conditions, but that they often do not have the time to provide adequate guidance, especially during the initial stages of scientific writing. My experience is that students encounter most of the problems – if not an outright writer's block – during the writing of their first "big" paper, hence tutoring at that stage is of utmost importance. Especially the fact that what is written seldomly is a literal account of the work done (see Sect. 2), is a serious obstacle. The beginning phase of a research project often is the most creative and intellectually most passionate – but also the most chaotic. Specifically, instruction and guidance are essential when writing down the first

[69]Fletcher (1982) was based on a manuscript autobiography, which Fletcher insisted should be published posthumously.

[70]Fletcher (1982): *"We were together nearly every afternoon for the next two years."*

[71]Note that this relatively long paper in *Science* was followed by a much longer paper, with exactly the same title, in *Physical Review* (Millikan 1911).

[72]Bible quote, Matthew 7:12.

[73]At both SWYA schools, but also on other occasions, see Sterken (2006).

results, and training in acquiring communication skills – by means of intensive reading combined with multiple passes of revision of whatever was written – are the vital first steps. In addition, a fast turnaround time of draft manuscripts is a most important prerequisite for the student to learn all aspects of the writing process in a smooth and progressive way.

From surveying the literature, it appears that the chief problems with research supervisors are related to the core essence of the research (Sect. 12.1), and, above all, to the various kinds of authorship issues (Sect. 12.2).

12.1 Creativity or routine?

Bronowski (2008) once told an audience of school children that the world would never change if they did not contradict their elders – only to run into outraged parents the next morning. Such kind of independence in observation and in thoughts he calls *the basis of the scientific method.* The trouble is to find the optimal balance between scientific creativity and work routine. Some research supervisors show similar reactions as Bronowski experienced, and harm their pupils. The problem, however, has two faces: my research supervisor once instructed me to *"stop making troubles, and just apply the methods that the other people apply"* (the method had systematic flaws and was abandoned soon after). Vice versa, I witnessed one of my own PhD students being attacked for just doing what I was ever told to do: apply methods that other people apply, in cases where the basic assumptions were not fulfilled.

The first point to recognize is the knowledge that routine is deadly: as Walter Baade wrote to Thackeray on 13 December 1949

> ... ROUTINE PROGRAMS HAVE A DEADENING EFFECT ONCE THEY GET THE UPPER HAND. WORST OF ALL, THEY ATTRACT THE WRONG KIND OF PEOPLE. ...[74]

Routine projects comprise – or quite often lead to – experimental imitation (for example, doing the same kind of investigation in just one other stellar cluster), and often lead to pathological or cargo cult science.

The second point, not recognizing true creativity, is deadly too. Take, as an example, Cecilia Payne's PhD thesis (Payne 1925a), which established the cosmic abundances of the chemical elements, and showed that the variety of spectral types result from temperature rather than from abundance differences. Her Table XXVIII gives the values of the relative abundance of the elements, with extreme values for the abundances of H and He. Her work has been called by Otto Struve "the most brilliant Ph. D. thesis ever written in astronomy",[75] but on the urging of Henry Norris Russell (her reviewer), she rejected this finding as spurious (Gingerich 1982), and in her PhD paper, she explained these super-outliers as follows:

[74] Cited by Feast (2000).

[75] Quoted from Gingerich (1982).

> HYDROGEN AND HELIUM ARE OMITTED FROM THE TABLE. THE STELLAR ABUNDANCE DEDUCED FOR THESE ELEMENTS IS IMPROBABLY HIGH, AND IS ALMOST CERTAINLY NOT REAL

(Payne 1925b). Four years later, Russell (1929) proposed the same finding without a single reference to Payne, in a 72-pages paper in *ApJ*, see also Hearnshaw (2010).

12.2 Authorship ethics

Wilcox (1998) assessed authorship disputes brought to the Ombuds Office of a consortium of medical schools, and listed the following typical authorship issues:[76]

- *though listed as an author, I never reviewed the article, saw the reviewers' feedback, or signed off on the final document*;
- *I was promised first authorship when I completed a project. When the work was done, and without informing me, the principal investigator added the work of someone else who was made first author*;
- *someone is demanding first author position but the contribution was primarily in implementing the experiments, not in creating the design or writing*, and
- *after I left my lab, I got no credit for the projects and articles that used my work.*

According to this author, issues of authorship (as percentage of total issues) increased from 2.3% to 10.7% over five years, and the problem continues to grow, especially in light of greater competition for fewer grant dollars (Wilcox 2010).

The last item listed above concerns the specific situation of "leaving the lab", *i.e.*, leaving your results behind, and losing credit for your work. Quitting the lab – quite often for no other reason than for not finding a postdoctoral fellowship – also may mean a total cut off from professional journals and other resources, and no more access to your data. One former PhD student told me that her latest manuscript (with over half a dozen consenting coauthors – among whom three supervisors and two native English speakers), dwelled almost 18 months in the hands of one of the senior coauthors, after which she was summoned to make major adjustments, including reconstruction of vast tabular material. In addition to the technical obstacles of upgrading a manuscript that was dormant for one and a half year, the question really is what these supervisors are up to: either the science in the paper is mediocre (and the manuscript should then not be submitted in the first place), or the science presented in the PhD was OK, and then it is most likely outdated after two years of gathering dust in one author's office.

Tarnow (1999) probed junior postdoctoral fellows in physics about their awareness of ethical principles on authorship, and reports that *"The supervisor, a joint author in 92% of all papers, is inappropriately given authorship in 14% of those papers. In 33% of the papers with additional authors other than the supervisor, one or more authors were perceived as inappropriately listed. In contrast, the postdoctoral fellow was an*

[76] Verbatim quotations of statements extracted from reports, with permission.

inappropriate author on only one percent of all papers."[77] Tarnow also describes that one postdoctoral fellow told him that his investigation was "offensive" – a reaction alike the one that I experienced already 20 years ago at my very first talk dealing with the ethics of authorship,[78] when an upset senior participant bluntly told me *"What you describe is like prostitution: we all know that it exists, but we do not talk about it"*. Incidentally, in 2009 I gave a lecture entitled *Integrity and misconduct in scientific communication: the role and impact of the thesis supervisor* (EASE 2009). After concluding my talk, a gentlemen stood up and said *"I have colleagues who have more than 80 students, and who do not have the time to read all their students' papers. Are you telling us that we should now disgrace those people?"* I answered – referring to what I was told 20 years ago – *"Professors with more than 80 students, who do not have the time to read the papers they coauthor, are the pimps of the system."* The audience – essentially Editors and Publishers and their staffs – was left aghast.

Also Anderson *et al.* (2007) give a realistic account of the perverse effects of authorship issues on scientists' work and relationships: *"I moved and I started some work. And the principal investigator says, 'This is your paper. You are doing this work, so you are the first author'. Okay, we get some good data, and it looks interesting. Then somebody else wants to be first author. And then the person claims, 'Oh, I did most of the writing'. And different people have different claims you know. And somebody can grab that paper and write and rewrite and rewrite and rewrite, and keep it, you know, for six months or one year, just to be able to make some claim that they did a lot of work on it"*. Is that not a fitting explanation for the leaving-the-lab syndrome exposed on the previous page?

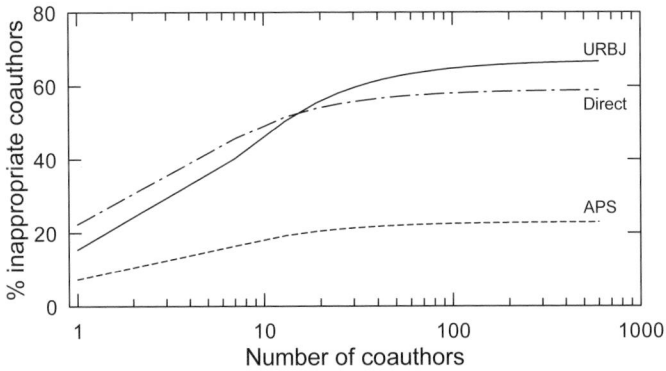

Fig. 11. Three functional relationships (Eq. (2) of Tarnow 2002) that describe the percentage of inappropriate coauthorship using APS, URBJ, or "direct contribution" criteria for coauthorship (see text).

In a large and detailed survey on the ethics of scientific coauthorship, members of the *American Physical Society* (APS) were asked to judge the number of appropriate coauthors on his or her last published paper. Tarnow (2002) presented a functional form

[77]This study does not include honorary authorship through lending the name by a celebrity (who always exits clean-handed when problems surface).

[78]Published in Sterken (1988).

that well describes the percentage of inappropriate coauthorship.[79] The 2-parameter fit shows that within the same criterion – *i.e.*, using the APS authorship guideline[80], the "direct contribution" approach, or the *Uniform Requirement for Manuscripts Submitted to Biomedical Journals*[81], there is apparently little difference in authorship assignment on papers with 10 coauthors or papers with 10^3 coauthors. The model shows that inappropriate coauthorship is strongly dependent on the criterion used, with an almost linear trend in inappropriateness that extends up to about a dozen authors. The asymptotic curve yields 20–60% inappropriate coauthors, depending on the coauthor assignment approach. Figure 11 shows the three functional relationships.

13 Dealing with the Press

When dealing with the Press (also called the "media"), the outcome is often quite different from what you initially expected. The first aspect to keep in mind is that the majority of press reporters are not trained scientists – *i.e.*, some reporters have obtained a degree in science, but have lost the practice of science (or never mastered it), or do not have decades-long experience in scientific analysis and communication of scientific results. Russell (2009) discusses the norms of professional science journalist behavior, and also **the difference between the ethos of science and the ethos of news**. He emphasizes that scientists and media people talk in different languages, and are focussed on different issues. Media people that entertain the public introduce "errors" that are, in fact, unavoidable. There is, of course, a huge difference between the quality newspaper science journalists, and some other publications that foster a more sensational way of presenting scientific issues.

A telling example is what happened to Holton's (1978) analysis of Millikan's laboratory notebooks, in which he pointed out that there was some degree of selectivity in the data that Millikan had picked out to publish. This led journalists Broad & Wade (1982) to the conclusion that Millikan had cooked his data by eliminating a number of oil-drop data, while in the Millikan (1913) paper the author had explicitly stated that he had included **all** data. Millikan thus appeared to have reported only those results that supported his hypothesis, and as such obtained tighter error margins than would otherwise be possible.

Goodstein (2010) also analyzed these laboratory notebooks, and Millikan's paper, and presents the opinion that the sentence on Millikan's (1913) page 133 "*... a complete summary of the results obtained on all of the 58 different drops upon which complete series of observations like the above were made ...*" and the sentence appearing on page 138 "*... this is not a selected group of drops but represents all of the drops experimented upon during 60 consecutive days ...*" reflect *only* on the confirmation of his presumed formula for correcting Stoke's law. Goodstein remarks that the four pages in between both statements contain only graphs and Tables, and he contends that it seems most likely that in Millikan's original typescript (which was lost), the second sentence would have

[79] The experimental data are shown in his Figure 6.

[80] `http://www.aps.org/policy/statements/02_2.cfm`

[81] URBJ, see `http://www.icmje.org/ethical_1author.html`

almost immediately followed the first. This corroborates the reflection that the thoughts of readers are guided by the way the texts are arranged and presented (Eisenstein 2005). In other words, **the printed layout changes the context**, and the apparent signs of misconduct is the consequence of *reading* a message that appears visually different to the reader as it appears to the writer. Hence the importance of Figure and Table placement, and the general layout, see the discussion on the typeset page as an image, in Section 10.3 of Paper II.

Broad & Wade's (1982) book *Betrayers of the Truth: Fraud and Deceit in the Halls of Science*, however, was not the product of malicious deceit – though it received a quite hostile reception at a meeting of the *American Association for the Advancement of Science* (David 1983). Quite the contrary happened with the reporting of John Herschel's astronomical expedition to South Africa, when Richard Locke, a *New York Sun*[82] reporter without scientific training, described Herschel's telescope lens as of being "24 foot diameter weighing 6700 kg" (25 August 1835). His stories, entitled *Great Astronomical Discoveries Lately Made By Sir John Herschel*, were accompanied by lithographic prints that sold very well. Figure 12 replicates one of these illustrations depicting natives of the Moon (*Vespertilio-Homo*), whose *"amusements would but ill comport with our terrestrial notions of decorum"*.[83]

Fig. 12. *Vespertilio-Homo*. Source: *New York Sun*, 1835. Courtesy Alex Boese.

[82] A newspaper that appeared from 1833 to 1950.
[83] Decorum means dignified behavior.

Whereas the *Great Moon Hoax of 1835* was a scientific impropriety, it was relatively harmless, and stories of this sort would be unthinkable these days. Thus, we live under the illusion that we can now easily distinguish between scientific journals and sensational journalism, although television stations faking scientific results – and in doing so endanger people's health – do exist. Cyranoski (2007), for example, reports that a television station faked scientific results[84] in one of its programmes, and *"used dubbing to put false words into the mouths of foreign researchers"* up to a content that was about 60% incorrect. This, obviously, is commerce-driven fraud.

Our modern press reporters commit "less unethical" transgressions, such as not verifying their sources, mounting dramatic sound tracks on images that were not captured at the same time, or giving the impression of *in situ*[85] reporting while being at a safe remote distance. The reasons for these practices mainly are:

1. commercialization of the media sector, with increasing pressure from commercial companies to inject commercial content in the news;
2. the force of *embedded journalism* – *i.e.*, news reporters attached to military units – rendering reporters transmitters of news rather than gatherers;
3. the fast advancement in communication and imaging technology;
4. overabundance of junk news, *i.e.*, trivial and insignificant news items;
5. deadly competition, and
6. too short *publication time*, causing sloppy work with no time to check the sources.

Van de Looverbosch & Deltour (2011) underline the need for strict separation of powers: rigorous detachment of commercial and editorial aspects of the media company, *i.e.*, in a structure that resembles the one depicted in Figure 13 below. These authors also see a direct relationship between the number of journalists (and their working conditions) on the one side, and the quality of their news product on the other side.

What is the lesson that we should learn from the Press? Well, the first lesson simply is: always verify every level of your work with the same rigor as you verify the proofs of a scientific paper delivered by the Publisher's prepress services. And transliterate the six points mentioned above into

1. separate Publisher and Editor, and see to it that Editors have total scientific freedom;
2. critically dissect the data you receive from others;
3. master the communication (writing and graphics) technology;
4. limit trivial and duplicate papers;
5. eschew harsh competition, and
6. be very careful when working under the pressure of time.

[84]On soya bean fermentation.
[85]Latin for *in the place*.

14 The signatures of scientific misconduct

A most common aspect of all forgeries is that the forger seems to have expert knowledge of observational or computational methods, that the data most often appear to excel in precision, but that it is the error pattern that reveals the forgery. And, very often, the forger refuses to render the raw data when asked for, or says that the data were destroyed, see the description of the D'Angos affair on page 196. As Sechrest (1987) puts it

> REANALYSIS OF OLD DATA (IF AVAILABLE AT ALL) IS OFTEN IMPOSSIBLE ... ALMOST ANYONE WHO HAS ATTEMPTED TO ASSEMBLE SETS OF DATA FOR REANALYSIS HAS PROBABLY BECOME SOMEWHAT CYNICAL, ... WHAT WITH ALL THE TALES OF FIRES, FILES LOST IN MOVING, VINDICTIVELY DESTRUCTIVE RESEARCH ASSISTANTS, AND SO ON.

In addition, forgers seem to be very prolific paper producers with excessive publication rates. Another sign is that their papers have vague and ambiguously formulated arguments with a high degree of "under-description" of the data, and, above all, that the scientific establishment accepts their data rather blindly.

Racker (1989) describes the common features of cheating scientists: they have an outstanding intellect, they recognize the missing links that – if solved – would represent a breakthrough, they are skilled experimenters, and are often headed by an investigator or supervisor with a reputation of integrity. His opinion is that scientific fraud, perpetrated by such "professional" cheaters, springs from an unbalanced, ill mind. He also emphasizes the need for careful supervision, because

1. cautious supervision detects experimental errors;
2. careful supervision enhances confidence, and
3. elaborate supervision greatly helps in the write-up of the papers.

A very remarkable element is Racker's (1989) first feature, *viz.*, that perpetrators seem not to be the kind of people who really need to behave badly, as is also suggested by the statement

> HE WAS A VERY GOOD SCIENTIST, BUT SOMETHING HAPPENED AND HE GOT INTO THIS BUSINESS OF DUPLICATING PAPERS.[86]

This Section closes with a long quote that summarizes about everything that was said about fraud, detecting fraud, not recognizing that published data are unacceptable, inadequate reporting, and even the bandwagon effect. It is the case of the so-called "Burt scandal", referring to the irregularities in the psychological intelligence-test data published by Cyril Burt[87] (1883–1971), so well described by Mackintosh (1980):

[86] Chris Pickles, quoted by Reich (2010).

[87] This person was also an Editor of an international journal, and he seems to have published several of his own unrefereed papers, and also added articles written by himself, but signed by fictitious names (Fancher 1985), see also Eysenck & Kamin (1981) and Rushton (2002).

> WHY WERE BURT'S DATA ACCEPTED FOR SO LONG? IGNORING THE QUESTION OF FRAUD, THE FACT OF THE MATTER IS THAT THE CRUCIAL EVIDENCE THAT HIS DATA ON IQ ARE SCIENTIFICALLY UNACCEPTABLE DOES NOT DEPEND ON ANY EXAMINATION OF BURT'S DIARIES OR CORRESPONDENCE. IT IS TO BE FOUND IN THE DATA THEMSELVES. THE EVIDENCE WAS THERE ... IN 1961. IT WAS, INDEED, CLEAR TO ANYONE WITH EYES TO SEE IN 1958. BUT IT WAS NOT SEEN UNTIL 1972, WHEN KAMIN FIRST POINTED TO BURT'S TOTALLY INADEQUATE REPORTING OF HIS DATA AND TO THE IMPOSSIBLE CONSISTENCIES IN HIS CORRELATION COEFFICIENTS. UNTIL THEN THE DATA WERE CITED WITH RESPECT BORDERING ON REVERENCE, AS THE MOST TELLING PROOF OF THE HERITABILITY OF IQ. IT IS A SORRY COMMENT ON THE WIDER SCIENTIFIC COMMUNITY THAT 'NUMBERS ... SIMPLY NOT WORTHY OF OUR CURRENT SCIENTIFIC ATTENTION' ... SHOULD HAVE ENTERED NEARLY EVERY PSYCHOLOGICAL TEXTBOOK, AND BEEN USED WITHOUT QUESTION IN EVERY DISCUSSION OF THIS ISSUE.

15 The causes of scientific misconduct

What are the deep causes of scientific misconduct? There are many, varying from sheer egotripping to hardcore ambition. This Section concentrates only on those aspects that distress scientists, *viz.*, administrative burdens, career and competition pressure, and the ever increasing "publish or perish" crushing force.

15.1 Increased administrative burdens

Kennedy (1985) formulates one reason for scientific misconduct:

> INCREASED ADMINISTRATIVE BURDENS ON PRINCIPAL INVESTIGATORS, ARISING IN PART OUT OF MORE ONEROUS GOVERNMENT REGULATION, MAY ALSO CONTRIBUTE – BY DECREASING THE TIME AVAILABLE FOR ACTIVE PARTICIPATION IN RESEARCH, JUST WHEN ITS IMPORTANCE TO ACADEMIC CAREERS IS INCREASING.

There is not much to add, except that in the quarter of decade since Kennedy's book, the detrimental effect of administrative loads has increased manifold.

15.2 Career pressure and competition

A second cause is career pressure, especially at the moment when important time deadlines are imposed, for example, the end of a contract, or a PhD that must be defended in order to get a contract renewal. But career pressure (*i.e.*, the urge to produce more papers, simply because the system expects you to add a dozen or more papers each year) changes the very essence of investigatory work – in just the same way as the system pressure on some police units to issue a minimum number of tickets each month modifies the essence of their activity: society expects that tickets are issued as a penalty, and not for the cheapest of motives of furthering careers or boasting "production".

The very well known "everybody does it" syndrome – *i.e.*, the idea that we are powerless to change anything – is an additional motive to succumb to career pressures that drive the "papering". Snow's novel (see page 200) very nicely illustrates the misguided attitude of his key character:

> I'M NOT INTERESTED IN ANY DAMN DISCOVERIES. ALL I'M INTERESTED IN IS COOKING UP A THESIS. THEN I CAN PUBLISH A PAPER OR TWO BY HOOK OR BY CROOK. THAT'S THE WAY EVERYONE'S PLAYING THIS GAME. AND I'M GOING TO PLAY THE SAME GAME TOO.

Competition among scientists is fierce – even within the same institute can teamwork be fraught with competition, jealousy and fear. Maddox (1993) is persuaded that

> COMPETITION TO SUCCEED IN DISCOVERY, OR AT LEAST **TO BE SEEN TO SUCCEED**, EXPLAINS THE RASH OF PERVERSIONS OF THE LITERATURE IN THE PAST 15 YEARS,[88]

while he laments the passing away of the more genteel decades. In the same vein, Maddox (1993) describes an example of what can happen when four articles announcing essentially the same results were published within one month in *Science* and *Nature* (one of the senior authors of one article even was the excluded referee of the other):

> SO MUCH ZEAL IS NOW INVESTED IN BEING THE FIRST, AND SO MUCH EMOTION IS SPENT IF SEEMING TO BE SECOND, THAT PEOPLE'S EFFECTIVENESS MUST BE UNDERMINED.

The situation has only become worse. Competition is and will always be part of the trade of science, for many reasons (some even as simple as the job market with far fewer jobs for a growing pool of people). But competition should remain healthy, and should be robust competition – not just any kind of competition – therefore Maddox's remark on **to be seen** to succeed is very important, especially in the light of Section 15.3. Unhealthy competition is often the consequence of wrongful mentoring, for example when a supervisor tells the students that in science, *"only first prize counts"*, it is not surprising that events as recounted on page 215 can unfold. The effect of such cases on the mutual collaboration, even more so on the victim's morale, is just devastating.

15.3 Misunderstanding bibliometry

Beginners in science, but also seasoned administrators, often have misguided views on bibliometrics, and make the following errors:

1. looking at the absolute value of one bibliometric parameter or indicator;
2. overlooking the factor of time in a bibliographic record;
3. overvalueing conference poster papers, and
4. sheer narcissism.

The last point refers to the feeling of optimism that some bibliometric scores suggest, what leads to the creation of false images of scientists, and of their institutions.

[88] *i.e.*, 1978–1993; emphasis mine.

Bibliometry is scientific scorekeeping, though with bibliometric scores that possess no absolute value whatsoever – in a similar way as basketball scores have no higher intrinsic value than soccer scores, despite the fact that they are numerically higher. But whereas sport scores are accurate, bibliometric scores are erratic and biased, and this wrongful system of bibliometric measurement and evaluation (see Sect. 18) is now one of the most relentless drivers of mediocre – if not pathological – science.

16 Cures for misconduct

This Section portrays three cures: the first one is on a personal level, the second is on the level of the scientific community, and the third deals with the responsibilities in the publishing structures. One model example of a misconduct policy is discussed in detail, and one example of a search for dual publication and plagiarism is worked out.

16.1 Accepting the limits of discovery

There are two severe misconceptions that the young scientist must abandon:

1. that your work was entirely achieved by yourself, without any help from others, and
2. the conviction that whatever you have discovered and written, is always totally new.

The first item is related to the fact that you always rely on the scientific results of others, even before you started your study. The second point is that it can happen that, after finishing a study, you discover that someone else has done a similar thing long before you, see the paragraph on innocent plagiarism on page 207. As said already, there is no other way out than to mention that reference, possibly adding a statement that you discovered this work only after you had written down your conclusions.

One part of the road towards accepting the limits of discovery – that is, learning to accept that strangers criticize your work – is perhaps one of the most difficult psychological aspects a young scientist must learn to live with: **someone, whom you may not know or whom you may not like, will take your work apart, and render some critical remarks**. And someone, sooner or later, will – totally independently – bring forth exactly the same bright ideas that you yourself ever came up with.

16.2 Research misconduct policies

Numerous universities and research institutions engage in developing and implementing research misconduct policies along well-defined procedures. Section 16.4 elaborates in detail one very specific misconduct policy: the *Caltech policy on research misconduct*.

Other efforts also develop in the community at large, for example, the *Scientific Red Cards* website[89] is a place where scientific misconduct can be reported. It inventorizes scientific publications for which research misconduct has been assessed – *viz.*, data-related misconduct, publication-related misconduct, and research practice misconduct – with the

[89] `http://www.scientificredcards.org/`

aim to promote a responsible conduct of research. Cases are classified into three categories with different color key: fraudulent data are flagged red, non-complying publications are indicated in blue, and unethical research practices are labeled in green. The website also provides a platform for discussing related questions, in particular via the FAQ Section. The site has a very interesting note about disclosure of misconduct (see also Sects. 10.6 and 17.3):

> PLEASE NOTE THAT THE AUTHORS OF THIS WEBSITE CANNOT BE ACCUSED OF WHISTLE-BLOWING, SINCE THEY CANNOT BE HOLD RESPONSIBLE FOR THE CONTENT OF LINKED PAGES ASSESSING MISCONDUCTS.

16.3 Mending the business model of journals

The business model of a journal is a crucial element in the journal's ethical universe. Diverse structures exist: from the smallest journals with no more than a couple of Editors, up to the larger journals that include editorial, management and publishing Boards.[90] The diversity of operational modes, and the basic function for which journals were created – *i.e.*, to communicate scientific results – make journals poor vehicles for tackling all sorts of scientific misconduct. Maddox (1989) explains that

> THERE IS A SCHOOL – CHIEFLY REPRESENTED BY THE EDITORS OF JOURNALS, WHICH HOLDS THAT SCIENTIFIC LITERATURE IS AND SHOULD BE A PASSIVE MEANS OF COMMUNICATION – IN OTHER WORDS, JOURNALS ARE PASSIVE ENTITIES WHOSE DUTY IS OVER ONCE THE PAPER HAS BEEN PUBLISHED. WHATEVER THE PASSIVITY, THERE IS AN ACTIVE FEEDBACK LINK BETWEEN JOURNALS AND THE CONDUCT OF RESEARCH.

That active link, however, means that when *Nature* or its sister journals receive serious allegations about wrongful data or about author misconduct, they do follow a clear procedure to work out whether the published record needs to be revised. Thus, in practice, there is no such thing as total passivity.[91]

Brown (2001) refers to the growing commercialization of campus life, with its ugly aspects of *"golden arches of fast-food outlets"*, and warns against other new features of commercialization that constitute a serious threat: privatization and the business model of a university are the potential menaces that admit an increased dependence on industry to operate the university. This includes, among others, **the commercial interests overriding the public's interest in free shared knowledge**, and attempts to run universities like a business that treats students and industry as clients, and academics as service providers with something to sell.

The literature on misconduct in the publishing sector vividly demonstrates that there is a need to separate the commercial and the academic branches of the publishing enterprise. We can take an example from one operational model based on the *separation*

[90] It is, though, not always very clear whether the refereeing is done by external experts, or by members of the editorial board, nor what are the ties with the commercial world.

[91] Editorial, April 29th, 2010.

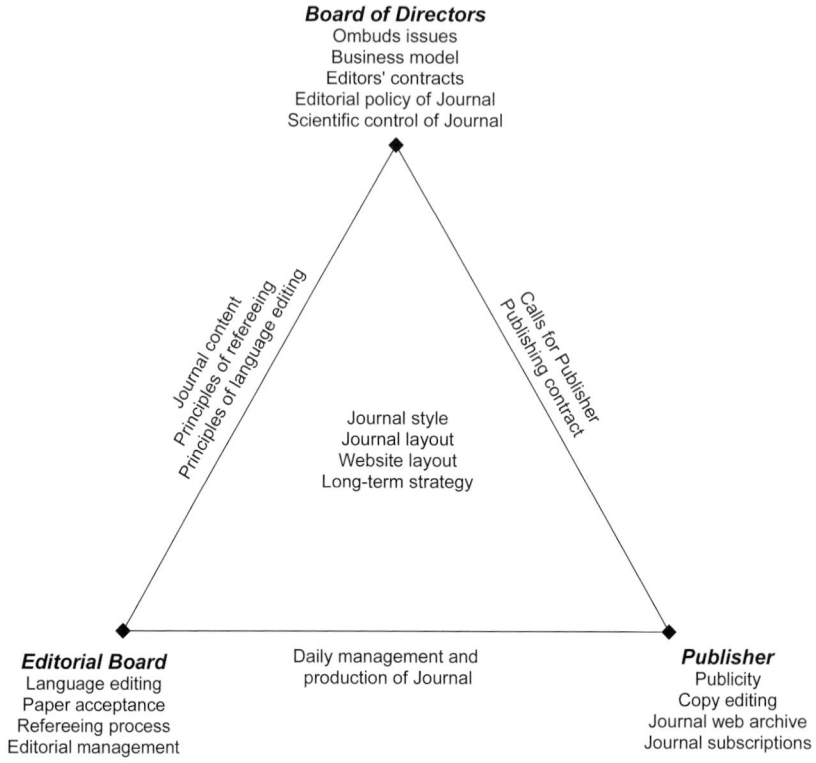

Fig. 13. Triangular (qualitative) diagram illustrating a business model of a journal. Each apex lists *some* basic functions that *solely* pertain to the associated body. Functions listed along the x, y and z axis only belong to the domain of the bodies on both sides of each axis. Functions listed inside the triangle are taken care of by the three parties in the set. The referees participate in the publishing process, but are not part of the formal journal structure.

of powers, a model of governance due to Charles de Montesquieu (1689–1755), that allows for independent branches with separate responsibilities – for example, an executive, a legislative, and a judiciary power. Figure 13 shows a qualitative ternary diagram that exemplifies such a structure, and illustrates the synergy between the three constituents.[92] The representation can be seen as an allegoric equivalent of a trichromatic color space that may become defective if

1. the top apex would disappear, so that all decisions and issues come near to, or on the base line, or

2. the bandwidths of the primaries become too wide: Editors never *only* edit – they are also authors – likewise, members of the external Board also are authors (and

[92]Ternary diagrams are explained in Section 4.3.7 of Paper II.

referees), and can even be authors of other publications marketed by the third party, the Publisher.[93]

So, as long as the distances are kept safe, and the bandwidths are narrow enough to avoid conflicts of interest and breaches of confidentiality, there should be no fear for a degenerate space leading to excesses as described in Sections 10 and 11. Preserving the distance between the Publisher, Editors and Board of Directors will also reduce the consequences of *groupthink*, as listed on page 204.

16.4 Establishing policies on research misconduct: an example

Caltech has developed a policy on research misconduct with the intent to act resolutely if a case of misconduct comes up, **at the same time providing maximum protection to the plaintiff and to the accused**. The policy covers the definition of research misconduct, the findings of an act of misconduct, the procedures of investigation, and the conclusions.

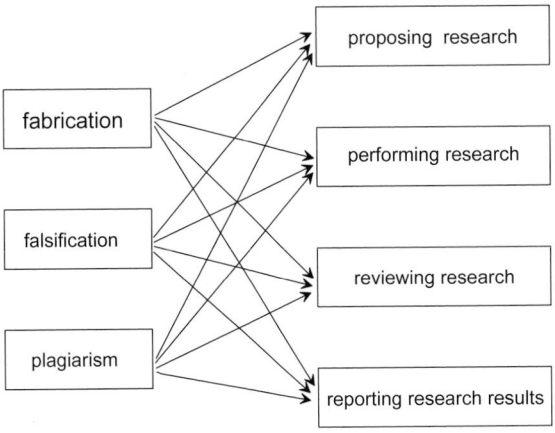

Fig. 14. Research misconduct, in the *Caltech policy*, is defined as fabrication, falsification, or plagiarism in proposing, performing, or reviewing research, or in reporting research results.

Definitions. The Caltech policy on research misconduct defines misconduct as *"fabrication, falsification, or plagiarism in proposing, performing, or reviewing research, or in reporting research results"* (Goodstein 2010). Figure 14 summarizes this definition in graphical form, and the list below clarifies the terms:

- *fabrication*: making up data or results and recording or reporting them;[94]
- *falsification*: manipulating research materials, equipment, or processes, or changing or omitting data or results such that the research is not accurately represented in the research record;[95]

[93]Therefore, members of external Boards should only be involved in general editorial policies, and should never be consulted on specific author issues – otherwise the principle of confidentiality will be transgressed (see Sect. 10.6).

[94]This definition covers what Babbage (1830) calls *hoaxing* and *forging*.

[95]This definition covers what Babbage (1830) calls *trimming* and *cooking*.

- *plagiarism*: appropriation of another person's ideas, processes, results, or words without giving appropriate credit.

Findings. A finding of research misconduct requires that:

- there be significant departure from accepted practices of the scientific community for maintaining the integrity of the research record;
- the misconduct be committed intentionally, or knowingly, or in reckless disregard of accepted practices, and
- the allegation be proven by evidence.

Procedure. The procedure involves *Inquiry*, *Investigation*, and *Judgment*.[96] Those responsible for conducting each phase should bear in mind the following important responsibilities:

1. the institute must vigorously pursue and resolve all charges of research misconduct;
2. all parties must be treated with justice and fairness;
3. confidentiality should be maintained to the maximum practical extent;
4. all semblance of conflict of interest must be avoided at all stages;
5. all stages of the procedure should be fully documented, and
6. unnecessary damage to the broad enterprise of academic research should be avoided.

i) **Inquiry.** The purpose of this stage is to determine, with minimum publicity and maximum confidentiality, whether there exists a sufficiently serious problem to warrant a formal investigation. Inquiry starts with confidential counsel, by the division chair, of those involved, and should be completed within 30 days of its initiation, with a judgment whether a formal investigation is warranted.

ii) **Investigation.** The investigation is carried out by an Investigation Committee, and should be completed within 120 days of its initiation. At this stage, the demands of confidentiality become secondary to the necessity that a vigorous investigation makes a conclusive determination of the facts. Nevertheless, every attempt is made to protect the reputations of all parties involved. A draft of the committee report is submitted to both complainant and respondent for comment before the final report is written.

iii) **Resolution and judgment.** The findings may be grouped into two broad categories:

1. no finding of research misconduct: the case is closed, but if the allegations are found to have been maliciously motivated, appropriate disciplinary action may be undertaken against the complainant;
2. finding of research misconduct: sanctions are considered, possibly including

[96]These stages are required by regulations issued by the federal government applicable to sponsored research.

- removal from the project;
- written reprimands;
- special monitoring of future work;
- probation or suspension;
- salary or rank reduction;
- termination of employment, and
- notification of all interested parties of the final outcome, *viz.*, the complainant, coauthors, collaborators, Editors of journals that have published compromised results, professional societies, other institutions that might consider employing the respondent, and the state justice system.

Summary. The *Caltech policy on research misconduct* covers a much wider field than the terms covered by the Babbage (1830) description: indeed, the composers of the policy have chosen the term *research misconduct* instead of the narrower *scientific misconduct*. On the other hand, the policy is not intended to deal with other problems – such as disputes over order of authorship – that the policy makers do not classify as research misconduct. Reading the complete document clearly exposes the rationale behind the policy:

1. when a complaint comes forth, it must be investigated;
2. if research misconduct has been committed, it must never be covered up;
3. if there is an indication that research misconduct has occurred, the case must be pursued, even in the absence of a formal allegation;
4. the case must be pursued to its conclusion, even if complainants and/or respondents resign from their positions at the institute;
5. the investigators' primary commitment is guarding the integrity of academic research;
6. maliciously motivated accusations cannot be tolerated, and
7. confidentiality, and avoidance of damage to people, are of prime importance.

Other Universities have published similar policies, see for example Stanford's *Research Policy Handbook* that lists the definition of research misconduct, and procedures for investigating and reporting allegations of misconduct.[97] Several of the above items should also belong to the rationale of any journal's editorial policy.

16.5 Detecting dual publication and plagiarism: an example

CrossCheck is an initiative started by CrossRef to help its members actively engage in efforts to prevent scholarly and professional plagiarism. Using the iThenticate software from iParadigms,[98] CrossCheck creates a continuously growing database of current and archival scholarly literature. As such, this service is a powerful tool in many Editors' continuous struggle with plagiarism.

[97] `http://rph.stanford.edu/2-5.html`
[98] `http://www.iparadigms.com`

Table 3. Similarity indices for papers in both Volumes of these Proceedings, as returned by iThenticate. The corrected index refers to the paper stripped of references and all quotations. The source for the highest similarity is listed, blank entries mean that the score is about 1%.

author	similarity index (%)	corrected index (%)	correction factor (%)	source of similarity
Bertout	37	36	2	Bertout & Schneider (2004)
Shore	3	1	66	
Henri	5	4	20	`www.edpsciences.org`
Adams	18	18	0	Adams *et al.* (2008)
Adams *et al.*	88	88	0	`www.aanda.org`
Grothkopf	12	8	33	`www.eso.org`
Cambrésy	19	17	11	`aanda.u-strasbg.fr`
Sterken (I)	28	23	18	Sterken (2006)
Sterken (II)	7	3	57	
Sterken (III)	22	8	64	Sterken (2006)

To illustrate its power, papers from both Volumes of these Proceedings have been screened with iThenticate, the CrossCheck plagiarism checker, and the result[99] is presented in Table 3. The principal outcome of the tool is the *similarity index*, *i.e.*, the percentage of similarity between a text submitted to the server, and information existing in the iThenticate databases selected as search targets. Beware that **this index is not a calibrated bibliometric index**, but a parameter that is the result of crosschecking multiple sources. And remember that **the index will change when the user changes the set of search databases**, or when content is added by CrossCheck members or iThenticate. Hence, comparing papers should always be done with the same set of reference databanks. Table 3 also provides a corrected index that refers to each paper after stripping the references and all quotations. The source for the highest similarity is listed, but sources that score below 1% are not given. The corrected similarity index for the complete Proceedings is < 10%.

The similarity index, evidently, should be seen as the result of a numerical experiment in a similar way as the citation and impact indices discussed in Section 18 were looked at – *i.e.*, by carefully monitoring biases and hidden "errors of measurement". Table 3 shows that

1. excluding references or quotes can decrease the similarity index by more than 60%;

2. this occurs for papers with a reference lists exceeding 10% of the paper length;

3. papers with many graphs or quotations (like Paper II) are affected likewise;

4. very short papers may yield similarity levels above 1%, because of the overheads (affiliations, publisher information, web addresses, etc.), and

[99]This exercise was concluded on 24 December 2010.

5. very large similarity indices can be misleading: the Adams *et al.* 88% score entirely comes from a match with just one internet source: this very manuscript was put online as a service to the readers of the journal, and exclusion of this source reduces the similarity index to 0%.

The conclusion of this exercise (see also Sect. 18.6) is that crossreferencing for plagiarism is a powerful method, and that technology is able to effectively bring down paper duplication. Indeed, Reich (2010) already sees a decline in the number of duplications as funding agencies tighten the rules – even refuse to subsidize duplicate articles in published conference proceedings and journals. The resulting similarity indices, though, should be used with care because of considerable margins of uncertainty, especially when taking the similarity index at face value in the framework of whistleblowing (see Sect. 17). Depending on the length of the article, I suppose that similarity indices of 1 to 4% – if not coming from one or two blocks of text – are unavoidable, and thus represent the threshold error level of the system.

17 Where to go for help: whistleblowing

A whistleblower is someone who exposes and reports wrongdoing on a specific case. Was Pasquich's assistant (see page 196) a whistleblower? Yes, certainly, but his accusations did not pass the verifications. The same applies to von Littrow (Sect. 17.2). Was Fletcher a whistleblower? In a sense no, and it was anyway too late, since his information was released posthumously. To understand Fletcher, one must see the situation in the context of his times, and weigh the circumstances of Millikan's support and assistance in his pupil's career development. But, as Goodstein (2010) says: *"ethical issues are not always black or white"*: it can be very difficult to prove the truth of your claims, and whistleblowing thus calls for a most prudent approach.

17.1 Motivations for whistleblowing

Whistleblowing should never be done with the purpose to take an advantage in ruthless competition, to bring benefit to your career, or in revenge. Nor should it be driven by any sort of *Robin Hood motive*, by malicious motivation, or by a conflict of interest. Whistleblowers should avoid crossing that ethical line, and they should be aware of the fact that bad motivations are very poor guides. Uncovering misconduct must be done in good faith and with great care, especially because in real life the case can be complicated by personal animosity.

Whistleblowing – like writing – is not just an act, it is a process, consisting of several phases. Sieber (1998) explains *attribution – i.e.*, the first step consisting of the perception that someone has engaged in a wilful act of deception. This first step is an act of interpretation of available facts or data, and must go unchallenged: is the whistleblower right for what concerns the data (observations, calculations, analysis, authorship, etc.) and, if so, is the perpetrator's misbehavior really deliberate (*i.e.*, can plain error be excluded)? Then comes the step to find others who could share your conclusions – where "others" are individuals, and not just the vaguely defined surrounding group comprising the everybody-knows-about-this-case population.

17.2 Whistleblowing: how to

Whistleblowing by exposition. Currie (1997) reports that, during his career, he personally experienced events that constituted fraud, misconduct, or simple unfairness to more junior staff. He exposes the following facts in some kind of public whistleblowing:

- a senior person submitting a dissertation for a higher degree, in which an appreciable part of the introduction had been copied verbatim from a published paper;
- manufacturing of important data (on two occasions);
- junior staff were omitted from published work;
- a junior researcher conceived a study and wrote the first draft of the proposal, but was excluded from the final submission that received substantial funding – the exclusion was considered to improve the chance of funding;
- token authorship;[100]
- token supervision of higher degrees;
- destruction of computerized data when a funding body decided to scrutinize the work more closely, and
- multiple applications for funding of essentially the same proposed work.

Whistleblowing by citation. A couple of years ago, an economist filed a *defamation lawsuit*[101] against an author who had written that someone's hypothesis does not seem to be true because other scholars could not replicate the results. The complainant defended that the term *replicate* means that other scholars have analyzed the same data, but could not reach the same result, which implies that the results were falsified (Shermer 2006).

The difference is very subtle, as can be easily demonstrated by an example from this very paper: someone may say that the results shown in Figures 5, 7–9, etc. cannot be replicated, although the database sources used are open to anyone (who pays) – thus implying that the evidence is fudged. So, why is exact duplication not possible? Simply because the graphs do not carry a date stamp, because both databases are continuously updated and corrected, because the selection criteria have not been rigorously described, and because the names of A, B, C and D have not been given in the first place.

The example shows that you must always be very careful with accusations in your printed work. Untested allegations expressed via websites that disclose scientific fraud can cause huge damage to individuals, teams and institutes, because no formal mechanisms of investigation are called for. But that also refers to referee reports: whenever you, as referee, have the impression that some kind of misconduct is involved, it is better to

[100] Insignificantly small share in the work.

[101] A suit against an abusive attack on a person's good name. It encompasses both *libel* (written defamation) and *slander* (oral defamation).

report it to the Editor by using the confidential box in the referee report form: there you can write your opinion with least risk to be sued, simply because the Editor is bound by the obligation of confidentiality (see Sect. 10.3). A special warning is due on national libel laws, for example English libel law (see also page 221) is much tougher than the United States libel law: the burden of proof in England is on the defendant, not on the plaintiff – just the opposite of US law.[102]

Whistleblowing by intervention. Resorting to *Scientific Red Cards*, Section 16.2, is one form of whistleblowing by intervention. A more straightforward approach is the direct procedure, which is not without pitfalls, especially if you are directly involved as one of the victims of misconduct. The problem with personal involvement is that it is very difficult to temper intense anger against mistreatment, and that you end up burning bridges, and giving your adversaries even more reasons to isolate you. The following routes are then open when you feel the need for blowing the whistle:

- tell your thesis or research supervisor(s);
- inform Human Resources Department at your university;
- transmit the facts to the Ombuds Office at your university;
- communicate with the ombudsperson at a scientific journal;
- or just start blowing if you are very sure that you have a solid case.

Gunsalus (1998) explains that filing charges of scientific misconduct can be a risky and dangerous endeavor, and concludes that it is possible to blow the whistle and still have a career afterwards, but that it takes a combination of common sense, prudence, and luck. But there are no guarantees. She presents rules of conduct to follow for reporting perceived misconduct, and a set of procedures that describe how to pursue responsible whistleblowing **once the decision to report misconduct has been made**. Her six rules and eight procedural steps are given below.

Rule 1 Consider alternative explanations (especially that you may be wrong), and re-examine your logic to see if it still holds up.

Rule 2 In the light of rule 1, ask questions, do not make charges. Here she explicitly refers to coauthorship issues (see also Paper I, Sects. 9.2 and 13.1), and to the fact that students are not always fully aware of contributions that were made by others in the laboratory, including the supervisor.

Rule 3 Figure out what documentation supports your concerns and where it is, and present yourself as professionally and unemotionally as possible, so that the focus is kept on the issues, and does not shift to the personality of the whistleblower (as happens all too often).

Rule 4 Separate your personal and professional concerns. Avoid being overwhelmed by anger, frustration, or resentment, and speak to real friends – *i.e.*, exclude the people you deal with professionally in this issue.

[102] Karla Adam, *Washington Post Foreign Service*, Wednesday, February 24, 2010.

Rule 5 Assess your goals: it is critical to know what you personally seek from the situation, and analyzing your goals can be very helpful in choosing the path to follow. Moreover, the question *"what are your goals"* will almost certainly be asked, and having a coherent and reasoned answer will have a constructive effect on the process.

Rule 6 Seek advice and listen to it: this is in fact the moment when you decide to stop, or to go on and file charges. From here on you follow Steps 1–8, taking into account that there is risk in seeking advice, and that you run the risk of losing control of the situation.

Step 1 Review your concerns with someone you trust, and who is in a position to evaluate your concerns. The emphasis here is on seeking **confidential** advice, and by all means avoiding someone with a close personal relationship with the subject of your concerns.

Step 2 Listen to what that person tells you: if that person agrees with you that there is a problem, ask what steps can be taken and who will take them; if that person disagrees with you, then evaluate whether the disagreement is objectively based on the facts you provided, or if the disagreement is a matter of personal predilection. In the latter case, proceed to Step 3, but with care.

Step 3 Get a second opinion and take that seriously. Remember that you keep focussed on the facts, and that you are posing questions, not lodging accusations.

Step 4 If you decide to initiate formal proceedings, seek strength in numbers. In other words, find out who will stay on your side, and also find out why some people do not want to support you (like you did in Step 2): having support may mean the difference between surviving the process with a career intact or not.

Step 5 Find the right place to file charges, and study the procedures. One mistake is to address journal editors and ask them to investigate claims made in papers submitted to them, rather than to the organization (or research team) for which the alleged offender works.

Step 6 Report your concerns and provide all the documented information you can. From now on the situation becomes stressful, and many people under stress do not hear or remember things as well as they might wish.

Step 7 Ask questions and keep notes of the answers you are given.

Step 8 Cultivate patience: this process always takes longer than will feel reasonable.

17.3 Whistleblowing and its aftermath

Sieber (1998) draws attention to the fact that beginning researchers[103] **have an inescapable obligation to act** – that is, to blow the whistle on misconduct – if they suspect a colleague of violating the ethical standards of the scientific community. But she judiciously

[103] If not all researchers.

adds that when the other side is powerful, the whistleblower hardly stands a chance of surviving the conflict unscathed, unless great sophistication or institutional wisdom accompanies the process. Gunsalus (1998) expressed the same concern (see previous Section), and Clouthier (2003) draws attention to the dangers for young people by saying

> IT IS UNLIKELY THAT SUFFICIENT PROTECTIONS FOR JUNIOR FACULTY MEMBERS, FELLOWS, OR STUDENTS CAN EVER EFFECTIVELY BE IMPLEMENTED INTO AN ACADEMIC SYSTEM SO PREDICATED ON THE POLITICS OF PUBLICATION.[104]

Swazey (1999) comments on the adverse consequences, especially that whistleblowers are judged to have violated the norms of loyalty to the team, in particular if they report outside the working group. She also indicates that it is an enormously stressful experience to be accused – or to be the accuser – and that the process may have negative effects on the mental health. Sieber refers to psychological shocks that are in store for the whistleblower, and adds *"... if one survives the experience intact, one is forever changed"*. One of her sources advises

> FORGET IT. BUT IF YOU CAN'T FORGET IT, THEN LEAK THE INFORMATION – MAKING SURE THAT YOUR NAME ISN'T ASSOCIATED WITH IT.

Blowing the whistle can also blow you out of business altogether, or at least blow you out of your field of research.

Unfortunately – as I found out myself – most whistleblowers are naive about the precautions they should take, and about the fact that the whistleblower typically operates in a hostile environment, and that virtually no one will be on their side once the case gets underway (especially during the phase of being banished from the society inside which the whistle is blown). Anyone who considers acting as whistleblower, should read Sieber (1998) and also Gunsalus (1998), and I deeply regret that no such papers were around – nor did I have the slightest clue – when I blew whistles two decades ago.

Over my career, I have been confronted with three severe cases of research and publishing misconduct, and in each of the cases I dealt with the situation in a different way. In the first case I discovered unacceptable data trimming and possible forging, but I was not personally implicated. Moreover, my whistleblowing was guided by very wise advice from a senior colleague from another institute,[105] and I could manage the process with minimal emotional involvement.[106] Nevertheless, I did not come out of the affair unscathed: my relationship with some senior colleagues was ruined forever. In a second case I was directly involved, and I lost control because I could not retain my *sang-froid*,[107] and the consequences were that I paid the bill. Emotion is a poor guide to obtaining your rights, because **emotion causes a very strong – even dominant – noise component in the chain of communication** as depicted in Figure 2 of Paper II. In the third case (explained in detail on page 215), I had learned that "they could break you", and I simply shut up.

---

[104] That is, instances of plagiarism perpetrated by senior faculty members against their graduate students.

[105] Dr. André B. Muller (1918–2006), see West (2006), and Sterken (2002).

[106] It is well known that emotional load can change your perception of the situation.

[107] *Sang-froid* or *sangfroid*, is the ability to stay calm under strain.

What is clear, is that the whistleblower – and (or) the victim – may never hope to receive an apology from the offender/wrongdoer – nor from "the system". And that whistleblowing does not come cost-free: there is always a price to pay – as Eldakar & Wilson (2008) demonstrate, *"punishment can be costly to the punisher"*.

18 On the biases of bibliometric indices

Bibliometric counts are measurements that – like any physical measurements – are beset with errors. The emanating investigations and their conclusions are thus, in turn, biased to some extent by database errors, and by interpretation errors. Database errors comprise errors of procedure, errors of confusion, errors of omission, and errors of content.[108] The analysis and conclusions are also affected by procedural errors, especially when the bibliometric indices are not properly calibrated, and – above all – when these indices are (by design) unreliable indicators of scientific value or quality.

Seglen (1992), about citation numbers as a useful parameter for evaluating scientists and scientific projects, remarks that citations represent a measure of (a limited kind of) *utility* rather than of quality. He broadly classifies problems into four major categories:

1. *Citation bias*, *i.e.*, citations reflect both the needs and the mannerisms of the citer, including such factors as utility, quality, availability, advertising (self-citation), collaboration or comraderie (in-house citation), chauvinism, mentoring, personal sympathies and antipathies, competition, neglect, argumentation, flattery, convention, reference copying, reviewing, secondary referencing, and so on;
2. *Incomplete journal coverage and other registration problems*: partial sampling of the total scientific literature by databases using a too narrow sampling window;
3. *Author variability*: the significant differences between authors, with the additional difficulty of how to grant credit to multiauthored papers, and how to take article length (and author name position) into account, and
4. *Field effects*: the difference in citation properties between different scientific fields.

Wright & Armstrong (2007) list three types of citation problems: omissions (failure to include relevant studies, see Sect. 7), incorrect references (see Sect. 9.12 in Paper I), and quotation errors (misquotation and misattribution, see Sect. 11.9 in Paper I).

In fact, citation counts can somehow be compared to television viewing figures, and we all know too well how easily these figures are influenced by adding a flavor of adventure, or a zest of erotics to an otherwise dull performance: popularity and high rankings are not necessarily tightly correlated with artistic excellence or deep insight. The same can also occur in astronomy: a dull star that you happened to observe in your student days may suddenly erupt, and receive short but intense observational attention with large telescopes, a situation that will undoubtedly boost a citation count that may perhaps have been dormant for more than a while. But also the publication of grave mistakes may boost

[108] See also Uta Grothkopf's paper in Part 1 of these Proceedings.

the citation index, in exactly the same way as some *mala fide*[109] webshops score high in web search machines, for no other reason than the many complaints that increase their scores.

Kennedy (1997) doubts the tightness of the correlation between citation index and excellence, though he sees a correlation between utility and citation: **papers that are almost never cited are likely not very influential**. He indicates that up to 75% of papers are never cited at all, though the several cases discussed in Figure 7 render a more optimistic view for astronomy papers: a mere 20–30% remains uncited.

Citation analysis can lead to completely wrong conclusions when done by non-experts, or when done improperly, and can severely harm those concerned. Many users of bibliometric databases are lay users, even those who do this job on a regular base for evaluating people and projects for the university or for a research foundation. I consider myself as non-expert – though not a novice – user.

The following Sections and paragraphs present some case studies that are based on data extracted from the ISI and ADS databases.

18.1 Errors of confusion by the database construction

Misattribution of scientific disciplines or subject areas. My non-expert ISI Web of Science analysis of my own publication record[110] reveals that

- the category Multidisciplinary Sciences contains one astronomy paper, and
- a paper on photometry of galaxies is listed under Geochemistry & Geophysics.

Flagging these categories on or off affects the Hirsch index: improper inclusion or omission of disciplines in someone's citation analysis may easily lead to an uncertainty of several units of h.

Confusion of the type of paper. The same analysis shows that

- a substantial number of the listed papers are outright or concealed unrefereed conference papers;
- several of these show up as genuine research papers, and can only be recognized as being non-research papers by someone who effectively consults those papers;
- unflagging the Conference Proceedings Citation Index option still returns various unrefereed conference papers – among others IAU Symposia, but also an "untitled" paper in an ISI-listed journal (introductory notes of a conference proceedings, Duerbeck *et al.* 2002), and a paper on the history of astronomy (Sterken 2002).

[109] From Latin *mala fides*: bad faith, deceptive.

[110] After removal of some papers in clinical neurology, which are due to a namesake.

18.2 Errors of confusion by the database user

The prime confusing elements are using the wrong type of database, mixing data of different persons with the same names and initials, and misattribution of subject areas.

The type of database. The nature of the database must be weighted: the ISI Web of Science, for example, is designed for scientometrics, whereas ADS has been set up for information retrieval. ADS covers a much wider selection of sources, whereas ISI is more restricted. Nevertheless, both databases are very useful and complementary. However, **when comparing people or projects, all information should be drawn from the same database, and at the same time**.

Authors with the same name and initials. Namesakes appear in disciplines as different as pediatry and astronomy, but also inside a same scientific discipline the problem occurs: some people simply have same surname and initials, and their bibliometric data cannot be automatically separated, unless their research fields are very different. Figure 15, for example, displays ADS-based statistics for two authors with same name and first initial: the senior author always uses only one initial, the junior author sometimes uses two initials. The upper forked curve (which bifurcates in 1993, the moment when the junior author started publishing) represents all ADS entries (as in Fig. 5), the lower forked curve refers to refereed papers only. The lower branch of each fork results from correcting the upper branch for the production by the junior author.

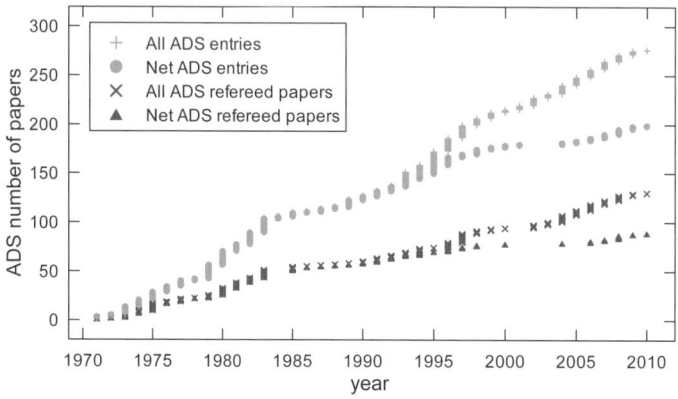

Fig. 15. ADS-based statistics on two authors with same name and first initial. Both curves bifurcate in 1993: the lower branch of the fork results from correcting the upper branch for the scientific production by the junior author.

Figure 16 illustrates the difference in the distribution of citedness of both authors: the junior author, despite a *scientific age*[111] difference of 25 years and a number of papers that is less by a factor of two, has a h index that is 4 units higher. Note also the contrast in citation counts for the top papers. Part of these differences can be attributed to the different nature of the field of research of these authors.

[111] Scientific age is a vaguely-defined concept, and can be somehow equated with the total length of a scientist's publishing career, or with post-doctoral age.

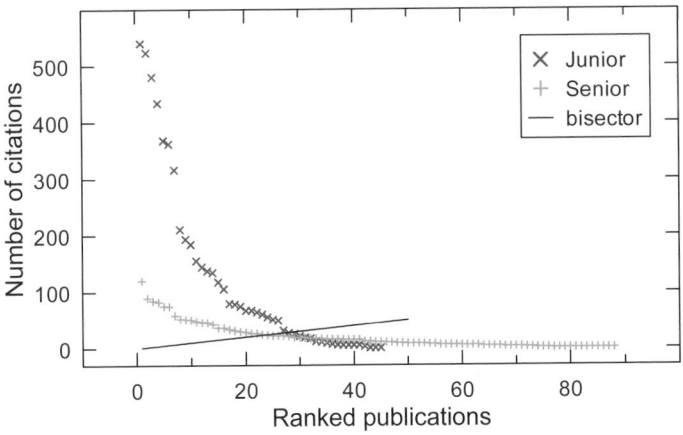

Fig. 16. Similar to Figure 7 for both authors referred to in Figure 15 (uncorrected ISI data). The junior author, with only half as many papers, has a much higher h. Note the contrast in citation counts for the top-score papers.

Figure 17 shows the ISI-listed subject areas and corresponding number of papers for another specific author name and initial. The filled bars indicate the top five discipline scores that appear by default when consulting the database. Note that both examples refer to not-so-common author names to which only a couple of persons correspond, and that a very common name will lead to many more problems. Take, as an example, astronomer Myron A. Smith, a stellar spectroscopist: the outcome is about 3000 papers in more than 100 ISI Subject Areas, with astronomy not even listed in the top five, with several papers unavoidably lost in areas like MULTIDISCIPLINARY SCIENCES.

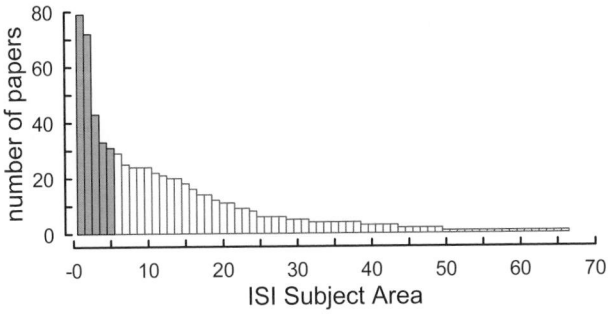

Fig. 17. ISI-listed subject areas and corresponding number of papers for one specific author name and initial. The filled bars indicate the top five discipline scores that appear by default, when consulting the database in default search mode settings.

18.3 Errors of omission and inclusion

The journal bibliometric observables feature several biases: the sample of journals that are included in the count, the width of the sampling window, and the definition of which articles are citable and which are not.

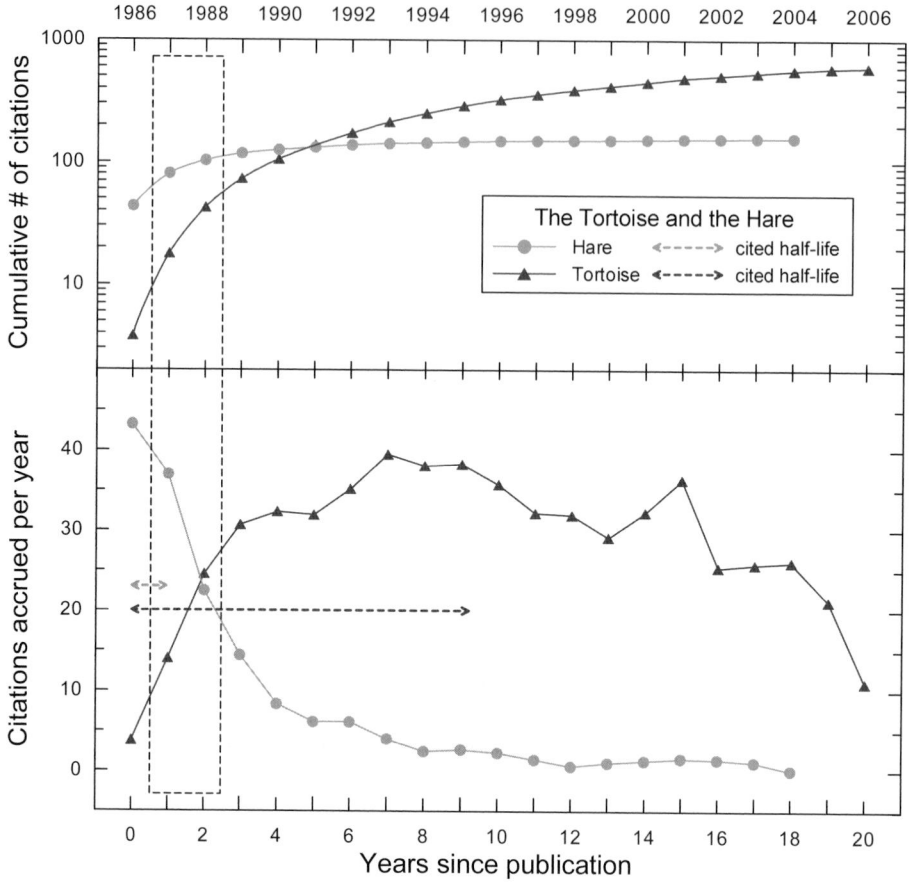

Fig. 18. The Tortoise and the Hare. *Top*: $\log_{10}$ of cumulative number of citations, as a function of calendar year. *Bottom*: citations accrued by each journal each year after publication. Based on data from Figure 8 of Vanclay (2009). The rectangle represents the sampling window used in calculating the journal impact factor (year 0 is the year of publication and is excluded from the calculation). The horizontal arrows indicate the cited half-life (the time interval that contains 50% of the citations).

Incomplete sample of journals. Vanclay (2009) presents statistics showing that the disproportionate sampling implies that the impact factor gives a misleading indication of the true impact of journals, and that it is biased in favor of journals that have a rapid rather than a prolonged impact. This author gathered citation rates of 14 titles, among which he selected two journals with a comparable number of citations in 1994, but with contrasting trends of citation accrual (denoted the Tortoise and the Hare, with impact factors 0.2 and 1.3, and cited half-lives of >10 and 2.4, respectively). Figure 18 shows the cumulative number of citations as a function of calendar year, and the citations accrued by each journal for each year after publication. The sampling window used in the calculation of the journal impact factor is shown. The Hare shows an abrupt culmination of the citation accrual pattern, whereas the Tortoise (and several other journals of the sample) shows a steady accumulation of citations to each year of publication, with little evidence of declining interest in earlier papers. Notwithstanding the manifest lasting value of papers in the latter journal, its impact factor is a factor of 6 smaller than the former. It is evident that the 2-year window used to calculate the impact factor creates a large distortion in favor of the Hare.

Bibliometric studies have revealed that the impact factor is severely affected by a small number of papers (*Nature* 2005, Vol. 435, p. 1003), and the inflated status and limited metric value of the impact factor is very well known.

Inadvertent omission of data subsets. Bibliometric databases sometimes inadvertently omit data subsets. The best way to evaluate this effect is to plot entries for two databases against each other. Figure 19 shows the relation between citation counts, as I extracted from ISI and ADS, for one selected senior author of intermediate scientific production (136 papers in refereed journals, common to both databases, not corrected for self-citation). The least-squares fit yields the relationship $ISI \approx (0.90 \pm 0.02) \times ADS$ with a correlation coefficient of 0.98. The graph shows that some papers yield markedly less citations from ISI than from ADS. Likewise, inspecting my personal citation index, I discovered that one review paper in an ISI-indexed journal had accrued only 1 (one) citation over the last 15 years, although more than 30 citations are listed by ADS.

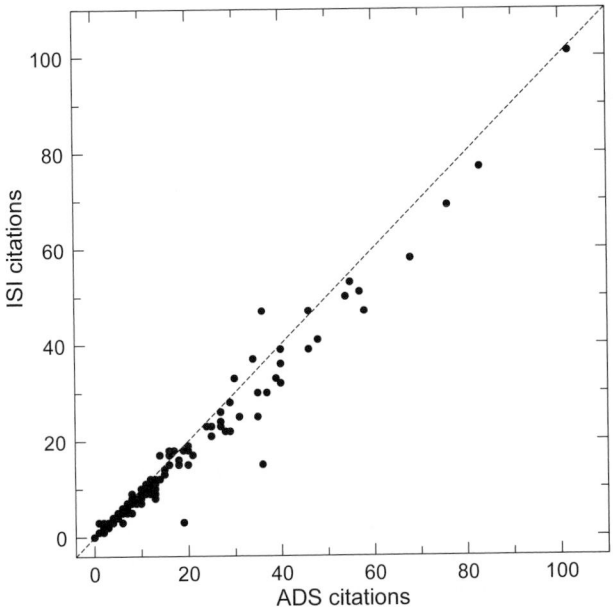

Fig. 19. Correlation between citations, as given by ISI and ADS, for one selected senior author of intermediate scientific production. The dashed line is the bisector.

Figure 20 is a similar graph for a set of authors, and shows the correlation between citation counts, as given by ISI and ADS, for a selection of top-score papers. The least-squares fit yields $ISI \approx (0.97 \pm 0.03) \times ADS$. The comparison reveals that a considerable number of citations have no counterpart in the ISI database. The $O - C$ amounts to several hundred units, with a standard deviation of 180, which illustrates the impact of the selected database. The inset shows the full range of all citations, and the outlier on the x-axis originates from an erroneous 0-citation caused by a misspelled first-author surname in the ISI database.

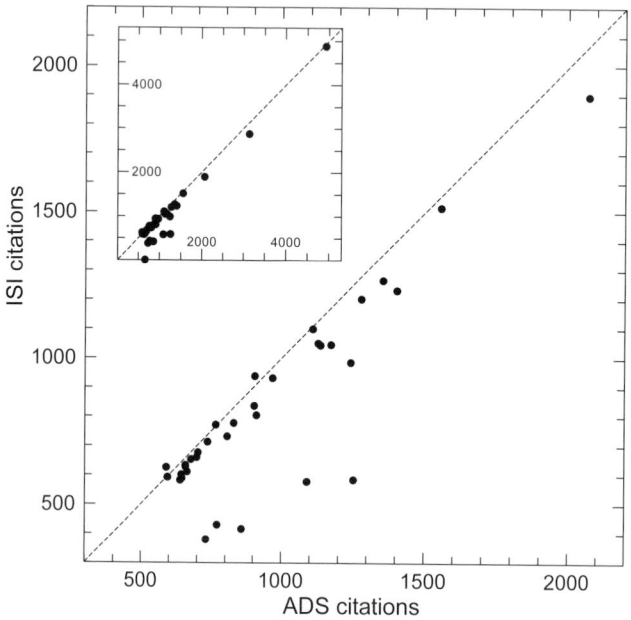

Fig. 20. Correlation between citation counts, as given by ISI and ADS, for a selection of authors of frequently cited papers. The inset shows the full range of all citations, the outlier on the x-axis originates from an erroneous 0-citation caused by a misspelled first-author surname in the ISI database (not visible in the main diagram because the axes have different zeropoints). The dashed line is the bisector.

These comparisons bring into light the impact of errors in the database. Tort & Amaral (2007) also question the accuracy of citation counts, because the references cited by online supplementary information in the journals are not being tracked as citations in databases like ISI (see also Seeber 2008).

18.4 Inherent bias

Inherent bias is implicit bias that is not readily apparent: it covers the effect of the type of publication, and of the research field.

The type of paper. Section 5 in Paper I lists categories of scientific papers, among which more than a dozen qualify to figure in bibliometric databases. Each of these types of papers generates citations in its own way. Letters to the Editor appear much faster than regular research papers, and are expected to yield higher citation rates. Review papers, as a rule, also generate lots of citations. With data papers, the situation is more ambiguous: some data papers generate citations only sporadically, but others – especially those papers that qualify as tools – are very popular, see the top scores in Figure 20, the discussion on page 188, and also Piwowar *et al.* (2007). Classical ground-based data, in particular, lead to less citations than data from large ground-based and space facilities – not in the least because most of these (raw) data are publicly accessible through data repositories. Also salami papers (Sect. 5.16 in Paper I) create citations in their own way via enhanced advertizing, as de Maupertuis (1698–1759) said[112] already 250 years ago:

> PUBLISH SMALL WORKS OFTEN, AND YOU WILL DOMINATE ALL OF LITERATURE.

What about conference papers? The Proceedings book of ASP Conference Series Vol. 349 (Sterken & Aerts 2006, a PhD school with a somewhat similar audience as SWYA), included 15 invited talks by senior scientists. The 11 scientific overview papers have, at this moment, an average citation rate of 3.4 citations per paper. Four other talks, dealing with databases and scientific communication, have received not one single citation so far. The outcome of a student poll at the end of that meeting, though, revealed that the paper on scientific writing (the precursor of Papers I–III in this Volume) was seen as the most useful lecture of the meeting. Notwithstanding the fact that this was the only paper of the book in *Open Access*, the paper has not rendered one single citation in more than four years. So, what is the lesson? The conclusion is that

1. the citation database is far from complete;
2. papers with 0 citations are not useless: some of these even contain most valuable information that is not found elsewhere, nor is widely taught, and
3. some journal and proceedings papers not taken into account by bibliometric databases, are effectively cited, and have an impact on science – but **they do not count**.

The *Publishing Research Consortium*[113] presented a global study (more than 3800 respondents) assessing the importance of, and ease of access to, professional and academic information (Publishing Research Consortium 2010). Figure 21 represents the responses for the subject area PHYSICS & ASTRONOMY to two questions:

1. *When it comes to information, how important is it that you have access to the types of information listed below?*[114]
2. *Of the various types which you consider important, how would you describe your level of access to them?*[115]

The study clearly indicates that of all types of papers, **original research articles in journals are ranked highest in terms of importance and in terms of ease of access**. Next come reference works, which also include review papers. It is thus not surprising that these kinds of papers also attract the highest citations. Beware, though, that this graph

[112] Quoted by Terrall (2002).

[113] PRC is a group of associations and publishers, which supports global research into scholarly communication in order to enable evidence-based discussion. The aim is to promote an understanding of the role of publishing and its impact on research and teaching. `http://www.publishingresearch.net/`

[114] The types given in Figure 21; scale 1 = not at all, 7 = very (only scales above 3 are shown in Fig. 21).

[115] Scale: very difficult, fairly difficult, fairly easy, very easy.

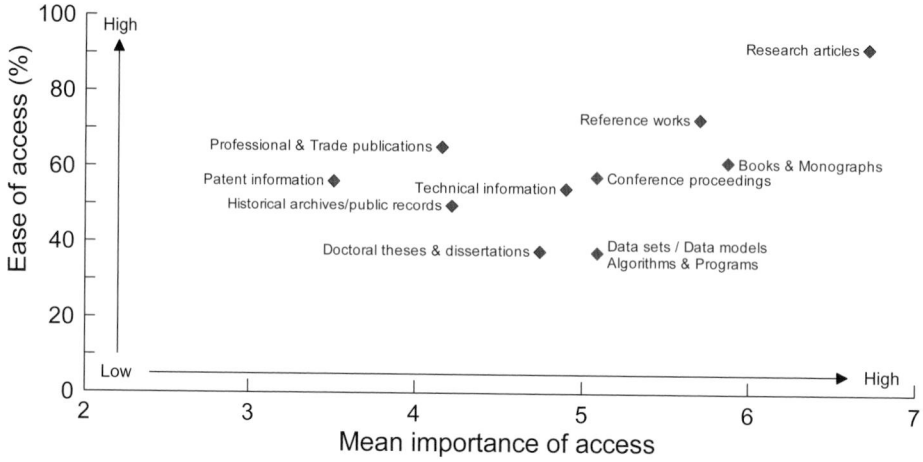

Fig. 21. Information types: importance versus ease of access. Based on a global study, PRC (2010).

should not lead to the widespread misinterpretation that unrefereed research papers – *i.e.*, everything with a mean importance of access below 6 in Figure 21 – may be composed in a sloppy way, simply because these works are not refereed. In my opinion, any of the other types of papers should be written with the same standards of rigor as any other refereed paper – and this is especially so for PhD thesis manuscripts.

The impact of the research field. The field of research is a quite dominant factor in citedness (for example, citations in theoretical stellar physics are not in step with citation counts in some branches of observational astronomy). The impact of the nature of the research field is obvious: it has long been known from biological sciences that studies on short-lived animals (for example, fruit flies) generate more citations than papers on long-lived organisms (*e.g.*, land tortoises). In a similar way, astronomical phenomena with short cycles (*e.g.*, δ Scuti stars on time scales of hours) will yield papers with more citations than studies where decades of data are needed (*e.g.*, luminous blue variables). Nor is this factor constant in time: some fields that were quite popular in the past (for example, observation and modeling of binaries) have now become much less citable, whereas exoplanetary and extragalactic science are much hotter topics now. The evolution, moreover, is not unidirectional: some disciplines (for example astrometry) have regained popularity because of large space projects. Habing (2009), for example, underlines that X-ray astronomy is almost completely dependent on satellite observations, which is not the case for infrared and visual-band astronomy. The high citation rates of X-ray papers reflect that these lead to rapid progress in our understanding of the cosmos, though these numbers also reflect the number of scientists working in this specific field: *"the community interested in cosmology is large, the satellites few"* (Habing 2009).

Two principal determinants of citedness can be recognized: the *scientific objective*, and the *scientific project*:

- **key scientific objectives** (for example searching for habitable planets of nearby stars) always attract many citations, and
- **key projects**, especially large and space-based (see the discussion related to Fig. 6) also attract many citations.

Figure 22 shows the evolution of the number of Letters per A&A Volume for the period 1996–2010. The impact of key projects on the number of Letters is evident, and because Letters have a higher citedness, the citation bias is thus driven by project as well as by type of publication.

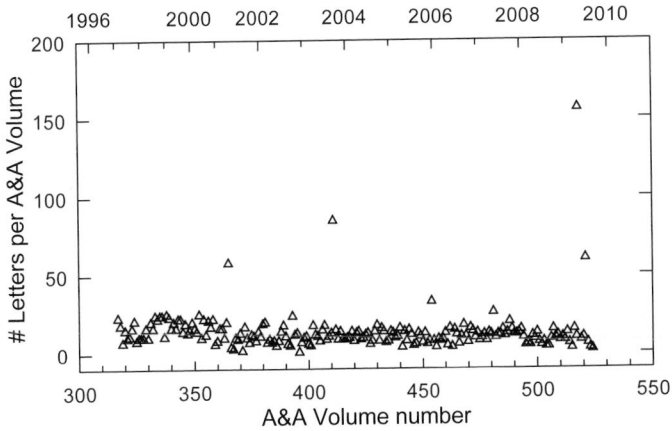

Fig. 22. Evolution of the number of Letters per A&A Volume. The peak Volumes are special Volumes dedicated to a project. The scale of the time-axis changes in 2001, when the number of Volumes per year was changed from 12 to 16.

18.5 Errors of procedure

The citation rate, impact factor, Hirsch index, etc. are indices, *i.e.*, numbers resulting from a count, or from a simple arithmetic operation using these counts. Most indices are just markers without any relationship to the quality of the measured quantity. An index may eventually become an indicator when it has been shown to really quantify a characteristic in an unbiased way. All bibliometric parameters discussed here, unfortunately, possess considerable biases, and, in addition, incorporate procedural faults.

The time bias. The citation-based indices are biased in the sense that they depend on scientific age, because an author's number of published papers and citations grows with time. A scientist's production is incremental, though sometimes discontinuous, including gaps of inactivity, and bursts of hyperactivity. These irregularities are sometimes due to family situations (for example pregnancies, or illness), or job situations (administration and teaching loads), which are not taken into account in any considered bibliometric parameter.

Some individual papers hit more citations than a single most productive scientist (*viz.*, cases A–D) accumulates in an entire professional life. On the other hand, it also happens that some very good papers attract very few citations, or that their visibility starts only after several years – for example, the case of Lemaître's (1927) paper discussed in Papers I and II – for reasons of omission of a particular element, like a graph. Another such case is Hoyle's (1954) fundamental paper that showed that all chemical elements were synthesized from hydrogen and helium by nucleosynthesis: Clayton (2007) attributes the undercited status of this paper to the omission of one single equation that is central to the theory, an equation that Hoyle, though, had described verbally.

The ratchet effect. Merton (1968) refers to a *ratchet effect* operating in the careers of scientists such that, once having achieved a particular degree of eminence, they do not later fall much below that level (although they may be outdistanced by newcomers, and so suffer a relative decline in prestige, as illustrated in Figs. 5 and 15). Merton finds that this pattern of recognition, skewed in favor of the established scientist, appears principally

1. in cases of collaboration;
2. in cases of independent multiple discoveries made by scientists of distinctly different rank, and
3. when a paper written by a comparatively unknown scientist is submitted instead by an eminent colleague, and thus stands a better chance of being published.

The ratchet effect is similar to the *Matthew effect*, defined by Merton (1968), as an adverse effect on recognition of the "lesser" authors in collaborations with more famous colleagues, resulting in the enhancement of the position of already eminent scientists who are given disproportionate credit in cases of collaboration or of independent discoveries. An enhanced ratchet effect can also be the result of publicity, and of a team's image-building policies, sometimes fostered by sending team members to many conferences.

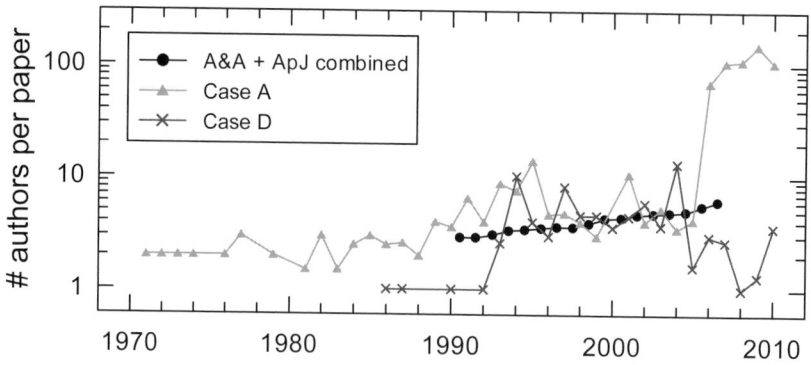

Fig. 23. Average number of authors per paper for cases A and D, and average number of authors per paper published in A&A and ApJ combined (• from Fig. 4), on a $\log_{10}$ scale.

The number of coauthors bias. It should be stressed, though, that there exists a serious effect of group size, as Crane (1965) concluded

SCIENTISTS AT MAJOR SCHOOLS ARE MORE LIKELY TO BE PRODUCTIVE AND TO WIN RECOGNITION THAN SCIENTISTS AT MINOR UNIVERSITIES, WHICH SUGGESTS THAT UNIVERSITIES PROVIDE DIFFERENT ENVIRONMENTS FOR SCIENTIFIC RESEARCH.

It is well known that the scientist's environment shapes citedness, as citation behavior is modulated by the size of the research group (members of large teams cite one another), and by the mobility of its members (especially in hyper-collaborative branches of science, like space research). This leads to a kind of limitation of view that favors research groups, some kind of group Matthew-effect, especially when authors recur to *"citing by example"*. Moreover, isolated researchers belonging to small research units are less well adapted to handle fluctuations, see also Section 5.2.3.

Collaboration is not obligatory, but it has become the norm, and this ever increasing specialization and synergism simply leads to larger outputs. This factor is not reflected in the citation analysis provided by bibliometric databases, nor is author escalation taken into account. Figures 23 and 24 show the impact of intense team work, and also illustrate that citation count and h are severely influenced by the number of coauthors: cases A and D have opposite behavior in number-of-authors policies, nevertheless $h_A \gg h_D$. In addition, next to faulty citation counts, there is also an effect caused by faulty citations (*i.e.*, erroneous references, see Sect. 9.12 in Paper I).

Figure 25 is a bar chart that shows the number of times each author shares authorship for a collection of about 30 refereed papers published in the same year, involving nearly 200 different author names. This set of papers will lead to almost 1500 entries in any bibliometric database. The inset shows the same data in a histogram of the number of times each author appears in the authorlist: about 40 scientists share authorship only once, but three authors share more than 85% of these papers. Both graphs display the

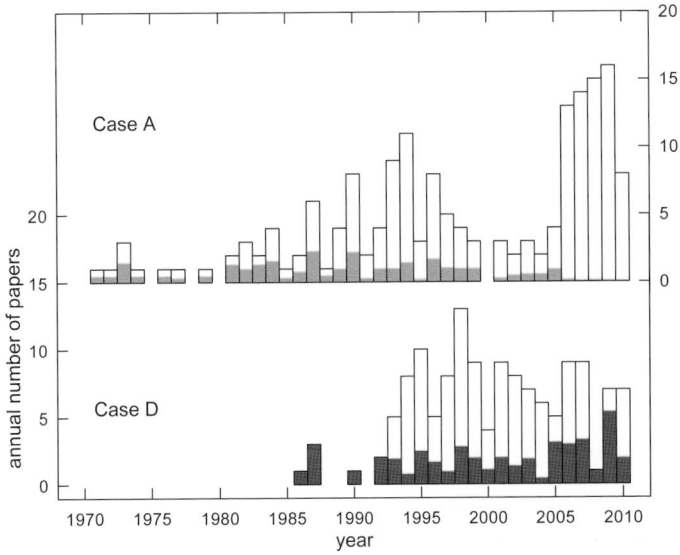

Fig. 24. Annual number of papers for cases A and D: annual number of papers in ISI-listed journals (white bars), and that number divided by the average annual number of authors (filled bars). Bars are not stacked, and should be read from the 0-line.

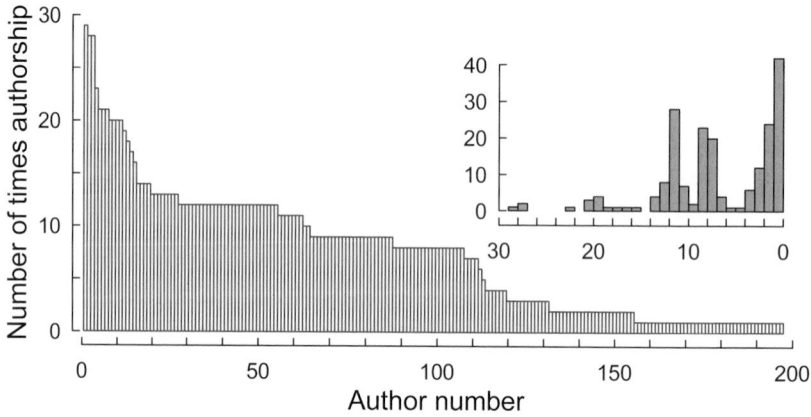

Fig. 25. Bar chart of number of times each author shares authorship for a set of 30 papers published in the same year. The inset (with reversed x-axis scale) shows the histogram of the distribution.

same information, but the histogram is helpful in assessing the role of some authors in this joint enterprise. The median number of authors per paper is 57, the median page per author is 0.07, and the average total accumulated page per author is less than 3 pages.

The question is how such a set of papers can be adequately and independently refereed: the pool of suitable referees for such a set of papers becomes really small, and much less independent as is the case for regular papers. This example case was not brought up for judging the appropriateness of these publications, but merely to ask the question **how to deal with such bibliometric data from the point of view of scientometry**.

Bias through self-citations. Self-citations – authors citing their own works – account for a significant portion of all citations. Fowler & Aksnes (2007) debate that self-references may result from the cumulative nature of individual research, from the need for personal gratification and self-agrandizement, or from the value of self-citation as a tactical tool in gaining more visibility. Anyway, they conclude that

THE MORE ONE CITES ONESELF THE MORE ONE IS CITED BY OTHER SCHOLARS.

Figure 26 is based on citations given by ADS, and illustrates the impact of self-citations on the h-index of one specific author in astronomy. The squares represent all citations to the ranked paper, and yield $h = 14$, the filled circles are the same data but with self-citations excluded, and lead to $h = 7$: obviously, the ranking sequence changes when self-citations are excluded. The inset shows the histogram of the percentage of references that count as self-citations, for the papers that lead to $h = 14$: self-citations count for 50–70% of all citations.

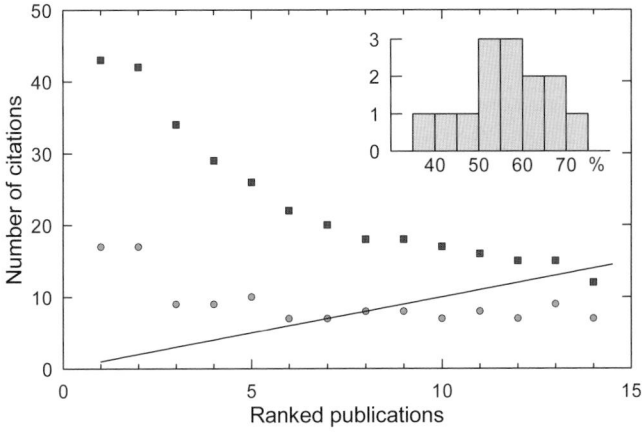

Fig. 26. Impact of self-citations on h for one specific author. ■: all citations; •: excluding self-citations. The line is the locus of papers with rank equal to the number of citations. Inset: histogram of references in these 14 papers that count as self-citations.

18.6 On rankings based on one single quantity

Measuring the value of a complex quantity using one single number or index – and applying this datum for the ranking of human beings or human enterprises – is a questionable practice. It is like assessing a person's health solely on the basis of the *Body Mass Index* (*BMI*), or a person's intelligence only with the *Intelligence Quotient* (*IQ*). The following paragraphs identify some biases of the widely applied *BMI* and *IQ* indicators, and compare the shortcomings with those of some bibliometric markers, in particular with the impact factor.

The *Body Mass Index*. The *BMI*, or Quetelet index,[116] proposed by Quetelet (1835), is a statistical measure of body weight that involves a person's body height, *viz.*, $BMI = m \times h^{-2}$, with mass and height expressed in units of kg and m.[117] The "ideal" range – that is, for the sample drawn by Quetelet – is $18.5 < BMI < 25$. The index is beset with various problems:

1. *BMI* is, first of all, a screening tool, and not a diagnostic tool;

2. *BMI* (along with h and *IQ*) is very easy to calculate, but delivers a false impression of accuracy;

3. just like the bibliometric indices discussed before equate citedness with value and impact, *BMI* equates height with body frame size, and equalizes muscles or fat with bone.[118] So, the more muscle (and bone) a person has, the more the numerator, and

[116] Adolphe Quetelet (1796–1874) was a Belgian astronomer, mathematician, statistician and sociologist, who was the first director of the Royal Observatory of Belgium. He invented the concept of *average man* to which human measures (and behavior) are referred to.

[117] The original formulation is *les poids chez les individus développés et de hauteur différentes, sont à peu près comme les carrés des tailles*, see also Quetelet (1842) for an English translation of this work.

[118] Human muscle has a density of about 900 gram per liter, fat has a similar specific weight, whereas bone density is twice as much.

hence the index, will increase – in fact, a situation comparable to the ill-defined denominator in the impact-factor formula;

4. as just one illustration of bias, consider Figure 27, which shows a histogram of the body circumference of a population of Scottish soldiers analyzed by Quetelet. The × represents the average of 10 undated measurements of my very own chest circumference. The conclusion is obvious: the statistics suggest an undersized person – although Quetelet's sample is incompatible with the measured subject, simply because the histogram data were sampled (almost 2 centuries ago) from a population of soldiers.

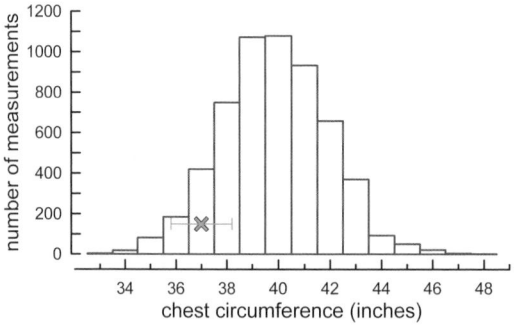

Fig. 27. Histogram of body circumference of a population of Scottish soldiers, based on data from Quetelet (1835). The × is the average of 10 undated measurements of the author's body circumference, the error bars represent the mean error resulting from a sequence of simple measurements, in the absence of any specification of the procedures of measurement.

The lesson, apart from the discussion of *BMI*, is that **you never ever should plot data from a catalogue without reading the background information about the samples, and without knowing the procedures of measurement that were followed**, see also Quetelet (1849) on the measurement of human chests.

The intelligence quotient. The concept of intelligence, and the question of its heritability was, for many decades, at the center of bitter debates. The early measurements of human intelligence date from Francis Galton (1822–1911), but the basis for the modern intelligence test was laid by Alfred Binet (1857–1911), whose tests had several defects.

Stern (1912) defined the concept *mental age*: a child of mental age[119] n means that, whatever its chronological age, its intelligence is equivalent to the average intelligence of the children of age n. The intelligence quotient, then, simply is the quotient of the mental age and the chronological age: $IQ = \frac{mental\ age}{chronological\ age} \times 100$ (where the factor 100 serves to get rid of the decimals). The IQ denominator is exact, but the numerator is a test result, with all its biases and errors.

High IQ is seen more as a signature of genius than as a marker of how well a person's educational and cultural background makes her or him fit for the tests. To the dismay of modern IQ-testers, inadequate intelligence tests were even suggested as a means of

[119]The original term was the German term *Intelligenzalter*.

deciding who could intellectually qualify for voting – even for producing offspring – and racial and national differences were even established based on IQ-scores lacking any significance. Binet (1900) was also the very first to warn against what he calls

> THE REPRESENTATION OF ALL OSCILLATIONS OF A THOUGHT BY A BRUTAL NUMBER THAT ONLY CAN HAVE AN ILLUSORY PRECISION.[120]

Unfortunately, biases that were known to Binet are now – 110 years later – present in our bibliometric indices, especially in the impact-factor index.

The impact factor again. The effect of replacing a complete distribution by one single bibliometric number is illustrated in Figure 28, which reproduces the principal elements of Figure 18 for the two journals. The dashed line is the average citation per year, calculated over a time span of about 20 years. The following conclusions are inevitable:

1. replacing a skewed distribution by a single number (the average) is unfounded, even more so if such a narrow impact-factor observing window is applied;
2. the factor of 6 by which the Hare wins from the Tortoise in the impact-factor race should be adjusted by a factor of 3.5 by which the Tortoise in average wins from the Hare in citations accrued;
3. the impact factor can fluctuate wildly if its denominator is a small number with large variation (as in mathematics papers and journals of modest size), and
4. there can be no doubt that this simple number could not possibly say anything about any individual paper in these journals.

There is another lesson to be taken in relation to the perceived third-decimal "accuracy" of the impact factor. Figure 26 illustrated the impact of self-citations on h for one specific astronomer. I analyzed for plagiarism (see page 240) all papers that were involved in the construction of Figure 26, and obtained an average similarity index of $41.6 \pm 1.7\%$, with an average journal impact factor $IF = 4.7 \pm 0.1$. The result is represented graphically in Figure 29, and leads to the following conclusions:

1. the box-whisker plot shows a very tight distribution, pointing to a systematic presence of identical passages in many papers;
2. it is obvious that we deal here with salami slicing in a narrow research niche, and
3. there appears a strong (negative) correlation between the average similarity index, and the 2009 journal impact factor.

[120] *Mais pour le moment, il est certain que l'on éprouve quelque embarras à exprimer par un chiffre brutal toutes les oscillations d'une pensée ; le chiffre ne peut avoir qu'une précision trompeuse ; comment en effet pourrait-il résumer ce qui aurait besoin de plusieurs pages de description ! Nous croyons nécessaire d'insister fortement sur cette question ; la suggestibilité d'une personne ne peut pas s'exprimer entièrement par un chiffre, alors même que ce chiffre correspondrait exactement au degré de sa suggestibilité : il faut en outre compléter ce chiffre par la description de tous les petits faits qui complètent la physionomie de l'expérience.*

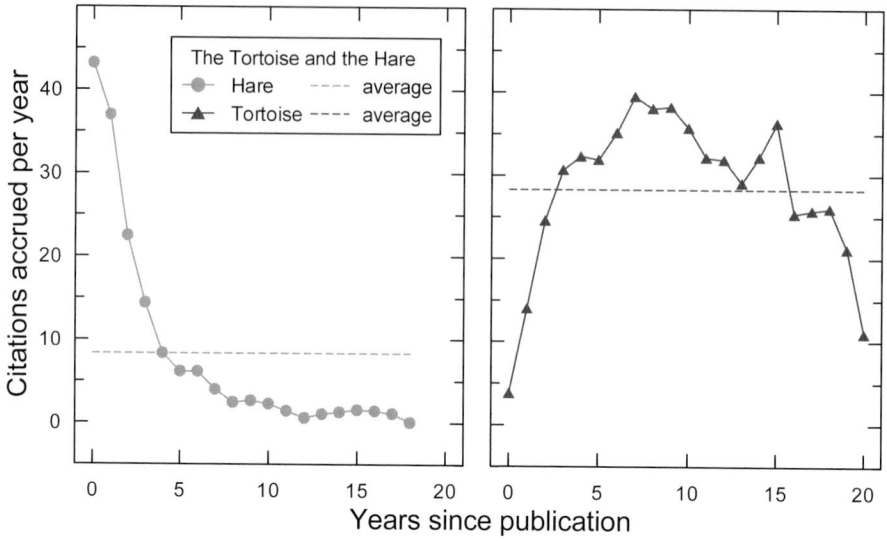

Fig. 28. The Tortoise and the Hare: distribution of citations accrued by each journal each year after publication. Based on data from Figure 8 of Vanclay (2009). The dashed lines indicate the average annual citations for each journal.

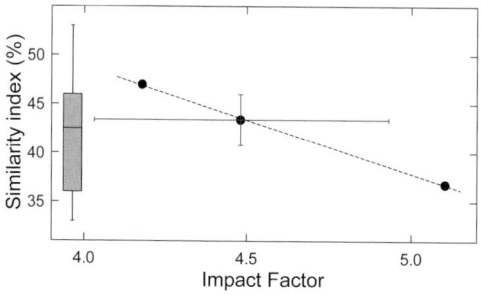

Fig. 29. Similarity index for a set of 14 papers. *Left*: box-whisker plot. *Right*: relationship between the *average* similarity index (per journal), and the 2009 journal impact factor. The vertical error bar is $\pm 1\sigma$, the horizontal error bar is $\pm 0.1 \times IF$. Error bars for the two other points are similar. The dashed line is the least-squares fit.

This is a very interesting graph, because it *suggests* a linear relationship – at least if you try not to see the error bars – between similarity index and impact factor, and merits a discussion in the light of nonsense correlations (see also Fig. 47 in Paper II). The vertical error bar in Figure 29 results from the "measurements", but formal error bars for the impact factor are not available.[121] If we would use the precision suggested by the third decimal in the published IF, no error bars would be visible at all. The symbol size used corresponds to $\pm 2\%$, and the horizontal error bar shown in Figure 29 is an optimistic

[121]The horizontal error bar was added on purpose to avoid unintentional misleading, in case this graph is reproduced out of context.

±10% – although 20% may perhaps be closer to the truth. Anyway, when using an error of 10%, it immediately follows that the relationship is a spurious correlation, the more so because the graph is really based on small-number statistics.

But the analysis also shows that the collection of similarity-index data could very well become a standard tool in assessment and evaluation of authors and projects, if done properly, of course. The drama of the data shown in Figure 29 is that all involved papers are refereed research papers in leading-IF astronomy journals as illustrated in Figure 1. By contrast, Table 3 contains only technical and educational papers of the type that is produced for conference proceedings. The differences in similarity indices speak for themselves.

18.7 On the use and abuse of bibliometric indices

The previous Sections demonstrated that, so far, we lack an exact framework for analysis of the value or worthiness of publications and of their authors. This reduces our ability to correctly analyze the value and judge the quality of scientific projects and their results.

1. One misuse of, for example, the impact factor, is to utilize its absolute value for taking decisions such as which journals a university library may purchase: if only the top ten journals – across all disciplines – may be purchased, there is a definite chance that most astronomy journals will not be among the selection.

2. Applying impact-factor statistics to single authors: the correlation between article citation rate and journal impact is quite weak (Seglen 1997), and such application is unfair towards some very productive scientists who happen not to have the financial means to pay the page charges of the higher impact-factor journals, and are thus limited to the journals that do not charge.

3. The unfounded belief in an absolute h-scale: for example, Kelly & Jennions (2006) give $h > 60$ for top physicists, but Table 1 shows that even in the same field (biological sciences) top scientists may differ by a factor of two in h.[122] One problem with this absolute interpretation (already mentioned on page 234) is the non-linearity of the index, another is the internal variation, which is index-value dependent.

4. Another misapplication is chasing after fashionable "hot" topics – the kind that get into high-impact journals, as well as squeezing research proposals within the boundaries of what is currently cited. In other words: using the bibliometric index as a driver of science.

5. Rewarding researchers with cash proportional to the journal impact factor of their publications, see Fuyuno & Cyranoski (2006), is a strategy that can lead to a dangerous fixation on a few indicators of scientific success, making scientists rush to publish, and ignore scientific and academic ethics. Or, as Lawrence (2003) put forward:

[122] Hirsch's (2005) Figure 2 – a histogram giving the number of Nobel prize recipients in physics in the last 20 years *versus* their h value – reveals that the peak is at $35 < h < 39$.

WHAT BEGAN AS SOMEONE ELSE'S MEASURE HAS BECOME OUR (OWN) GOAL.

What about the predictive power of bibliometric indices: is it possible to use, for example, h as a predictor, *i.e.*, does a high h mean high academic ability that makes certain that the hired researcher will eventually produce top-class science? Not at all. First of all, h always increases with time, hence it cannot reveal a decline in the performance of a scientist (for example, the case of the senior author in Fig. 15). Second, – in parallel with Herrnstein's (1973) conclusion that *"a low IQ predicts poor performance more reliably than a high one predicts good performance"* – a low[123] h will be a good predictor of poor performance, whereas a high h value is a necessary but not a sufficient condition for guaranteeing future high performance, simply because the uncertainties of the "measurement" grow with its level. Figure 30 is a rendering of Herrnstein's (1973) Figure 2, a much-idealized representation of how school grades may relate to IQ. The shaded region shows the spread; the dashed line is the average. The graph shows that **the regions with lowest risk to misjudge are the highest and the lowest, and that the domain with lowest predictive power is the medium and above-medium region**.

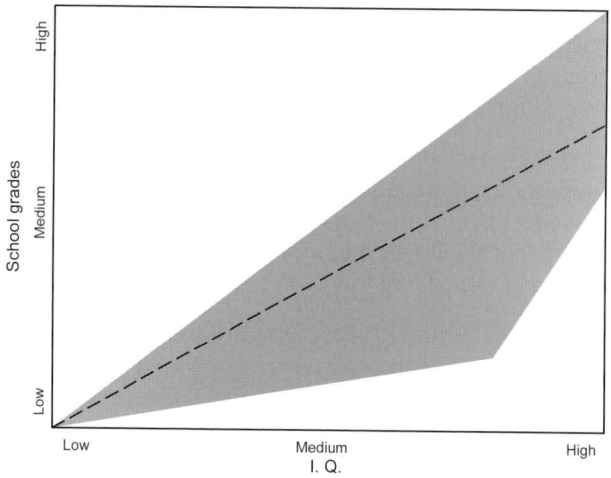

Fig. 30. Rendering of Figure 2 of Herrnstein (1973): an impressionistic, idealized representation of how school grades may relate to IQ. The shaded region shows the spread; the dashed line shows the average.

18.8 Some serious warnings

Young scientists should be warned that the citation index, combined with the impact factor, are tools developed *by and for the scientific administration* for *statistical* assessment of the outcome of scientific projects and funding schemes. These tools should be used with cautious restraint: these parameters are not even calibrated over a modest range of disciplines inside one science, not to speak of comparing citation counts in mathematics, physics and chemistry. I dare say:

[123]That is, low in relation to others in the same research discipline.

> YOU SHOULD NEVER EVER ATTEMPT TO MEASURE YOUR OWN SCIENTIFIC VALUE ONLY BY THE OUTCOME OF SOME KIND OF CITATION COUNT: YOU WILL EITHER END UP WITH OVERWHELMING UNHAPPINESS, OR WITH A MISTAKEN SENSE OF SELF-IMPORTANCE.

The index with the lowest meaning and the highest degree of ill-treatment, no doubt, is the sacred impact factor. Polderman (2007) expresses his deep concern that impact factors are increasingly used to evaluate the quality of single papers and of individual researchers, and he vividly protests this inappropriate practice: *"Keep your hands off our impact factor"*. These warnings are not only coming from individuals, but more and more emanate from organizations and scientific journals. The *European Association of Science Editors*[124] (EASE) recommends that journal impact factors are used only – and cautiously – for measuring and comparing the influence of entire journals, but not for the assessment of single papers, and certainly not for the assessment of researchers or research programmes, either directly or as a surrogate. The *World Association of Medical Editors* (WAME)[125] posted a strong warning that the impact factor can be affected by **a number of aspects unrelated to journal quality**, including self-citation by a journal, publication timing, and types of articles published. As a result, the WAME Board makes the following recommendations to WAME members:

1. More research is needed to evaluate the impact factor and other measures of journal and article quality.
2. Journal Editors should look beyond the impact factor as a summary statistics, and present other indicators of journal visibility, such as circulation, number of published articles, and the distribution of the citations. Such demographics of a journal should be regularly published to inform journal readers and authors.
3. Journal Editors have the responsibility to educate their readers, authors, administrators, and their scientific community in general about the impact factor and its relevance, as well as about other measures of journal and article quality.

Even if you forget these warnings, then simply just remember:

JOURNAL STATISTICS $\not\equiv$ INDIVIDUAL AUTHOR STATISTICS

CITATION RATE $\not\equiv$ VALUE OF A PAPER OR OF A SCIENTIST

JOURNAL IMPACT FACTOR $\not\equiv$ SCIENTIFIC IMPACT OF A PAPER

CITATION RATES DETERMINE THE IMPACT FACTOR, NOT VICE VERSA

18.9 Summary of bias factors

In the following item list I summarize most of the bias factors of previously discussed bibliometric indices currently in use (see also Seglen 1997):

[124] `http://www.ease.org.uk/`

[125] `http://www.wame.org/resources/policies#impact`

- some types of research are just not cited much;
- the impact factor favors English-language and US journals, and as the US community is larger, this makes these impact factors to grow;
- the impact factor is favorably influenced by journals that publish fast: a Letter journal publishes faster and has a higher impact factor, but a shorter cited half-life, whereas a review journal has a longer cited half-life, and a higher impact factor because it generates more citations than does a journal with only research papers;
- the impact factor strongly depends on the dynamics (expansion or contraction) of the research field: rapidly expanding literature that uses many references per article will produce a higher impact factor;
- the numerator and denominator in some indices are defined differently;
- impact factors are not statistically representative of individual journal articles;
- impact factors correlate poorly with actual citations of individual articles;
- the impact factor is a function of the number of references per paper in the research field;
- uncited articles get credit through the impact factor because of a few highly cited papers: the impact factor is strongly influenced by a minority of papers, and the decimals suggest false precision, whereas its accuracy is rather in the 10–20% range;
- the impact factor can fluctuate wildly for small journals, especially when the term in the denominator is small;
- authors may use criteria other than impact when submitting to journals;
- citations to "non-citable" items are sometimes included in the database;
- self-citations are not corrected for;
- review articles are heavily cited and inflate the impact factor;
- long articles collect many citations and yield high impact factors;
- articles tend to preferentially cite other articles in the same journal;[126]
- small research fields tend to lack journals with high impact;
- books are not included in the database as a source for citations;
- the set of journals in the database may vary from year to year;
- research fields with literature that rapidly becomes obsolete are favored;
- citation patterns change with time, *e.g.*, because of electronic document retrieval;
- all bibliometric indices are on a non-linear scale;
- bibliometric indices are not corrected for the number of authors per paper, and
- there is a substantial body of science beyond the literature listed in the ISI-database.

[126]This is called *journal self-citation*.

18.10 Repairing bibliometrics

The commerce in impact factors and related indices, obviously, is a blooming research industry. Should we simply follow Lippmann's (1922) suggestion to drown all the intelligence testers and all their questionnaires?[127] Certainly not: we should work to revise and refine the bibliometric methods. Tierney (2005) hits the nail on the head:

> IT IS CLEAR THAT THERE IS SOME ARBITRARINESS ABOUT WHAT TYPES OF ARTICLES IT [ISI] COUNTS IN THE DENOMINATOR AND WHAT TYPES OF CITATIONS IT COUNTS IN THE NUMERATOR. THERE IS NO WAY AROUND THIS, AND WE ARE STUCK. THEY ARE A FOR-PROFIT COMPANY, AND NOBODY ELSE IS DOING THIS.

Scientists, but also supervisors and students seem to be convinced that there is no escape from this increasing pressure. Lawrence (2003), though, has a different opinion:

> EVALUATIONS OF SCIENTISTS DEPEND ON NUMBERS OF PAPERS, POSITIONS IN LISTS OF AUTHORS, AND JOURNALS' IMPACT FACTORS. IN JAPAN, SPAIN AND ELSEWHERE, SUCH ASSESSMENTS HAVE REACHED FORMULAIC PRECISION. BUT BUREAUCRATS ARE NOT WHOLLY RESPONSIBLE FOR THESE CHANGES – WE SCIENTISTS HAVE ENTHUSIASTICALLY COLLUDED.

The literature contains suggestions for repair. Garfield (2006), for example, proposes to repair bibliometric indices by adjusting all citation studies to account for variables such as specialty, citation density, and half-life. Burton (2007), on the other hand, introduces *effective contribution per author* as the average number of pages divided by the average number of authors per paper published. Some of the following points may be considered.

1. Eliminate author name ambiguities documented in Section 18.2. The *Open Researcher & Contributor ID* initiative[128] aims to solve the author/contributor name ambiguity problem in scholarly communications by creating a central registry of unique identifiers for individual researchers (in a way similar to the *Digital Object Identifier*). This should lead to accurate identification of researchers and their work, and solve once and forever the confusions and interferences caused by the commonality of name, as illustrated in Figures 15 and 16.
2. Standardize the subject areas over all journals in a same scientific discipline by introducing a subject area or discipline list, to which authors can link through their ORCID number. This would allow a same paper to be subject-sorted according to each author's individual preferences: for example, if I coauthor a paper together with an aerospace engineer, I would flag my paper as an astronomy paper, whereas the engineer could list that very same paper under space technology. After all, I am evaluated on the basis of my scientific contributions to astronomy, whereas engineers are assessed on their contributions to engineering science.

[127] Lippmann (1922): *If, for example, the impression takes root that these tests really measure intelligence, that they constitute a sort of last judgment on the child's capacity, that they reveal 'scientifically' his predestined ability, then it would be a thousand times better if all the intelligence testers and all their questionnaires were sunk without warning in the Sargasso Sea.*

[128] ORCID Inc. is a non-profit organization, `http://www.orcid.org`

3. Make the transition to *Open Access*, and explain the system to authors. Salager-Meyer (2009) strongly advocates that all scientific journals become *Open Access*, and that all research conducted with public funding be posted either as preprints or e-prints on institutional repositories.
4. Correct the bibliometric parameters for *scientific age*. This age should be expressed relative to age group, something analogous to Wechsler's 1955 *deviation quotient* (WAIS)[129]. Note that the error on scientific age is proportionally largest for young careers, and that when this parameter is entered in the denominator of an index, the result becomes very uncertain, especially for the very youngest among us.
5. Remove the false precision of the impact factor by replacing it by a classification such as the credit rating system used by banks (see the top diagram in Fig. 31). A similar rating scheme for researchers is already in use in some countries, but it is only applicable to the local market.
6. Promote independent impact factor and citation raters on the market.

The last two points in the list may, though, lead to disfunction as well. Take as an example the *credit ratings* in economics. A credit rating is an evaluation that estimates the credit worthiness of an individual, or of a corporation, based on the subject's credit history. The recent financial crisis revealed that several rating agencies overvalued some subjects, even to the extent that it was exactly the bankruptcy of some top-rated institutions that set off the 2008 global credit crisis. How was it possible to assign such a high rating to dubious companies? Well, credit raters are paid by those corporations that issue the bonds, so they make self-fulfilling prophecies and participate in indirect market manipulation. That rating companies are paid by the firms whose bonds they rate is absurd: rating agencies act as referees in an environment invoking strong conflicts of interest, and they get money from the companies who are rated – it is just like paying your own referee. In our business, the raters are not paid by the companies (*i.e.*, the journals), but by those who consult the database.

One way out of this scheme would be to use a free journal-ranking tool as, for example, described by Butler (2008). Tierney's (2005) statement on page 267 refers to the fact that our bibliometric standards are set by just a few for-profit companies, so perhaps there should be more competition on the market? Introducing more competition in the ratings market will not necessarily improve the reliability of the ratings, at least as long as the follies of the system are not removed. This point is corroborated by Becker & Milbourn (2009), who conclude that competition leads to lower quality in the ratings market: the current agencies produce more issuer-friendly and less informative ratings when competition is stronger – in other words, **their reports do not satisfy the basic requirements of scientific writing**. Figure 31 shows the distribution of credit ratings in low- and high-competition credit-rating environments (absolute and differential data). The vertical arrow indicates the transition from the so-called *junk bonds* to *investment grade bonds*: by convention, anything above that line is seen as good quality, everything below that line is considered as rubble. The metaphor can be used to illustrate the parallel with high impact

[129] *Wechsler Adult Intelligence Scale*: `en.wikipedia.org/wiki/Wechsler_Adult_Intelligence_Scale`

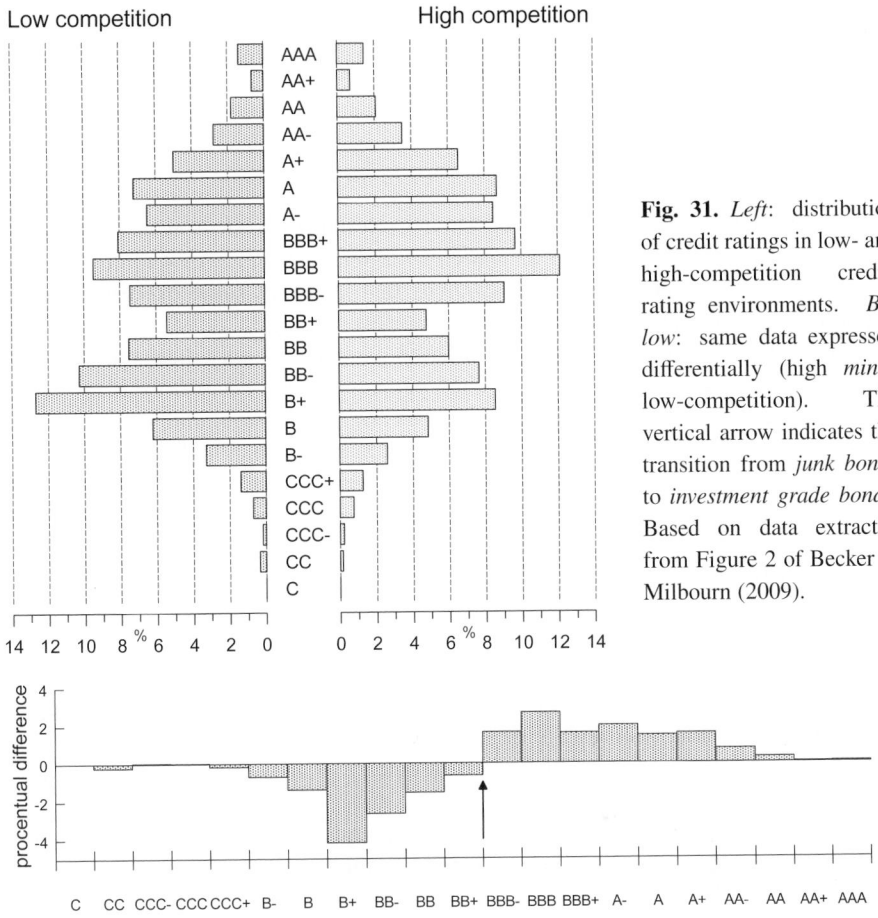

Fig. 31. *Left*: distribution of credit ratings in low- and high-competition credit-rating environments. *Below*: same data expressed differentially (high *minus* low-competition). The vertical arrow indicates the transition from *junk bonds* to *investment grade bonds*. Based on data extracted from Figure 2 of Becker & Milbourn (2009).

factor journals and lower-rated publications, and high-rated scientists and more average colleagues, and leads to the following observations:

1. the systematics are obvious: the best and the worst bonds are beset with the smallest variance – a situation similar to the one shown in Figure 30;
2. the middle-class bonds are either systematically overrated or underrated;
3. the transition between underrating and overrating appears at the level between BB+ and BBB–: the data analysis exactly reproduces the model agreed by convention;
4. the rating system is not an absolute set of standardized values with a rigid set of metrological rules, but is a relative scale, *i.e.*, the highest class will always be AAA, and the lowest will always be C.

Would such a scheme work for bibliometrics? Perhaps, if it is constructed in a serious way, using bibliometric indices supplemented by some form of personal assessment of the scientific production worthiness of a researcher (or of a team of scientists). Such ratings should be "watertight" – *i.e.*, **accurate and reliable, constant in time, and have a predictive property**.

18.11 On the true meaning of bibliometric indicators

So, after all, what is the Hirsch index? The Hirsch index is exactly what Hirsch (2005) described: *"an index to quantify an individual's scientific research output"* – in other words: h is a single entity called *productivity* that measures **some overall productive capacity** of an individual in order to allow her or his ranking.

But a citation – and thus any citation-based parameter – only means that the citing author has consulted (or heard of) the cited paper, what not necessarily is the same as having read the paper: ***cited* is not the same as *read*** – moreover citations depend **only** on crossreferencing objects, without any quality control. Also, a high h may possibly be an indicator of speedy production rather than of research depth: quick and shallow may boost citedness, whereas depth may result in an apparent decrease in productivity, and hence citedness, see case D in Figures 5, 8 and 24.

These defects in an indicator may be as harmful when comparing scientists from different cultural and institutional backgrounds, as when using IQ for gauging cultural and racial variations in different populations. BMI, IQ, h and IF are based on statistical information that reveals basically correct empirical relationships, but their interpretation – equating IF with quality of an individual paper, IQ with innate intelligence, BMI with health, and h with quality of researcher – matches more Feynman's cargo cult science (see page 177) than it touches on calibrated measurement.

The point simply is that any kind of **accurate and reliable, constant and coherent bibliometric index should at the start define exactly what it measures**, as so well called into question by Kelly & Jennions (2006): *"Finally, what do h indices really tell us?"*[130] Well, Boring (1923) posed the equivalent question on intelligence, and his answer was one of the first operational definition of intelligence (Watson & Campbell 1963):

> INTELLIGENCE IS WHAT THE TESTS TEST.

Boring's (1923) concept is that intelligence is like *power*, as the physicist uses the word: the amount of work that can be done in a given time – because most intelligence tests are speed tests. Hence, the quantity suggested by the impact factor – an average citation score per unit of time[131] – can thus be called a measure of *citation power* of a journal. In other words, **the impact factor is a measure of the average citation power of the papers in a journal, not of the journal's impact on the progression of scientific knowledge, nor of the quality of all the papers that this journal accepts for publication**. In the same vein, a scientist's citation index or any derivative (like h), if normalized by scientific age, could then be seen as **a measure of the citation power of that scientist** – though, again,

[130] See the full quote on page 272.

[131] The unit of time is the impact-factor window: two years.

it remains a manifestation of citation, not of scientific value of the person or of her/his work. Provided, of course, that all previously discussed biases and errors of measurement are remediated. That is exactly where our current citation metrics stand today.

We astronomers can so easily counter the craze of using one single number to qualify a complex phenomenon (the human body, human intelligence, or human scientific value): we know since long that stellar spectral energy distributions can be represented by one single number – a color index – **if and only if the star radiates like a black body** (see Fig. 19 in Paper II).[132] So, if a government agency would force us to persistently represent all spectral energy distributions by just one single low-precision number, then most of us astronomers would run out of business in no time. But would we comply with that folly? No doubt, we would rise up. Tuchman (1984) specifies *folly* through an if/then construction:[133]

> IF PURSUING DISADVANTAGE AFTER THE DISADVANTAGE HAS BECOME OBVIOUS IS IRRATIONAL, THEN REJECTION OF REASON IS THE PRIME CHARACTERISTIC OF FOLLY.

Why, then, do we hang on to folly when it comes to measuring ourselves and our journals?

THE CONCLUSION IS THAT WE MUST RECOGNIZE THAT, AT THIS MOMENT, WE HAVE NO RELIABLE DIAGNOSTIC SYSTEM, AND THAT IT IS TIME TO ABANDON THE TRADITIONAL BIBLIOMETRIC QUANTIFICATIONS, BECAUSE THEY ARE SCIENTIFICALLY UNFOUNDED IN THE SENSE THAT THEY NEGLECT ALL PRINCIPLES OF SCIENTIFIC DATA GATHERING AND STATISTICAL COHERENCE.

IF WE PROVE TO BE UNABLE TO TURN THE TIDE, AND IF WE GO ON BLOWING BUBBLES IN COLLECTIVE MADNESS, OUR PAPERS-CITATION-IMPACT FACTOR DOMINATED UNIVERSE WILL SOON BECOME AS VOID AS THE SOAP BUBBLE THAT LED TO THE FINANCIAL CRISIS.

19 How can truth and value ever survive?

The rate of productivity of scientists steadily grows: the number of papers, and the number of authors per paper increase in step, but the percentage of researchers completing a life-long career is on the decrease. So, if science is the survival of the fittest[134] – implying a harsh struggle for existence that includes clawing each other's papers, and dog-eats-dog rules with free play for the strongest – how can truth and value ever survive?

[132] Thus, real spectral energy distributions necessarily need one additional color index for every other deviation from the Planck curve.

[133] An if/then construction expresses that one part of a sentence is only true if a condition expressed in the second part of that sentence is met, and the proper use is to separate the sentences by "then" – just like in computer programming.

[134] In scientific *production*.

That question, in evolutionary terms, was exactly Charles Darwin's problem: why do so many species have non-reproductive members, and why can non-*reproduction* persist in the first place? The answer simply is that nature provides room for non-reproductive workers who provide resources to the reproductive castes.[135]

In the context of science, we do not deal with *reproduction*, but with *production*, yet the scientific community also counts on "non-productive" elements,[136] *viz.*: the many work-for-pay helpers, some of whom we acknowledge at the end of our papers. But there is much more than that, or at least there was a time when each institute or each team had a couple of – often senior – colleagues who helped the others out of *altruistic*[137] motives through personal effort that benefits the other members of the group. Parents are generally altruistic to their offspring, and a similar parental care may be expected from thesis supervisors and mentors, who provide unselfish support, a feature needed for the thriving of any scientific endeavor. Kelly & Jennions (2006) put their fingers on the wound of the complex endeavor that education in science really is:

> FINALLY, WHAT DO h INDICES REALLY TELL US? OF COURSE, IT IS DISTASTEFUL TO REDUCE A LIFETIME'S WORK TO A NUMBER. SOME SCIENTISTS MAKE HUGE CONTRIBUTIONS THROUGH THEIR MENTORING AND GENEROSITY WITH IDEAS, SKILLS AND TIME. WITHOUT THEM, ACADEMIA WOULD COLLAPSE: A DEPARTMENT SOLELY COMPRISED OF RELENTLESS PUBLISHERS WOULD BE A JOYLESS PLACE FOR EAGER STUDENTS.

But helping others without direct reward (for example, coauthorship) appears to give no benefit to the helper in the first place. I am convinced that this is not the case, although I must admit that the benefits may not be directly and immediately visible. Refereeing, for example, makes both parties gain, because refereeing involves research efforts that are mutually beneficial. Even proofreading someone else's manuscript does yield at least the minimum profit of expecting the same return service eventually.

And what about the cheaters in the system? Eldakar & Wilson (2008), in their studies of animal behavior, dealt with a problem that is also faced by scientific institutes, where individuals also have almost unlimited access to limited resources. Altruists benefit the group through sharing the resource, while selfish individuals undermine altruism through overconsuming the resource at the expense of the group. Eldakar & Wilson propose that selfish strategies, through eliminating other selfish strategies, cause selfishness to be a self-limiting strategy, enabling altruists to coexist in a stable equilibrium. This model identifies a strategy that involves behaving selfishly in first-order interactions, and altruistically in second-order interactions by punishing other selfish individuals. Or, much better expressed by Krakovsky (2009): cheaters help to sustain altruism by punishing other cheaters – only for motives of their own – because a group with too many cheaters does not have enough altruists to exploit. As Eldakar puts it,[138]

---

[135] Bees, wasps, ants, termites, ...

[136] In terms of writing scientific papers – not "unproductive" in its more general meaning.

[137] *Altruism* is selflessness, the opposite of competition and egoism.

[138] In his interview with Krakovsky.

> IF YOU'RE A SINGLE SELFISH INDIVIDUAL IN A GROUP OF ALTRUISTS, THE BEST THING YOU CAN DO EVOLUTIONARILY IS TO MAKE SURE NOBODY ELSE BECOMES SELFISH – MAKE SURE YOU'RE THE ONLY ONE.

Science can prosper only in an atmosphere of complete freedom, free of devastating publish–or–perish pressures. Lawrence (2003) clearly points out who are the people that have to safeguard this freedom:

> IT IS WE OLDER, WELL-ESTABLISHED SCIENTISTS WHO HAVE TO ACT TO CHANGE THINGS. WE SHOULD MAKE THESE POINTS ON COMMITTEES FOR GRANTS AND JOBS, AND SHOULD NOT BE SO DESPERATE TO PUSH OUR PAPERS INTO THE LEADING JOURNALS. WE CANNOT EXPECT YOUNGER SCIENTISTS TO ENDANGER THEIR FUTURE BY MAKING SACRIFICES FOR THE COMMON GOOD, AT LEAST NOT BEFORE WE DO.

I conclude this paper by expressing a hope, the hope of Richard Feynman (1974):

> **SO I HAVE JUST ONE WISH FOR YOU –**
> **THE GOOD LUCK TO BE SOMEWHERE WHERE YOU ARE FREE**
> **TO MAINTAIN THE KIND OF INTEGRITY I HAVE DESCRIBED,**
> **AND WHERE YOU DO NOT FEEL FORCED BY A NEED**
> **TO MAINTAIN YOUR POSITION IN THE ORGANIZATION,**
> **OR FINANCIAL SUPPORT, OR SO ON, TO LOSE YOUR INTEGRITY.**
> **MAY YOU HAVE THAT FREEDOM.**

Epilog: the road to wisdom

An epilog is a piece of writing at the end of a work, to bring it to closure.

Science is a creative, logical and coherent form of knowledge – nevertheless, the aspect of beauty is also of high value: **beauty in science means elegance, simplicity of expression, and clarity**. A proof of a theorem, for example, can be more elegant than another proof when it involves a simpler and clearer train of arguments. Paul Dirac (1902–1984) insisted on the role of beauty (Dirac 1963):

> IT IS MORE IMPORTANT TO HAVE BEAUTY IN ONE'S EQUATIONS THAN TO HAVE THEM FIT EXPERIMENT.

Dirac's precept – if quoted out of context, like here – is to be taken with a heavy dose of salt: the full story is about Erwin Schrödinger (1887–1961) who should have been more confident in the very beautiful equation he had derived, but that did not fit the experiment – simply because at that time it was not known that the electron has a spin. Chandrasekhar (1987) also stressed the role of esthetics, and the relation between the beauty of an idea and the power of this idea, to explain a physical phenomenon. Chandrasekhar, in his *Truth and Beauty*, also refers to Heisenberg's (1970) vision on beauty.

Papers I–III in this book emphasized the need for truthful attitudes in collecting and analyzing the data, and in reporting the results. In addition, the need for ethics and esthetics was repeatedly accentuated. But you should never forget: **first the truth, then the beauty**, even more so **first the truth, then the honor**. And you should always remember that notions of beauty and honor are not calibrated, and change with time. And now I close with my best advice, so beautifully worded by Piet Hein:[139]

The road to wisdom?
Well, it's plain
and simple to express:
err
and err
and err again,
but less
and less
and less.

Thanks to Dr. Ed Stolper (Caltech Provost) for giving permission to reproduce some verbatim statements from the Caltech policy on research misconduct, to Dr. Linda J. Wilcox for giving permission to reproduce the statements on page 226, and to Dr. Carol Tenopir for permission to reproduce the key points listed in Section 5.3. Ms. Kirsty Meddings, Product Manager at CrossRef, is acknowledged for her assistance with iThenticate.

I am grateful to Dr. Peter Brosche for providing access to his private copy of Volume 4 of *Correspondance Astronomique, Géographique, Hydrographique et Statistique du Baron de Zach*, to Dr. Hilmar W. Duerbeck for acquiring a copy of *Leipziger Magazin für reine und angewandte Mathematik* from the Universitätsbibliothek Leipzig, and to Mr. Alex Boese for providing the high-resolution digital image reproduced in Figure 12. I express my appreciation to Dr. Myron A. Smith for his permission to mention his name in the context of confusion of namesakes in bibliometric databases (on page 249).

I am very much indebted to Dr. Hilmar W. Duerbeck, Dr. Chris Koen, and Dr. Laszlo Szabados for deep proofreading of, and criticism, discussion and advice on my three papers included in this book.

I acknowledge financial support from the Vrije Universiteit Brussel, and I express gratitude to my Employer, the Belgian Fund for Scientific Research (FWO), for supporting several research projects that, over the years, gave me the opportunity to acquire the specialized knowledge presented in these papers.

[139] 1905–1996, Danish scientist, mathematician, inventor, designer, author, and poet, who studied at the Institute for Theoretical Physics of the University of Copenhagen (later to become the Niels Bohr Institute).

References

Abt, H.A., & Garfield, E., 2002, Is the Relationship Between Numbers of References and Paper Lengths the Same for All Sciences?, J. Amer. Soc. Inform. Sci. Technol., 53, 1106

Adams, J., Halliday, C., Mednick, A., Usdin, M., Bertout, C., & Walmsley, M., 2008, English language editing at A&A, A&A, 490, 19

Agassi, J., 1981, The Problems of Scientific Validation, in On scientific discovery: the Erice lectures 1977, ed. M.D. Grmek, R.S. Cohen & G. Cimino (Dordrecht: D. Reidel), p. 103

Anderson, M.S., Ronning, E.A., De Vries, R., & Martinsons, B.C., 2007, The Perverse Effects of Competition on Scientists' Work and Relationships, Sci. Eng. Ethics, 13, 437

Angier, N., 2000, William Hamilton, 63, Dies; An Evolutionary Biologist (New York Times), March 10

Babbage, C., 1830, Reflections on the Decline of Science in England, and on Some of its Causes (London: Fellowes)

Barber, T.X., 1974, Pitfalls In Human Research. Ten Pivotal Points (Oxford: Pergamon Press Inc.)

Becker, B., & Milbourn, T., 2009, Reputation and competition: evidence from the credit rating industry (Harvard Business School Working Papers), 09-051

Belton, M.J.S., Thomas, P., Veverka, J., *et al.*, 2007, The internal structure of Jupiter family cometary nuclei from Deep Impact observations, Icarus, 187, 332

Berger, P.A., Ginsburg, R.A., Barchas, J.D., *et al.*, 1978, Platelet monoamine oxidase in chronic schizophrenic patients, Amer. J. Psych., 135, 95

Berry III, J.N., 2000, The Real Purpose of Copyright, Lib. J., 125, 6

Bertout, C., & Schneider, P., 2004, Editorship and peer-review at A&A, A&A, 420, E1

Bhandari, M., Einhorn, T.A., Swiontkowski, M.F., & Heckman, J.D., 2003, Who Did What? (Mis)Perceptions About Authors' Contributions to Scientific Articles Based on Order of Authorship, J. Bone & Joint Surg., Amer., 85, 1605

Binet, A., 1900, La Suggestibilité (Paris: Schleicher Frères)

Boring, E.G., 1923, Intelligence as the Tests Test It, The New Republic, 34, 35 (reproduced in History, Psychology, and Science: selected papers, ed. R.I. Watson & D.T. Campbell, 1963 (New York: Wiley)

Boring, E.G., 1962, Newton and the Spectral Lines, Science, 136, 600

Bošnjak, L., Puljak, L., Vukojević, K., & Marušić, A., 2011, Analysis of a number and type of publications that editors publish in their own journals, Scientometrics, 86, 227

Botez, E., 2004, Maximilian Hell and the Northernmost Transit of Venus Expedition of 1769, J. Astron. Data, 10, 7

Bozzi, M.L., 2003, Truth and science: Bill Hamilton's legacy, in Origin of HIV and emerging persistent viruses, Atti dei Convegni Lincei, ed. Acc. Naz. dei Lincei, 187, 21

Broad, W.J., 1981, The Publishing Game: Getting More for Less, Science, 211, 1137

Broad, W., & Wade, N., 1982, Betrayers of the Truth: Fraud and Deceit in the Halls of Science (New York: Simon & Schuster)

Bronowski, J., 2008, Science and Human Values (London: Faber and Faber)

Brown, J.R., 2001, Who Rules in Science (Cambridge, Mass.: Harvard University Press)

Butler, D., 2008, Free journal-ranking tool enters citation market, Nature, 451, 6

Burton, W.B., 2007, in Future Professional Communication in Astronomy, ed. A. Heck & L. Houziaux (Brussels: Académie Royale de Belgique), 47

Campanario, J.M., 1996, The Competition for Journal Space among Referees, Editors, and Other Authors and Its Influence on Journals' Impact Factors, J. Amer. Soc. Inform. Sci., 47, 184

Chandrasekhar, S., 1987, Truth and Beauty (Chicago: University of Chicago Press)

Chew, M., Villanueva, E.V., & Van Der Weyden, M.B., 2007, Life and times of the impact factor: retrospective analysis of trends for seven medical journals (1994–2005) and their Editors' views, J. Royal Soc. Med., 100, 142

Clayton, D.D., 2007, Hoyle's Equation, Science, 318, 1876

Clouthier, S.G., 2003, Plagiarism Erodes Scientific Integrity, Science, 301, 311

Condon, J.J., 2009, No new observations, no new theories, enduring scientific value, A&A, 500, 219

Crane, D., 1965, Scientists at Major and Minor Universities: A Study of Productivity and Recognition, Amer. Sociol. Rev., 699

Currie, C., 1997, Author saw fraud, misconduct, and unfairness to more junior staff, BMJ, 315, 744

Cyranoski, D., 2007, Japanese TV show admits faking science, Nature, 445, 804

David, P., 1983, Scientific fraud: The system defends itself, Nature, 303, 369

Delambre, J., & Méchain, P., 1806, Base du Système Métrique Décimal, Discours Préliminaire, p. 158–159

Dewatripont, M., Ginsburgh, V., Legros, P., *et al.*, 2006, Study on the Economic and Technical Evolution of the Scientific Publication Markets in Europe, European Commission, Directorate-General for Research, Brussels, `http://ec.europa.eu/research/science-society/pdf/scientific-publication-study_en.pdf`

Dirac, P.A.M., 1963, The Evolution of the Physicist's Picture of Nature, Scient. Amer., 208 (5)

Draaisma, D., 2005, Het verdriet van de kosmopoliet, `www.douwedraaisma.nl/Lezing/Oratie.html`

Duerbeck, H.W., Pustylnik, I., Steinle, H., & Sterken, C., 2002, Astron. Nachr., 323, 523

Duerbeck, H.W., 2011, personal communication

Dugatkin, L.A., 2006, The Altruism Equation. Seven Scientists Search for the Origins of Goodness (Princeton University Press)

EASE, 2007, EASE statement on inappropriate use of impact factors, European Science Editing, 33, 99

EASE, 2009, Integrity in Science Communication, Tenth EASE General Assembly and Conference, 16–19 September 2009, Pisa, Italy

Eldakar, O.T., & Wilson, D.S., 2008, Selfishness as second-order altruism, Proc. Nation. Acad. Sci., 105, 6982

Eisenstein, E., 2005, The Printing Revolution in Early Modern Europe, 2nd edition (Cambridge University Press)

Encke, J.F., 1820, Correspond. Astron., Géog., Hydrog. Stat., 4, 456

Eysenck, H.J., & Kamin, L.J., 1981, The Intelligence Controversy (New York: Wiley)

Fancher, R.E., 1985, The Intelligence Men (New York: W.W. Norton & Co)

Feast, M., 2000, Stellar populations and the distance scale: the Baade-Thackeray correspondence, J. Hist. Astron., 31, 29

Feyerabend, P., 1987, Farewell to Reason (London: Verso)

Feynman, R.P., 1974, Cargo Cult Science, Eng. Sci., 37, 10

Fletcher, H., 1982, My work with Millikan on the oil-drop experiment, Physics Today, June 1982, 43–47

Fowler D.H., & Aksnes D.W., 2007, Does self-citation pay?, Scientometrics, 72, 427

Franklin, A., 1984, Forging, cooking trimming, and riding on the bandwagon, Amer. J. Phys., 52, 786

Franklin, A., 1989, The Neglect of Experiment (Cambridge University Press)

Fuyuno, I., & Cyranoski, D., 2006, Nature, 441, 792

Garfield, E., 2006, The history and meaning of the journal impact factor, J. Amer. Med. Ass., 295, 93

Garry, R.F., 2010, Retraction: Influenza or not influenza: Analysis of a case of high fever that happened 2000 years ago in Biblical time, Virol. J., 7, 190

Gauss, C.F., 1824, Ehrenrettung, Astron. Nach., 3, 77

Gee, H., 2006, A day in the life of Henry Gee, Nature/Nurture, autumn, p. 3

Gingerich, O., 1982, Obituary – Payne-Gaposchkin Cecilia, Quart. J. Royal Astron. Soc., 23, 450

Goodstein, D., 2002, Doing Business The Caltech Way: An Ethics Handbook, California Institute of Technology `http://asic.caltech.edu/documents/2-doing_business_the_caltech_way.pdf`

Goodstein, D., 2010, On Fact and Fraud (Princeton University Press)

Grant, B., 2009a, OA publisher accepts fake paper, The Scientist, 29 June 2009, `http://www.the-scientist.com/blog/display/55756/`

Grant, B., 2009b, Merck published fake journal, `www.the-scientist.com/blog/display/55671/`

Grothkopf, U., & Erdmann, C., 2008, Open Access – State of the Art, IAU Inf. Bull., 102, 64

Gunsalus, C.K., 1998, How to Blow the Whistle and Still Have a Career Afterwards, Sci. Eng. Ethics, 4, 51

Habing, H.J., 2009, Citation counts: trick or treat?, A&A, 500, 499

Hearnshaw, J.B., 2010, Auguste Comte's blunder: an account of the first century of stellar spectroscopy and how it took one hundred years to prove that Comte was wrong!, J. Astron. Hist. Herit., 13, 90

Heisenberg, W., 1970, The Meaning of Beauty in the Exact Sciences, Munich

Herrnstein, R.J., 1973, I. Q. in the Meritocracy (Atlantic Monthly Press)

Hetherington, N.S., 1972, Adriaan van Maanen and Internal Motions in Spiral Nebulae: a Historical Review, Quart. J. Royal Astron. Soc., 13, 25

Hetherington, N.S., 1983, Just how objective is science?, Nature, 306, 727

Heywood, E., 2008, Confidentiality, libel, peer review and the law, The Write Stuff, 17, 168

Hindenburg, F., (Ed.), 1786, Leipziger Magazin für reine und angewandte Mathematik (Leipzig: Müller), 132

Hirsch, J.E., 2005, An index to quantify an individual's scientific research output, Proc. Nat. Acad. Sci., 102, 16569

Holmes, R., 1666, A Narrative Concerning the Success of Pendulum-Watches at Sea for the Longitudes, Phil. Trans. Royal Society of London, 1, 13

Holton, G., 1978, The Scientific Imagination: Case Studies (Cambridge: Cambridge University Press)

Hon, K.L., Ng, P.C., & Leung, T.F., 2010, Influenza or not influenza: Analysis of a case of high fever that happened 2000 years ago in Biblical time, Virol. J., 7, 169

Horton, R., 1998, The Journal Ombudsperson: A Step Toward Scientific Press Oversight, JAMA, 280, 298

Hoyle, F., 1954, On Nuclear Reactions Occuring in Very Hot Stars. I. The Synthesis of Elements from Carbon to Nickel, ApJS, 1, 121

Janis, I.L., 1972, Victims of Groupthink: A Psychological Study of Foreign-policy Decisions and Fiascoes (Houghton Mifflin Company)

Jardine, L., 2010, Accidental Anglo-Dutch Collaborations: Seventeenth-Century Science in London and The Hague, Sarton Chair Lect. Univ. Gent, Sartoniana, 23, 15

Jones, K.G., 1991, Messier's Nebulae and Star Clusters (Cambridge University Press)

Kaiser, J., 2004, HHS Proposes Research Misconduct Rules, Science, 304, 501

Kale, R., 2006, Changing China, Brit. Med. J., 333, 19 August

Kelly, C.D., & Jennions, M.D., 2006, The *h* index and career assessment by numbers, Trends Ecol. Evol., 21, 167

Kelly, C.D., & Jennions, M.D., 2007, H-index: age and sex make it unreliable, Nature, 449, 403

Kennedy, D., 1985, On Academic Authorship, `http://rph.stanford.edu/2-8.html`

Kennedy, D., 1997, Academic Duty (Cambridge, Mass.: Harvard University Press)

Koen, C., & Lombard, F., 1993, The Analysis of Indexed Astronomical Time Series – Part One – Basic Methods, MNRAS, 263, 287

Kozak, M., 2010, personal communication

Krakovsky, M., 2009, Thriving on Selfishness: Why it Pays for Cheaters to Punish Other Cheaters, Scient. Amer., 300, 14

Kühr, H., Witzel, A., Pauliny-Toth, I.I.K., *et al.*, 1981, A catalogue of extragalactic radio sources having flux densities greater than 1 Jy at 5 GHz, A&AS, 45, 367

Langdon-Neuner, A.E., 2010, EASE forum, `http://www.ease.org.uk/forum/index.shtml`

Langmuir, I., 1954, Colloquium on Pathological Science, (transcribed from a tape recording, of the lecture on Pathological Science that Langmuir gave on December 18, 1953), `http://www.cs.princeton.edu/~ken/Langmuir/langmuir.htm`

Lawrence, P.A., 2003, The politics of publication, Nature, 422, 259

Lee, A., Dennis, C., & Campbell, P., 2007, Nature's guide for mentors, Nature, 447, 791

Lemaître, G., 1927, Un Univers homogène de masse constante et de rayon croissant, rendant compte de la vitesse radiale des nébuleuses extra-galactiques, Ann. Soc. Scient. Brux., XLVII, 49

Lippmann, W., 1922, The Abuse of the Tests, The New Republic, 32, `http://www.disabilitymuseum.org/lib/docs/1726.htm`

Lombard, F., & Koen, C., 1993, The Analysis of Indexed Astronomical Time Series – Part Two – the $O - C$ / Observed – Calculated / Technique Reconsidered, MNRAS, 263, 309

Luyten, W.J., 1974, Stellar Motion Survey by Automation, Science, 185, 351

Luyten, W.J., 1981, Plagiarism in astronomy, Proper motion survey with the 48-inch Schmidt telescope. LVIII – University of Minnesota, Minneapolis, MN

Mackintosh, N.J., 1980, Book review of Cyril Burt: Psychologist, Brit. J. Psychol., 71, 174

Maddox, J., 1984, Privacy and the peer-review system, Nature, 312, 497

Maddox, J., 1989, Can journals influence science?, Nature, 339, 657

Maddox, J., 1993, Competition and the death of science, Nature, 363, 667

Marcovitch, H., & Williamson, A., 2003, Editorial boards, European Ass. of Science Editors, June 2003

May, R.M., 2001, Memorial to Bill Hamilton, Phil. Trans. Royal Soc. London, Biol. Sci., 356, 785

McSherry, C., 2001, Who Owns Academic Work? Battling for Control of Intellectual Property (Cambridge, Mass.: Harvard University Press)

Medawar, P., 1990, Pluto's Republic (Oxford University Press)

Merton, R.K., 1968, The Matthew Effect in Science, Science, 159, 56

Millikan, R.A., 1910, The Isolation of an Ion, a Precision Measurement of its Charge, and the Correction of Stokes' Law, Science, 30 September 1910, 436–448

Millikan, R.A., 1911, The Isolation of an Ion, a Precision Measurement of its Charge, and the Correction of Stokes' Law, Phys. Rev. 32, 349–397

Millikan, R.A., 1913, On the Elementary Electrical Charge and the Avogadro Constant, Phys. Rev., 2, 109

Mitroff, I.I., 1974, Studying the lunar-rock scientist, Saturday Rev. World, Nov. 2, 1974, 64–65

Mommonier, M., 1996, How to Lie with Maps, 2nd edition (Chicago: University Of Chicago Press)

Monastersky, R., 2005, The Number that's Devouring Science, Chron. High. Educ., 52, A12

Newcomb, S., 1883, On Hell's alleged falsification of his observations of the transit of Venus in 1769, MNRAS, 43, 371

Öpik, E.J., 1973, The Optical Oblateness of Mars, Irish Astron. J., 11, 1

Öpik, E.J., 1977, About Dogma in Science, and Other Recollections of an Astronomer, Ann. Rev. Astron. Astrophys. 15, 1

Parrish, D., & Noonan, B., 2009, Image Manipulation as Research Misconduct, Sc. Eng. Ethics, 15, 161

Patkós, L., 2004, The Pasquich affair, Acta Hist. Astron., 24, 182

Payne, C.H., 1925a, Stellar Atmospheres; a Contribution to the Observational Study of High Temperature in the Reversing Layers of Stars, Ph.D. Thesis, Radcliffe College

Payne, C.H., 1925b, Astrophysical Data Bearing on the Relative Abundance of the Elements, Proc. Nation. Acad. Sci. USA, 11, 192

Pierson, D.J., 2003, Padding the Impact Factor, `http://www.wame.org/wame-listserve-discussions/impact-factor/`

Piwowar, H.A., Day, R.S., & Fridsma, D.B., 2007, Sharing Detailed Research Data Is Associated with Increased Citation Rate, PLoS ONE `www.plosone.org`, doi:10.1371/journal.pone.0000308

Polderman, A.K.S., 2007, Keep your hands off our impact factor, Euro. Sci. Edit., 33, 98

Popper, K., 1996, In Search of a Better World, Routledge

Potkin, S.G., Cannon, H.E., Murphy, D.L., & Wyatt, R.J., 1978, Are Paranoid Schizophrenics Biologically Different from Other Schizophrenics?, New Engl. J. Med., 298, 61

Publishing Research Consortium, 2010, Access vs. Importance. A global study assessing the importance of and ease of access to professional and academic information (Phase I Results), `http://www.publishingresearch.net/documents/PRCAccessvsImportanceGlobalNov2010_000.pdf`

Quetelet, A., 1835, Sur l'homme et le développement de ses facultés (Paris: Bachelier)

Quetelet, A., 1842, A Treatise on Man and the Development of His Faculties (Edinburgh: W. & R. Chambers)

Quetelet, A., 1849, Letters addressed to H.R.H. the Grand Duke of Saxe Coburg and Gotha, on the Theory of Probabilities, Translated by O.G. Downes, London: Charles & Edwin Layton, Letter XX, Note p. 276

Racker, E., 1989, A view of misconduct in science, Nature, 339, 91

Reich E.S., 2010, Self-plagiarism case prompts calls for agencies to tighten rules, Nature, 468, 745

Rennie, D., Flanagin, A., & Yank, V., 2000, The Contributions of Authors, JAMA, 284, 89

Rosenthal, R., 1976, Experimenter effects in behavioral research (New York: Irvington Publishers), p. 31

Rossner, M., & Yamada, K.M., 2004, What's in a picture? The temptation of image manipulation, J. Cell Biol., 166, 11

Rushton, J.P., 2002, New evidence on Sir Cyril Burt: His 1964 Speech to the Association of Educational Psychologists, Intelligence, 30, 555

Russell, H.N., 1929, On the Composition of the Sun's Atmosphere, ApJ, 70, 11

Russell, N., 2009, Communicating Science: Professional, Popular, Literary (Cambridge University Press)

Salager-Meyer, F., 2009, Academic equality and cooperative justice, TESOL Quart., 43, 703–709

Schaefer, B.E., 2008, A Problem with the Clustering of Recent Measures of the Distance to the Large Magellanic Cloud, AJ, 135, 112

Schreiber, M., 1997, A case study of the Hirsch index for 26 non-prominent physicists, Ann. Phys. 16, 640

Schumacher, H.C., 1824, Ehrenrettung Pasquich's. Geschichtserzählung, Astron. Nachr., 3, 65

Scott, T., 1997, Changing authorship system might be counterproductive, Brit. Med. J., 315, 744

Sechrest, L.B., 1987, Approaches to Ensuring Quality of Data and Performance: Lessons for Science?, in Scientific Excellence, ed. D.N. Jackson & J.P. Rushton (Sage Publications), p. 253

Seeber, F., 2008, Citations in supplementary information are invisible, Nature, 2008, 887

Seglen, P.O., 1992, The Skewness of Science, J. Am. Soc. Inf. Sci., 43, 628

Seglen, P.O., 1997, Why the impact factor of journals should not be used for evaluating research, Brit. Med. J., 314, 497

Shermer, M., 2006, Fake, Mistake, Replicate, Scient. Amer., 295, 40

Sieber, J.E., 1998, The Psychology of Whistleblowing, Sci. Eng. Ethics, 4, 7

Skármeta, A., 1985, Ardiente Paciencia (Santiago de Chile: Pehuén)

Smith R.S., 1997, Peer review: reform or revolution? Time to open up the black box of peer review, Brit. Med. J., 315, 759

Snow, C.P., 1959, The Two Cultures and the Scientific Revolution (Cambridge: Cambridge University Press)

Snow, C.P., 1960, The Affair (A Strangers and Brothers novel, Thirsk: House of Stratus)

Sokal, A., 1996, Transgressing the Boundaries: Towards a Transformative Hermeneutics of Quantum Gravity (Social Text, Duke University Press), 46/47, 217

Stearns, L., 1992, Copy Wrong: Plagiarism, Process, Property, and the Law, Calif. Law Rev., 80, 513

Sterken, C., 1988, in Coordination of Observational Projects in Astronomy, ed. C. Jaschek & C. Sterken (Cambridge: Cambridge University Press)

Sterken, C., 2002, ESO and European astronomy: four decades of reciprocity, Astron. Nachr., 323, 555

Sterken, C., 2005, The Light-Time Effect in Astrophysics: Causes and Cures of the $O-C$ Diagram, ed. Sterken, C. (San Francisco: Astronomical Society of the Pacific), ASP Conf. Ser., 335, 178

Sterken, C., 2006, Advice on Writing a Scientific Paper, in Astrophysics of Variable Stars, ed. Sterken, C. and Aerts, C. (San Francisco: Astronomical Society of the Pacific), ASP Conf. Ser., 349, 445, `http://www.aspbooks.org/publications/349-0445.pdf`

Sterken, C., 2011, Just one new measurement of the B[e] supergiant Hen-S22, IBVS, 6000

Sterken, C., & Aerts, C., 2006, Astrophysics of Variable Stars, ed. Sterken, C. and Aerts, C. (San Francisco: Astronomical Society of the Pacific), ASP Conf. Ser., 349

Stern, W., 1912, The Psychological Methods of Intelligence Testing, (Translated by G. Whipple) (Baltimore: Warwick and York)

Sterne, T.E., 1934, Appendix B: The Method of Computing e and s from the Means of Dates of Maximum, Distributed Regularly or Irregularly in Time, Harvard Coll. Obs. Circ., 386, 32

Swazey, J.P., 1999, Commentary on Lubalin and Matheson's "The Fallout: What Happens to Whistleblowers...", Sci. Eng. Ethics, 5, 251

Tarnow, E., 1999, The Authorship List in Science: Junior Physicists' Perceptions of Who Appears and Why, Sci. Eng. Ethics, 5, 73

Tarnow, E., 2002, Coauthorship in Physics, Sci. Eng. Ethics, 8, 175

Tennant D., 2006, Editorial integrity, Computerworld, October 9

Tenopir, C., 2010, personal communication

Tenopir, C., Allard, S., Bates, B., *et al.*, 2010, Research Publication Characteristics and Their Relative Values: A Report for the Publishing Research Consortium, Center for Information and Communication Studies University of Tennessee, `www.publishingresearch.net/PRCTenopiretalWord2010ResearchPublicationCharacteristics_000.docx`

Terrall, M., 2002, The Man Who Flattened the Earth: Maupertuis and the Sciences in the Enlightenment (Chicago: University Of Chicago Press)

Tierney, W.M., 2005, Impact Factor and Journal Importance, `http://www.wame.org/wame-listserve-discussions/impact-factor-and-journal-importance`

Tort, A.B.L., & Amaral, O.B., 2007, Measures for measures, but are we measuring right?, Nature/Nurture, Spring 2007

Toy, J., 2002, The Ingelfinger Rule: Franz Ingelfinger at the New England Journal of Medicine 1967–77, Science Editor, 25, 195

Tuchman, B.W., 1984, The March of Folly: From Troy to Vietnam (New York: Ballantine)

Vanclay, J.K., 2009, Bias in the journal impact factor, Scientometrics, 78, 3

Van de Looverbosch, M., & Deltour, P., 2011, Media onder vuur, De Standaard 19 January 2011, p. 22

van Maanen, A., 1916, Preliminary evidence of internal motion in the spiral nebula Messier 101, ApJ, 44, 210

Watson, J.D., 2007, Avoid Boring People (Oxford: Oxford University Press)

Watson, R.I., & Campbell, D.T., (Eds.), 1963, History, Psychology, and Science: selected papers (New York: Wiley)

Weil, V., 2001, Mentoring: Some Ethical Considerations, Sci. Eng. Ethics, 7, 471

Wen, Q., & Gao, Y., 2007, Dual publication and academic inequality, Internat. J. Appl. Lingu., 17, 221

West, R.M., 2006, The Messenger, 124, 37

Wilcox, L.J., 1998, Authorship: The Coin of the Realm, The Source of Complaints, JAMA, 280, 216

Wilcox, L.J., 2010, personal communication

Wilson, J.R., 2002, Responsible Authorship and Peer Review, Sci. Eng. Ethics, 8, 155

Wright, M., & Armstrong, J.S., 2007, Verification of Citations: Fawlty Towers of Knowledge?, Munich Personal RePEc Archive, 4149, `mpra.ub.uni-muenchen.de/`

Some additional readings:

Committee on Science, Engineering, and Public Policy, 2009, On Being a Scientist: A Guide to Responsible Conduct in Research, Third Edition, Nat. Acad. Sci., `http://www.nap.edu/catalog.php?record_id=12192`

Kohn, A., 1986, False Prophets (Oxford: Basil Blackwell)

LaFollette, M.C., 1996, Stealing into Print: Fraud, Plagiarism, and Misconduct in Scientific Publishing (University of California Press)

Mallon, T., 2001, Stolen Words – The Classic Book on Plagiarism (Orlando: Harcourt)

SUBJECT INDEX

NAME INDEX

Subject Index

Astronomy and Astrophysics
- anniversary issue, 101, 187
- business model, 236
- colors, 133
- Comments, 6
- copyright holder, 39
- country membership, 76
- economic model, 235
- journal, 2
- language editing, 11
- Letters, 6, 185, 255
- refereeing, 213
- Sections, 7
- structured abstracts, 23
- typefaces and fonts, 99
- website, 5

astrophysics
- Balmer discontinuity, 109, 154
- Cepheid period-distance relation, 67
- color index, 50, 67, 102, 109
- color-magnitude diagram, 104
- Comet 9P/Tempel 1, 113
- distance modulus, 200
- Doppler effect, 72
- effective temperature, 50
- ephemeris, 158
- evolution tracks, 104
- Fraunhofer lines, 129
- Hertzsprung–Russell diagram, 67, 104
- Hubble constant, 71, 155
- interstellar reddening, 104
- isochrones, 104
- Large Magellanic Cloud, 19, 152, 199
- Leonid meteor storm, 73
- luminous blue variable, 152
- metallicity, 50
- MK spectral types, 161
- Moon, 73
- open cluster, 104
- photometric system
  - Johnson *UBV*, 49, 104, 129
  - Strömgren *uvby*, 49
  - Walraven *VBLUW*, 50
- Planck radiation formula, 108
- planetesimal, 144
- planets
  - Earth, 157
  - Jupiter, 36, 158
  - Mars, 154, 158
  - Mercury, 157
  - Neptune, 36, 158
  - Saturn, 157
  - Uranus, 157
  - Venus, 158
  - Venus (transit), 196
- sodium spectrum, 129
- spiral nebulae, 175
- Sun
  - butterfly diagram, 84, 86
  - sunspot activity, 116
  - sunspot areas, 84
- theory of General Relativity, 71
- Titius-Bode law, 157

authorship, 40
- author order, 42
- Copenhagen compromise, 40
- corresponding author, 43
- credit, 3
- honorary author, 41
- Vancouver guidelines, 40

bibliometrics, 182
- abstracts of meetings, 182
- abuses, 263
- ADS, 247
- biases, 246, 256
- bibliometry, 27
- Body Mass Index, 259
- citations, 37, 187

citable items, 182
citation bias, 246
citation counts, 187, 190
citation index, 187
citation power, 270
citation rates, 180, 182
cited half-life, 184
deviation quotient, 268
editorials, 182
errata, 182
errors of confusion, 247
errors of procedure, 255
impact factor, 261
impact of research field, 254
Intelligence Quotient, 260
Journal Citation Reports, 182
key projects, 255
key scientific objectives, 255
Matthew effect, 256
Nobel Prize, 153, 189, 224
ORCID initiative, 267
parameters
Hirsch *h* index, 188–190
immediacy index, 184
impact factor, 180, 182
indicators, 180
paper counts, 184
similarity index, 240
peer-review process, 180
productivity, 186
ratchet effect, 256
repairing bibliometrics, 267
research papers, 182
review articles, 182
scientific age, 248
scientometry, 27
self-citations bias, 258
similarity index, 48, 261
threshold level, 189
time bias, 255
type of paper bias, 252
where to publish, 193

color
as decoration, 125
as didactic tool, 125
as identifier, 125
as organizer, 125
as quantitative information, 125
as symbol, 125
calibration, 140
changing color space, 142
chromaticity diagram, 129, 131
color balance, 139
color blindness, 126, 197
color circle, 126
color gamut, 130
color management, 138
color palette, 130, 136, 139
color picker, 139
color spaces, 83, 129, 130
color-defective eye, 127
color-mixing experiments, 127
Daltonism, 136
degenerate color spaces, 135, 236
deuteranopia, 135
device dependent, 130, 134
device independent, 135
dichromat, 135
false color, 148
gamma correction, 138
gray tonality, 139
in graphics, 124
indexed color, 92
lightfastness, 134
mixing dyes, 128
mixing light rays, 128
models, 130
photography, 125
primary colors (primaries), 128
protanopia, 136
purple line, 131
roygbiv, 126
secondary colors, 132
spectral response curves, 129
transformations, 140
trichromacy, 128
trichromatic model, 129
tritanopia, 136
Websafe palette, 136

workflow, 138

color space
CMYK, 95, 132
RGB, 129, 132, 140
XYZ, 131, 140
sRGB, 132
xyY, 131
xyz, 131
degenerate color spaces, 135
human observer, 131
spot colors, 135
video screen, 131

communication
entropy, 67
graphical content, 66
literate mode, 67
medium, 68
metaphor, 67
noise, 67, 68
oral mode, 67
textual mode, 66, 67
truthful communication, 174
visual mode, 67

copyright, 39
assignment, 39
Copyright Clearance Center, 48
copyright holder, 39
principle of fair use, 48, 207
rules, 15

data
accuracy, 103, 155
binning, 79, 85, 103
clipping, 84
cooking, 197
decorrelation, 159
estimate, 99
Fourier analysis, 89
Gaussian fit, 103
interquartile range, 153
invariance, 83
linear correlation, 154
metadata, 91
missing values, 84
nonsense correlation, 158, 262, 263
normally distributed, 112
observed – calculated ($O - C$), 158
outliers, 79, 84, 127, 151, 153
parameter, 99
precision, 103, 155
quartiles, 77, 153
random, 112
regression of x on y, 154
regression of y on x, 154
residuals, 79
robust regression, 155
small-number statistics, 263
spurious correlation, 159, 263
standard deviation, 79
symmetries, 79
tri-variate, 89
trimming, 197

databases and repositories
ADS, 3, 10, 12, 20, 45, 54, 85, 114, 185, 192, 247, 248, 251, 258
ISI, 182, 185, 187, 189, 192, 202, 220, 247–249, 251, 253, 257, 266
ORCID, 267
SIMBAD, 186
thesaurus, 15

error
blunder, 176
communication error, 177
conceptual error, 177
documentation error, 177
errors of science, 177
fooling others, 177
fooling yourself, 177
fruitful error, 177
interpretation error, 177
mistake, 176
negative knowledge, 176
numerical magnification error, 177
observational error, 177
personal error, 177

plug and play, 177
poor knowledge of instrument, 177
procedural error, 177
rounding error, 177
sampling error, 177
software error, 177
standard error of the mean, 17
statistical error, 177
textbook errors, 35

ethics, 194
abusing anonymity, 213
abusing the referee, 204
altruism, 272
amplification of work, 194
arm wrestling, 212
author misconduct, 201
authorship order, 202
bandwagon effect, 199
breaching confidentiality, 211, 221
Burt scandal, 231
bypassing refereeing process, 219
cargo cult science, 177, 181, 270
cascade submissions, 202
cheating, 174
citation practice, 204
codes of conduct, 173
competition, 232
confidentiality, 237
conflict of interest, 54, 214, 237
copyright infringement, 205
copyrights, 195
cryptomnesy, 48, 208
cures for misconduct, 234
data
data cooking, 146, 174
data fabrication, 237
data falsification, 237
data forging, 174, 195
data manipulation, 199
data trimming, 174
dataright infringement, 206
fabrication, 216
falsification, 216
defamation, 242
dishonesty, 174
duplicate papers, 209
editor misconduct, 217
embedded journalism, 230
ethos of news, 228
ethos of science, 228
forging referee reports, 219
fraud, 174, 231
freedom, 273
groupthink syndrome, 203
hoaxing, 174, 205
honesty, 177
honorary authorship, 203
image forensic software, 201
image manipulation, 200
improper attribution, 204
inappropriate authorship, 226, 227
Ingelfinger rule, 209
innocent plagiarism, 207
integrity, 177, 273
libel, 242
long revision times, 212
lying, 201
lying with graphs, 148
manipulating impact factor, 220, 223
manipulating journal content, 222
Matthew effect, 204
misconduct, 34
misconduct policies, 234
misquoting, 36, 204
modifications of manuscripts, 218
non-disclosure agreement, 195
non-responsive Editors, 217
noncontributing authors, 203
omitting authors, 202
paper slicing, 223
paraphrasing, 208
pathological science, 178
plagiarism, 206, 216, 238
plagiarism hunting, 207
Press, 228
pseudo-mentors, 223
publisher mismanagement, 222
redundant publications, 209
referee misconduct, 210, 216

refusing responsibility, 203
remaining impartial, 215
research misconduct policy, 237
reviewer misconduct, 216
secretiveness, 195
self-censorship, 204
self-plagiarism, 208
service observing, 41
sexism, 33
shortcutting refereeing process, 219
signatures of misconduct, 231
silencing dissent, 212
slander, 242
superficial revision, 212
supervisor misconduct, 223
survival of the fittest, 271
tacit knowledge, 48, 208
taking revenge, 212
text recycling, 208
translations, 210
trust, 214
truth, 4, 174, 180, 182, 271
value, 179, 182, 271
whistleblowing, 241, 242, 244

file conversions
16 bit to 8 bit, 94
8 bit to 16 bit, 94
raster to vector, 91
vector to raster, 91

file formats
BMP, 92
EPS, 93
FITS, 93
GIF, 93
JPEG, 93
JPEG2000, 93
PDF, 93
PICT, 93
PNG, 93
RAW, 93
TIFF, 93

graphics
animations, 70
artwork, 70
duotones, 90
full color images, 90
grayscale images, 90
halftones, 70
line art, 70, 90
movies, 70
multitone images, 90
Postscript bounding box, 119
resolution, 15, 95

graphs
3-D surfaces, 83
depth, 84
lighting angle, 84
abscissa, 102
autoclipping, 112
autoscaling, 112
axis, 102
direction, 102
labels, 104, 108
linear scale, 108
logarithmic scale, 108
orientation, 102
origin, 102
tickmarks, 102
title, 104
type, 106
units, 104
Bézier curves, 90, 160
backgrounds, 84, 110
bar graphs, 80
barycentric coordinates, 82
bitmap images, 90
box-whisker plots, 77, 153, 155, 198
bubble graphs, 82
cartoons, 87
color models, 160
composition, 117
contour maps, 82
curve fitting, 159
data density, 143
data units, 104
data-ink ratio, 143
decoration, 116

design type, 88
distortions, 83
dot joining, 153
drawings, 75, 87
duplicate axes, 102
error bars, 112
Euclidean rulers, 83
Figures
  captions, 73, 119
  numbering, 119
  placement, 119
framing, 110
galleries, 117
grids, 76, 115
Hermann grid illusion, 117
histograms, 78
insets, 114
inverted tickmarks, 102
isograms, 82
isolines, 82
key, 76, 113
labels, 81, 113
least-squares fitting, 154
legend, 76, 113
lie factor, 81
line graphs, 76
line weights, 98
linear correlation, 155
linearity, 154
linearization, 109
logos, 90
mesh plots, 84
nonsense correlations, 158
optical illusion, 144
ordinate, 102
orientation, 114
overshoots, 159
phase diagrams, 76, 106
pictographs, 82
pie charts, explosion factor, 81
pie charts, perception of surface, 80
plot, 75
plot symbols, 112
point-spread functions, 83
proofreading, 144
pseudocolor maps, 82
raster images, 90
retracing graphs, 149
revision, 144
scale factors, 114
scatter plots, 75, 82
SI units, 105, 107
size ratios, 118
sky maps, 82
spline curves, 160
spurious correlations, 83
stacked histogram, 79
step plots, 80
surface plots, 78
ternary diagrams, 82, 89
tickmarks spacing, 103
timelines, 77
title, 101
tracing, 91
triangular diagrams, 82
vector images, 90
vector plots, 76
viewing conditions, 110

images
  bit depth, 90
  capturing, 95
  color, 76, 90
  cropping, 95
  density, 139
  descreening, 96
  dynamic range, 139
  grayscale, 90
  halftone screen angle, 95
  halftones, 94, 121
  information content, 94
  lookup table, 76
  millefleur, 73
  Moiré patterns, 95
  monochromatic, 76, 90
  pseudo color, 76
  quality, 120
  rasterization, 91, 97
  reflectivity, 139
  rendering, 95

resolution, 74, 98, 120
dpi, 121
lpi, 121
pixel, 74, 121
ppi, 121
rotation, 121
sampling frequency, 122
scanning, 122
storage depth, 94
trimming, 95

information
channel, 67
content, 4, 15, 67
destination, 67
extraction, 68
meaning, 68
overload, 51
receiver, 67
reconstruction, 68
signal, 67
source, 67
theory, 67
transmitter, 67

journals
business model, 235
cited half-life, 184
data-protection policy, 221
editorials, 182
email policy, 221
evaluation, 182
high-impact journals, 31
house style, 15, 37, 119
immediacy index, 184
impact factor, 182
journal self-citation, 266
libel policy, 221
ombudsperson, 218
Open Access, 180, 219, 268
page charges, 7
refereed journal, 5
speed of publication, 7
the Hare and the Tortoise, 249

language, 68
daemon, 13
homophones, 17
if/then construction, 271
instinct, 11
jargon, 32
metaphor, 73
narrative, 2
nomenclature, 32
parser, 13, 33, 35
Pinyin, 12
proofreading, 29
reading process, 13, 69
stress languages, 12
syntax, 11
tautology, 14
terminology, 32
tonal/non-tonal languages, 12
transliteration, 12
writer's block, 51
writing process, 15, 69

legal aspects
copyright term, 205
copyrights, 148
data rights, 195
libel law, 221
non-disclosure agreement, 195
Principle of Fair Use, 47, 207
public domain, 205
trade secret law, 195

Open Access, 219, 268
Organisations and Institutions
HST, 199

page layout
aspect ratios, 147
composition, 147
copyrights, 148
lie factor, 150
margin justification, 147
orphans, 147
widows, 147

papers
structure
appendix, 28

abbreviations of given names, 21
acronyms, 14
authors and gender, 21
authorship, 20, 85
book reviews, 9
capitalization, 32
comments, 6
composer, 85
conclusions, 18
contributed conference paper, 8
data paper, 7, 51
discussion, 18
division of labor, 66
DOI, 18
duplicate papers, 10
epilog, 273
errata, 6
essays, 8
etiquette, 30
footnotes, 33
grandiloquence, 15
grey literature, 10
guidelines, 37
hoax articles, 10
IMRAD structure, 19, 127
information bulletins, 7
instructions, 37
instrument designs, 7
introduction, 18
invited conference talk, 8
Letter to the Editor, 6
manuscript, 18
methods, 18
misciting, 34
misquoting, 34
netiquette, 30
nomenclature, 33
PhD thesis book, 7
poster paper, 4, 8
postscript, 59
press releases, 11
proofreading, 14, 28
publicity papers, 9
quoting out of context, 34
refereed paper, 5
Research Note, 6
research paper, 5
results, 18
review paper, 6
salami paper, 9, 253, 261
sexism, 33
structure, 18
  abstract, 22
  acknowledgment, 25
  analysis, 24
  appendix, 28
  author affiliation, 22
  authors and coauthors, 20
  citations, 27
  conclusions, 24
  dedication, 25, 26
  discussion, 24
  Glossary, 28
  index, 28
  introduction, 23
  key words, 23
  methods, 24
  observations, 24
  postscript, 28
  References, 27
  results, 24
  running title, 19
  subtitle, 19
  title, 19
structured abstract, 127
style, 147
submitting, 28
telegrams, 7
ticket papers, 9

printing
addressability, 121
color, 69
dye-based inks, 134
dyeing, 125
grayscale, 69
imagesetter, 120
painting, 125
pigment-based inks, 134
weaving, 125

publishing
continuity editor, 14
copy editor, 14
DOI, 18
prepress, 92, 139
prepress staff, 140
production editor, 38
submitting image files, 93

referees
refereeing, 54
report, 3, 31, 47

references
format, 28
Harvard system, 27
list, 37
typos, 28
Vancouver style, 27

scientific
approximation to truth, 176
beauty, 180
fallibility, 176
imagination, 175
replication, 175
Reward System, 181
validation, 175

Tables
lines, 163
machine readable, 163
missing values, 163
OCR, 146
rules, 163

time systems
Anno Domini, 77
Anno Hegirae, 73
BC, BCE, 77
civil year, 77
Coordinated Universal Time, 110
Gregorian calendar, 74
heliocentric correction, 153
International Atomic Time, 110
Islamic calendar, 73
Julian Calendar, 73
Julian Date, 105
Julian Day, 77
Modified Julian Date, 105
Universal Time, 105

typesetting
acronyms, 14, 31
font sizes, 99
fonts, 38, 97
OpenType fonts, 98
picas, 98
pixels, 98
points, 98
PostScript Type 1 fonts, 98
PostScript Type 3 fonts, 98
TrueType fonts, 98
typeface, 38, 97

words
absolute, 15
acceptable, 15
accuracy, 15, 25
advise, 15
affirm, 15
algorithm, 15
amplitude, 15
analyze, 15
anticipate, 15
apprehend, 15
assert, 15
assumption, 15
authenticity, 15
average, 15
avow, 15
believe, 15, 17
bias, 15
big, 15
blunder, 15
boss, 15
bug, 17
calculate, 15
callname, 12
chance, 15
chaos, 15

cite, 15
claim, 15
class, 15
classic, 15
coach, 15
code, 15
colloquium, 15
color, 15
color index, 15
comparison star, 15
complete, 15
comprehend, 15
compute, 15
conclude, 15
conference, 15
confidence, 15
confirm, 15
conjecture, 15
consideration, 15
construe, 15
contend, 15
contrive, 15
copyright infringement, 15
correlation, 15
corroborate, 15
count, 15
default, 15
defect, 15
define, 15
demonstrate, 15
dependable, 15
devise, 15
differential, 15
dimension, 15
discern, 15
discovery, 15
dispersion, 15
distinguish, 15
doctrine, 15
duplication, 15
elegant, 15
elementary, 15
emulate, 15
entail, 15
entire, 15
entropy, 15
error, 15
establish, 15
estimate, 15
eventually, 17
example, 15
excellent, 15
expect, 15
explain, 15
extension, 15
extent, 15
fact, 15
factor, 15
false friends, 17
filter, 15
fit, 15
flaw, 15
flux, 15
font, 15
forecast, 15
foresee, 15
formulate, 15
frequency, 15
good, 15
grapheme, 12
great, 15
guess, 15
guide, 15
guidelines, 15
harmonic, 15
homophones, 17
honesty, 15
hunch, 15
hypothesis, 15
imagine, 15
implicate, 15
imply, 15
impose, 15
impossible, 15
inconceivable, 15
indefinite, 15
induce, 15
infinite, 15
innumerable, 15
insight, 15

instructions, 15
integrity, 15
intensity, 15
intercept, 15
interpret, 15
invention, 15
isotropy, 15
jargon, 15
large, 15
Latin words
  Anno Domini, 14, 77
  Anno Hegirae, 73
  e.g. (exempli gratia), 31
  a fortiori, 31
  ad hoc, 31
  caveat, 31
  et al. (et alii), 31
  etcetera, 31
  i.e. (id est), 31
  ibidem, 35
  imprimatur, 30, 181
  in extenso, 31
  in situ, 230
  indicum, 126
  lapsus, 138
  mala fide, 247
  minus, 31, 77
  nihil, 150
  nihil obstat, 181
  op. cit. (opus citatum), 35
  opus, 212
  post scriptum, 28
  sic, 36
  solidus, 106
  sphaera, 72
  thesaurus, 15
  verbatim, 36, 48, 206
  versus, 3, 17, 31, 191
  via, 120, 253
  vice versa, 27
  viz. (vidēre licet), 31
lecture, 15
likely, 17
look, 17
lower bound, 15
luck, 15
luminosity, 15
maintain, 15
manual, 15
manuscript, 15
maximum, 15
mean, 15
measurement, 15
memoir, 15
mention, 15
mentor, 15
metaphor, 15
minimum, 15
mistake, 15
mode, 15
model, 15
nickname, 12
noise, 15
nomenclature, 15
normal, 15
observable, 15
observation, 15
overthrow, 15
paper, 15
parameter, 15
paraphrase, 15
passband, 15
patron, 15
pattern, 15
perhaps, 17
period, 15
pleasing, 15
postulate, 15
power, 15
precision, 15, 25
predict, 15
premiss, 15
presentation, 15
probability, 15
probably, 17
process, 15
program, 15
proposition, 15
prove, 15

quote, 15
random error, 15
randomness, 15
range, 15
reckon, 15
recognize, 15
reduce, 15
redundancy, 15
refer, 15
referee, 15
refute, 15
regression, 15
relative, 15
relevant, 15
repeat, 15
replicate, 15
requirements, 15
residual, 15
resolution, 15
resolving power, 15
retention, 15
review, 15
revise, 15
robust, 15
rules, 15
satisfactory, 15
scatter, 15
scattering, 15
scintillation, 15
scrutinize, 15
secretiveness, 15
seeing, 15
seem, 17
self-plagiarism, 15
seminar, 15
serendipity, 15
significance, 15
significant, 15
simple, 15
simulate, 15
size, 15
sound, 15
speculate, 15
sponsor, 15
stable, 15
standard, 15
standard deviation, 17
standard star, 15
suggest, 15
supervisor, 15
support, 15
suppose, 15
supposition, 15
syllable, 12
symmetry, 15
symposium, 15
systematic error, 15
talk, 15
terminology, 15
text, 15
theory, 15
think, 15, 17
thought, 15
treatise, 15
trend, 15
tutorial, 15
type, 15
typeface, 15
uncertainty, 15
understand, 15
upper bound, 15
validate, 15
variable, 15
verify, 15
whole, 15
zero point, 15

writing
advancing insight, 3, 67
law of diminishing returns, 29
sparring partner, 29
writer's block, 15

Name Index

Anderson, Edgar, 26
Arago, François, 198
Archimedes, 35
Armstrong, Neil, 33

Baade, Walter, 225
Babbage, Charles, 174, 195, 197, 205
Baretti, Giuseppe, 35, 49
Batman, 31
Bayer, Johann, 49
Berg, Håkan, 72
Bernoulli, Daniel, 151
Bessel, Friedrich Wilhelm, 196
Bézier, Pierre, 90, 160
Binet, Alfred, 260
Bohr, Niels, 274
Burt, Cyril, 231

Cassini, Giovanni Domenico, 39
Chandrasekhar, Subramanyan, 273
Chatterley, Clifford, 36
Christie, William H. M., 85
Curie, Pierre, 20

Dalton, John, 135, 147
D'Angos, Jean Auguste, 196
Darwin, Charles, 272
Delambre, Jean-Baptiste, 151, 197
de Maupertuis, Pierre-Louis Moreau, 253
de Montesquieu, Charles, 236
de Sacrobosco, Johannes (John of Holywood), 72
Dirac, Paul, 273
Dürer, Albrecht, 74

Eco, Umberto, 3
Eddington, Arthur Stanley, 72, 210
Einstein, Albert, 71
Encke, Johann Franz, 196

Feynman, Richard, 24, 177, 181, 199, 273
Fisher, Ronald, 69, 148
Fletcher, Harvey, 199, 224, 241

Galilei, Galileo, 35
Galton, Francis, 260
Gauss, Carl Friedrich, 196

Hamilton, William, 189
Hein, Piet, 274
Heisenberg, Werner Karl, 274
Hell, Maximilian, 196
Hermann, Ludimar, 117
Herschel, Frederick William, 157
Herschel, John, 229
Hlava, Jaroslav, 35
Holmes, Robert, 198
Hoyle, Fred, 190, 256
Hubble, Edwin, 59, 71, 155
Hungerford, Margaret, 130
Huygens, Christiaan, 198

Ingelfinger, Franz Joseph, 209

Johnson, Harold, 49, 129
Joliot, Jean Frédéric, 20
Joliot-Curie, Irène, 20

Lawrence, David Herbert, 36
Le Blon, James, 125
Lemaître, Georges, 19, 71, 155, 210, 256
Lockyer, William, 85

Malthus, Thomas Robert, 29
Matzinger, Polly, 10
Maunder, Edward Walter, 21, 164
Maxwell, James Clerk, 127
Mentor, 223
Merton, Robert King, 256
Messier, Charles, 196
Méchain, Pierre, 151
Millikan, Robert, 146, 199, 209, 224, 241

Mirkwood, Galadriel, 10
Muller, André B., 245

Newcomb, Simon, 197
Newton, Isaac, 126, 129
Nyquist, Harry, 122

Odysseus, 223
Olbers, Heinrich Wilhelm, 196
Öpik, Ernst Julius, 212

Parra, Violeta, 4, 13
Pasquich, Johann, 196
Payne, Cecilia, 225
Pilate, Pontius, 175
Planck, Max, 108
Poggendorff, Johann Christian, 129
Popper, Karl, 176, 179
Poquelin, Jean-Baptiste (Molière), 60
Potter, Harry, 14
Prendergast, Kevin, 26
Ptolemy, 72
Purkinje, Jan Evangelista, 110

Quetelet, Adolphe, 259

Remi, Georges (Hergé), 33
Rodham, Hillary (Clinton), 20
Rømer, Ole, 35, 39, 103
Russell, Annie, 21
Russell, Henry Norris, 225

Sabine, Edward, 197
Safka, Melanie, 52, 215
Schiaparelli, Giovanni, 154
Schrödinger, Erwin, 273
Schumacher, Heinrich Christian, 196
Schwarzschild, Karl, 38
Shakespeare, William, 60
Shannon, Claude Elwood, 67, 122
Skármeta, Antonio, 206
Skłodowska-Curie, Marie, 20
Snow, Charles Percy, 200, 233
Stone, Lucy, 21
Strömgren, Bengt, 49
Struve, Otto, 225

Telemachus, 223
Thackeray, Andrew David, 225
Thomas, Dorothy, 21
Thomas, William, 21
Tintin, 33
Tolkien, John Ronald Reuel, 10
Tukey, John, 153
Tyndall, John, 117

Uplavici, Hlava, 35

van Maanen, Adriaan, 175, 181
von Fraunhofer, Joseph, 129
von Helmholtz, Hermann, 129
von Littrow, Carl Ludwig, 196, 241
von Zach, Franz Xaver, 20, 196

Walraven, Theodore, 49
Watson, James, 189
Winsor, Charlie, 26

Young, Thomas, 127, 129

EAS Publications Series

Volume 1 AGN in their Cosmic Environment (JENAM'99)
September 7-9, 1999 • **30 €**

Volume 2 GAIA: A European Space Project
May 14-18, 2001 • **60 €**

Volume 3 Star Formation and the Physics of Young Stars
(Summer School on Stellar Physics X)
September 18-22, 2000 • **40 €**

Volume 4 Infrared and Submillimeter Space Astronomy
June 11-13, 2001 • **65 €**

Volume 5 Radiative Transfer and Hydrodynamics in Astrophysics (GRETA)
November 28, 2001 • **25 €**

Volume 6 Observing with the VLT Interferometer (EuroWinter School)
February 3-8, 2002 • **47 €**

Volume 7 Final Stages of Stellar Evolution (Summer School on Stellar Physics XI)
September 16-21, 2001 • **52 €**

Volume 8 Astronomy with High Contrast Imaging
May 13-16, 2002 • **56 €**

Volume 9 Magnetism and Activity of the Sun and Stars
September 17-21, 2002 • **56 €**

Volume 10 Galactic & Stellar Dynamics (JENAM 2002)
September 3-6, 2002 • **36 €**

Volume 11 The Future Astronuclear Physics
August 20-22, 2003 • **47.50 €**

Volume 12 Astronomy with High Contrast Imaging II
October 6-10, 2003 • **56 €**

Volume 13 Evolution of Massive Stars, Mass Loss and Winds
(Summer School on Stellar Physics XII-XIII)
October 13-18, 2002 and *October 6-10, 2003* • **50 €**

Volume 14 Dome C: Astronomy and Astrophysics Meeting
June 28-30, 2004 • **63 €**

Volume 15 Radio Astronomy from Karl Jansky to Microjansky (JENAM 2003)
August 27-30, 2003 • **72 €**

Volume 16 Teaching and Communicating Astronomy (JENAM 2004)
September 13-17, 2004 • **44 €**

Volume 17 Element Stratification in Stars: 40 Years of Atomic Diffusion Meeting in Honour of Georges Michaud
June 6-10, 2005 • **60 €**

Volume 18 Radiative Transfer and Applications to Very Large Telescopes (GRETA)
May 11-13, 2005 • **47 €**

Volume 19 Stars and Nuclei: A Tribute to Manuel Forestini
March 4-5, 2004 • **36 €**

Volume 20 Mass Profiles and Shapes of Cosmological Structures (IAP 2005)
July 4-9, 2005 • **47 €**

Volume 21 Stellar Fluid Dynamics and Numerical Simulations: From the Sun to Neutron Stars
September, 2004 and May, 2005 • **55 €**

Volume 22 Astronomy with High Contrast Imaging III: Instrumental Techniques, Modeling and data Processing
May 16-19, 2005 • **55 €**

Volume 23 Sky Polarisation at Far-Infrared to Radio Wavelengths: The Galactic Screen before the Cosmic Microwave Background
September 12-15, 2005 • **39 €**

Volume 24 CRAL-2006. Chemodynamics: From First Starts to Local Galaxies
July 10-14, 2006 • **47 €**

Volume 25 1st ARENA Conference on "Large Astronomical Infrastructures at CONCORDIA, prospects and constraints for Antarctic Optical/IR Astronomy"
October 16-19, 2006 • **47 €**

Volume 26 Stellar Evolution and Seismic Tools for Asteroseismology
November 20-23, 2006 • **32 €**

Volume 27 The Third European Summer School on Experimental Nuclear Astrophysics
October 2-9, 2005 • **39 €**

Volume 28 Perspectives in Radiative Transfer and Interferometry
May 14-16, 2007 • **25 €**

Volume 29 Tidal Effects in stars, Planets and Disks
September 5-9, 2005 • **47 €**

Volume 30 Spanish Relativity Meeting – Encuentros Relativistas Españoles
ERE2007
September 10-14, 2007 • **63 €**

Volume 31 Far-Infrared Workshop 2007
November 5-7, 2007 • **39 €**

Volume 32 Stellar Nucleosynthesis 50 Years after B^2FH
December 4-8, 2006 • **63 €**

Volume 33 2nd ARENA Conference on "The Astrophysical Science Cases at Dome C"
September 17-21, 2007 • **55 €**

Volume 34 Astronomy in the Submillimeter and Far Infrared Domains
with the Herschel Space Observatory
April 23-May 4, 2007 • **47 €**

Volume 35 Interstellar Dust from Astronomical Observations to Fundamental Studies
May 1st-5, 2006 • **47 €**

Volume 36 Dark Energy and Dark Matter
and Theories
July 7-11, 2008 • **55 €**

Volume 37 Astrophysics Detector Workshop 2008
November 17-20, 2008 • **63 €**

Volume 38 Nonlinear Pulsations and Hydrodynamics of Cepheids
July 9-13, 2007 • **32 €**

Volume 39 Stellar Magnetism
September 24-28, 2007 • **39 €**

Volume 40 3rd ARENA Conference on "An Astronomical Observatory at
CONCORDIA (Dome C, Antarctica)"
May 11-15, 2009 • **87 €**

Volume 41 Physics and Astrophysics of Planetary Systems
February 18-29, 2008 • **95 €**

Volume 42 Extrasolar Planets in Multi-Body Systems: Theory and Observations
August 25-29, 2008 • **79 €**

Volume 43 Non-LTE Line Formation for Trace Elements in Stellar Atmospheres
July 30 – August 03, 2007 • **43 €**

Volume 44 JENAM 2008: Grand Challenges in Computational Astrophysics
September 8-11, 2008 • **26€**

Volume 45 GAIA: At the Frontiers of Astrometry
June 7-11, 2010 • **78€**

Volume 46 PAHs and the Universe: A Symposium to Celebrate the 25th Anniversary of the PAH Hypothesis
May 31-June 04, 2010 • **69€**

Volume 47 Astronomical Spectrography for Amateurs
May 06-11, 2003 • **35€**

Volume 48 CRAL-2010: A Universe of Dwarf Galaxies
June 14-18, 2010 • **87€**

Volume 49 Scientific Writing for Young Astronomers – Part 1
2008-2009 • **35€**

Order directly to EDP Sciences
17 av. du Hoggar • BP. 112 • P.A. de Courtabœuf • 91944 Les Ulis Cedex A • France
Tel.: 33 (0)1 69 18 75 75 • Fax: 33 (0)1 69 86 06 78 • books@edpsciences.org

Astronomy & Astrophysics

Aiming High

weekly international journal which publishes papers
n all aspects of astronomy and astrophysics (theoretical,
bservational, and instrumental) independently of the
echniques used to obtain the results.

www.aanda.org

EAS Publications Series

An internationa
publications serie

Proceedings of international colloquia, workshops and schools, in the field of astronomy & astrophysics

European Astronomical Society
Publications Series

EDP SCIENCES

eas

www.eas-journal.org